SOLUTIONS MANUAL

High School | Geometry

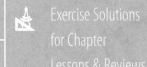

Exercise Solutions
for Chapter
Lessons & Reviews

Answers to Exercises For

GEOMETRY

Seeing, Doing, Understanding

Harold R. Jacobs

MasterBooks® CURRICULUM

Author: Harold R. Jacobs

Master Books Creative Team:

Editor: Laura Welch

Cover Design: Diana Bogardus

Copy Editors:
Craig Froman
Willow Meek

Curriculum Review:
Kristen Pratt
Laura Welch
Diana Bogardus

Master Books® Edition
First printing: February 2017
Eighth printing: October 2022

Master Books, P.O. Box 726, Green Forest, AR 72638

Master Books® is a division of the New Leaf Publishing Group, Inc.

ISBN: 978-1-68344-021-5
ISBN: 978-1-61458-093-5 (digital)

Unless otherwise noted, Scripture quotations are from the New King James Version of the Bible.

Printed in the United States of America

Please visit our website for other great titles:
www.masterbooks.com

HAROLD R. JACOBS is a teacher of mathematics and science, writer, and well-respected speaker. He received his B.A. from U.C.L.A. and his M.A.L.S from Wesleyan University. His other publications include *Mathematics: A Human Endeavor, Elementary Algebra*, and articles for *The Mathematics Teacher* and the *Encyclopedia Britannica*. Mr. Jacobs has received the Most Outstanding High School Mathematics Teacher in Los Angeles award, the Presidential Award for Excellence in Mathematics Teaching, and was featured in the book *101 Careers In Mathematics* published by the Mathematical Association of America.

Publisher's Note:
Additional reference sources have been included in some of the answers in this solution manual to offer learning extensions for students if they wish to explore more on specific topics. The books represent a variety of worldviews, scientific disciplines, and perspectives. Use of these additional resources is not required for the course. Educators should use their discretion on incorporating these additional resources per the needs and requirements of their students and education program.

ANSWERS TO EXERCISES

Chapter 1, Lesson 1

Set I (pages 10–11)

Ch'ang-an, now called Sian, was one of two great capital cities of ancient China. It was built on the bank of the Wei River on a main trade route to Central Asia.

The city measured 5 miles from north to south and 6 miles from east to west and was surrounded by a wall 17.5 feet high. Its population in the eighth century is estimated to have been 2 million.

More about Ch'ang-an can be found in chapter 5 of *Ancient China,* by Edward H. Schafer (*Great Ages of Man* series, Time-Life Books, 1967).

1. Collinear.
•2. 12 cm.
3. 6 cm.
4. 4 cm.
•5. About 1.3 cm.
6. 10 cm.
•7. 3 mi. ($\frac{6 \text{ mi}}{2}$ = 3 mi.)
8. 0.5 mi. ($\frac{6 \text{ mi}}{12 \text{ cm}} = \frac{0.5 \text{ mi}}{\text{cm}}$.)
9. 5 mi. (10 cm × $\frac{0.5 \text{ mi}}{\text{cm}}$ = 5 mi.)
10. No, because its sides are not all equal.
11. The colored regions (the Palace and Imperial Cities, the two markets, and the park).
12. 22 mi. (44 cm × $\frac{0.5 \text{ mi}}{\text{cm}}$ = 22 mi.)
13. Coplanar.
•14. Line segments.
15. A line segment is part of a line and has a measurable length. A line is infinite in extent.
16. 25.

•Answers to exercises marked with a bullet are also given in the textbook.

Set II (page 11)

The figure in this exercise set is a "nine-three" configuration, one of three such configurations in which nine lines and nine points are arranged so that there are three lines through every point and there are three points on every line. It and other configurations of projective geometry are discussed in chapter 3 of *Geometry and the Imagination,* by David Hilbert and Stefan Cohn-Vossen (A.M.S. Chelsea Publishing, 1990).

17. Nine.
18. Points that lie on a line.
19. Nine. They are E-F-I, A-B-G, A-D-I, G-H-I, B-C-D, B-E-H, D-F-H, A-C-F, C-E-G.
20. Lines that contain the same point.
•21. Three. (AG, AF, AI.)
22. Three. (AG, BD, BH.)
23. Three. (AF, BD, CG.)
24. Three. (Three lines are concurrent at every lettered point of the figure.)
•25. Six; AB, AG, AC, AF, AD, AI.
26. Six; BA, BC, BD, BE, BH, BG.
27. Six; CB, CD, CE, CG, CF, CA.

Set III (pages 11–12)

The first three figures in this exercise illustrate a special case of the dual of Pappus's theorem. As long as two sets of three lines in the figure are parallel, as in the first three figures, the added sets of lines are concurrent. A figure such as the last one in the exercise suggests that, when the lines are not parallel, the concurrence is lost. However, the two sets of three segments in the original figure must not necessarily belong to parallel lines. If they belong to concurrent sets of lines, the added sets of lines also are concurrent.

In the exercises, three sets of concurrent lines are named: AI, CF, DG; AF, CI, EH; and BG, CH, EI. There are three more sets that are not mentioned, of which one is AD, BE, GI. The other two sets, AE, BD, FH and BH, CG, DF, are more elusive (1) because, for certain spacings of the lines in the original figure, they are parallel rather than concurrent and (2) because, for the cases in which they *are* concurrent, the points of concurrency lie outside the figure.

1.

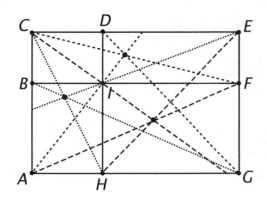

2. They seem to be concurrent.

3. These lines also seem to be concurrent.

4. These lines also seem to be concurrent.

5. *Example figure:*

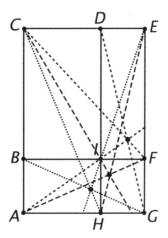

6. The three sets of lines again seem to be concurrent.

7. *Example figures:*

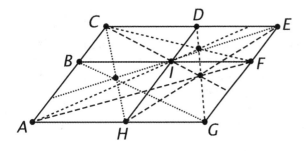

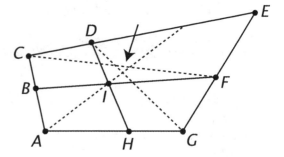

(Just one set of the three lines is shown for the figure above, and the lines are clearly not concurrent.)

8. As long as the line segments in the original figure consist of two parallel (or concurrent) sets, the three sets of lines seem to be concurrent. Otherwise, they do not.

Chapter 1, Lesson 2

Set I (pages 15–16)

In the problems about the constellations, it is interesting to realize that, although the stars that form them give the illusion of being both coplanar and close together, they are certainly not coplanar and not necessarily near one another. The angles that are measured in the exercises are merely the angles between the lines in our map of the sky and not the actual angles between the lines in space.

Triangle Measurements.

1. AB.

•2. ∠C.

3. AC.

4. ∠B.

5. AB = 5.1 cm, AC = 2.6 cm, BC = 3.7 cm.

•6. Rays AC and AB.

7. ∠A = 45°.

8. *(Student drawing.)*

9. Rays BA and BC.

•10. ∠B = 30°.

11. Rays CA and CB.

12. ∠C = 105°.

Constellation Angles.

Set II (pages 16–17)

According to *Rules of Thumb,* by Tom Parker (Houghton Mifflin, 1983), "the base of a ladder should be 30 percent of its height from the base of the wall it is leaning against."

The Sliding Ladder.

24.

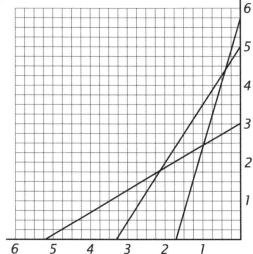

•25. Approximately $1\frac{3}{4}$ ft.

26. 74°.

27. 16°.

•28. Approximately $3\frac{1}{4}$ ft.

29. 56°.

•30. 34°.

31. Approximately $5\frac{1}{4}$ ft.

32. 30°.

33. 60°.

34. 45°. (The sum of the two angles seems to always be 90°.)

Set III (page 17)

More information on speculations about Mars can be found in the entertaining chapter titled "The Martian Dream" in *Pictorial Astronomy,* by Dinsmore Alter, Clarence H. Cleminshaw, and John G. Philips (Crowell, 1974).

Measuring Mars.

Dividing 24° into 360°, we get 15; so the distance around Mars is 15 times the distance between the two locations:

15 × 880 miles = 13,200 miles.

Chapter 1, Lesson 3

Set I (page 21)

Pentominoes were invented by Solomon Golomb, a long-time professor of mathematics and electrical engineering at the University of Southern California, who introduced them in a talk in 1953 at the Harvard Mathematics Club. His book titled *Polyominoes* (Princeton University Press, 1994) is the definitive book on the part of combinatorial geometry included in recreational mathematics. It is interesting to note that Arthur C. Clarke, the author of *2001—A Space Odyssey,* originally intended to use Pentominoes as the game played with the computer HAL before chess was substituted for it instead.

The problems about the Texas ranches were inspired by the fact that people sometimes naively assume that the size of a property can be measured by the time that it takes to walk around it. Proclus wrote about this assumption in 450 A.D. in his commentary on the first book of Euclid.

The Perimeter and Area of a Square.

1.

- •2. 9 in².
- •3. 12 in.
- •4. 36 in².
- •5. 24 in.
- 6. 1 ft².
- 7. 4 ft.
- 8. 144 in².
- 9. 48 in.
- 10. x^2 square units.
- 11. $4x$ units.

Pentominoes.

- 12. A polygon is a two-dimensional figure that is bounded by line segments.
- •13. Piece I because it has four sides.
- 14. 5 square units.
- 15. Piece P. (It has a perimeter of 10 units; all the other pieces have perimeters of 12 units.)

Texas Ranches.

- •16. Ranch A. (Ranch A needs 36 miles of fencing, whereas ranch B needs 32 miles).
- 17. Ranch B. (Ranch B has an area of 60 square miles, whereas ranch A has an area of 56 square miles).
- 18. Area.

Set II (page 22)

A book filled with fascinating information about the Great Pyramid is *Secrets of the Great Pyramid,* by Peter Tompkins (Harper & Row, 1971).

A Model of the Great Pyramid.

- 19. 64°.
- 20. Yes. (The sides opposite the 58° angles appear to be about 9.4 cm each.)
- 21. *(Model of Great Pyramid.)*
- 22. *Example figure:*

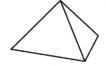

- •23. Polygons.
- 24. A polyhedron.
- •25. Eight.
- 26. Five.
- 27. Four.
- •28. Three.
- 29. Five.
- •30. A line.
- 31. *(Student estimate.)* (Calculations using trigonometry give a height of about 6.25 cm.)
- •32. 75.6 ft.
- 33. *(Student estimate.)* (6.25 × 75.6 gives a height of approximately 470 ft.)
- 34. A view from directly overhead.

Set III (page 23)

The square-to-triangle puzzle was created, as the footnote on page 23 states, by Henry Dudeney, England's greatest inventor of mathematical puzzles. In *The Canterbury Puzzles,* he told of exhibiting it to the Royal Society of London in 1905 in the form of a model made of four mahogany pieces hinged together in a chain. When closed in one direction, the pieces formed the triangle and, when closed in the other direction, they formed the square. The version illustrated on page 23 of the text is slightly altered to make it easier for beginning geometry students to draw. Martin Gardner gives directions for constructing the figure exactly on pages 33 and 34 of *The 2nd Scientific American Book of Mathematical Puzzles and Diversions* (Simon & Schuster, 1961).

A Square Puzzle.

- 1. 60°.
- 2. ∠BFP = 79° and ∠AEH = 49°.
- 3. ∠BEH = 131° and ∠EHA = 41°.
- 4. PF and PG.

5.

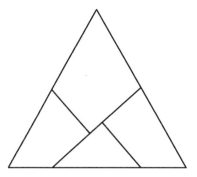

6. About 21.3 cm.

7. 60°.

8. The area of the triangle is equal to that of the square because they are made of the same pieces.

9. 196 cm².

10. It is longer. (The perimeter of the triangle is about 64 cm, whereas the perimeter of the square is 56 cm.)

Chapter 1, Lesson 4

An excellent history and mathematical treatment of sundials is *Sundials—Their Theory and Construction,* by Albert E. Waugh (Dover, 1973).

 The most common type of sundial is the horizontal one and is the kind depicted on the Kellogg's Raisin Bran box. For this type of sundial to tell time accurately, the angle of its gnomon must be equal to the latitude of the location in which the dial is used. The angle of the gnomon printed on the Kellogg box has a measure of about 39°, evidently chosen because that is a good approximation of the average latitude of the United States. Seattle has a latitude of almost 48°, whereas Miami's latitude is less than 26°; in locations such as these, the claim that "now you can tell time outdoors" with the cereal box sundial is less than accurate!

Set I (pages 26–27)

Finding North by Shadows.

1.

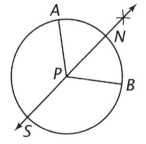

2. *(Student answer.)* (∠APB should be twice ∠APN.)

3.

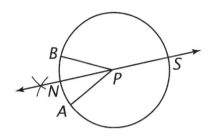

4. *(Student answer.)* (∠APB should be twice ∠APN.)

•5. Noncollinear.

6. Coplanar.

7.

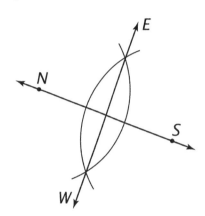

8. *(Student answer.)* (The angles formed by NS and EW have measures of 90°.)

9. *(Student answer.)* (The sum of all four angles is 360°.)

Bisectors in a Triangle.

10.

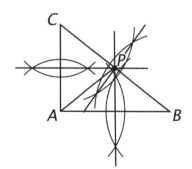

•11. The bisecting lines seem to be concurrent.

12. They seem to meet on side CB.

13. The distances seem to be equal.

14.

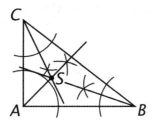

15. They seem to be concurrent.

•16. No.

Set II (pages 27–29)

About Sundials.

•17. West.

18. North.

19. South.

20. Noon.

•21. About 6 A.M. and 6 P.M.

22. ∠IOJ.

•23. OI bisects ∠HOJ.

24. No. ∠POQ and ∠QOR do not appear to be equal.

•25. ∠DOB and ∠EOA.

26. ∠POL.

27. ∠COB, ∠PON and ∠POQ.

28. ∠EOM.

29. *Example figure:*

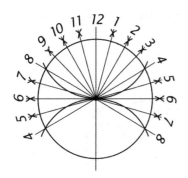

30. Collinear means that there is a line that contains the points.

•31. 60°.

32. 30°.

•33. 180°.

34. 90°.

35. 15°.

36. 4 minutes. ($\dfrac{60 \text{ minutes}}{15°} = \dfrac{4 \text{ minutes}}{1°}$.)

Set III (page 29)

This discovery exercise is adapted from a proof exercise (3.5.18) in *Geometry Civilized,* by J. L. Heilbron (Oxford University Press, 1998). Every geometry teacher should have a copy of this beautifully written and illustrated book. Dr. Heilbron was formerly Professor of History and the History of Science and Vice Chancellor of the University of California at Berkeley and is currently Senior Research Fellow at Worcester College, Oxford University, and at the Oxford Museum for the History of Science.

Angle Bisectors in a Rectangle.

1.

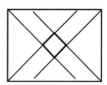

2. A square.

3.

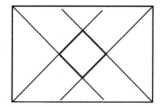

Again, the lines seem to form a square.

4.

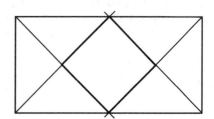

The lines seem to form a square.

5.

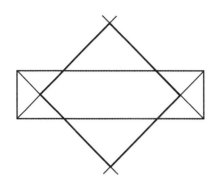

The lines still seem to form a square.

6. Yes. The more elongated the rectangle, the larger the square.

7. If the original rectangle were square in shape, the polygon would shrink to a point.

Chapter 1, Lesson 5

Set I (pages 31–32)

The incorrect Egyptian formula for finding the area of a quadrilateral was evidently based on the notion of multiplying the averages of the lengths of the opposite sides. It is easy to show (as the students will be asked to do later in the course) that this method gives the correct answer if the quadrilateral is a rectangle. For all other quadrilaterals, the method gives an area that is too large.

The Egyptian Tax Assessor.

• 1. 100 square units. ($\frac{1}{4}20 \cdot 20 = 100$.)

2. 88 square units. ($\frac{1}{4}16 \cdot 22 = 88$.)

3. 98 square units. ($\frac{1}{4}14 \cdot 28 = 98$.)

4. Ramses.

5. Yes. ($11 \cdot 8 = 88$.)

• 6. It contains 80 square units. ($10 \cdot 8 = 80$.)

7. It contains 108 square units. ($9 \cdot 12 = 108$.)

• 8. 54 square units. ($\frac{1}{2}108 = 54$.)

9. It contains 60 square units. ($5 \cdot 12 = 60$.)

10. 30 square units. ($\frac{1}{2}60 = 30$.)

• 11. It contains 84 square units. ($54 + 30 = 84$.)

12. Ramses is cheated the most. He pays the highest tax, yet actually has the smallest property. (See the chart below.)

	Assessment Area	Actual Area
Ramses	100	80
Cheops	88	88
Ptahotep	98	84

Set II (pages 32–34)

We will return to the trisection of line segments and angles later in the course. The method for trisecting an angle presented here is surely the first one to be thought of by most people encountering the problem for the first time. For small angles, it is a fairly good approximation but, for an angle of 60°, the error just begins to become detectable through measurement and, for obtuse angles, its failure becomes very obvious.

The classic book on angle trisection for many years was *The Trisection Problem,* by Robert C. Yates, originally published in 1942 and reprinted in 1971 by N.C.T.M. as a volume in the *Classics in Mathematics Education* series. The definitive book is currently *A Budget of Trisections,* by Underwood Dudley (Springer Verlag, 1987). Dr. Dudley includes some additional information on the subject in "Trisection of the Angle" in his book titled *Mathematical Cranks* (M.A.A., 1992). All make fascinating reading.

Trisection Problems.

13.

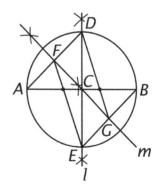

14.

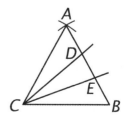

15. An accurate drawing and careful measurements suggest that ∠ACD, ∠DCE, and ∠ECB are about 19°, 22°, and 19°, respectively.

Set III (page 34)

This problem was discovered by Leo Moser, who, in a letter to Martin Gardner, said that he thinks it was first published in about 1950. The "obvious" formula for the number of regions, $y = 2^{n-1}$, where n is the number of points, is wrong. The correct formula,

$$y = \frac{1}{24}(n^4 - 6n^3 + 23n^2 - 18n + 24)$$

is surprisingly complicated. For more on this problem, see the chapter titled "Simplicity" in *Mathematical Circus* by Martin Gardner (Knopf, 1979).

Cutting Up a Circle.

1. *Example figure:*

2. 16.

3. 2.

4. 4.

5. No. of points connected   2   3   4   5
   No. of regions formed   2   4   8   16

6. It doubles.

7. 32.

8. 64.

9. *Example figures:*

10. No. (Six points result in at most 31 regions, and seven points in at most 57 regions.)

Chapter 1, Review

Set I (pages 36–37)

The transparent view of the cube used in exercises 15 through 17 is named after psychologist Hertha Kopfermann, who studied it in 1930. An interesting discussion of our perception of this figure appears on pages 27 and 28 of *Visual Intelligence,* by Donald D. Hoffman (Norton, 1998).

1. Line segments.

•2. Polygons (or squares).

3. A polyhedron.

4. 3.

•5. 12.

6. 3.

7. 6.

•8. Coplanar.

9. 9.

•10. PQ.

11. NO.

12. OP.

•13. T.

14. U and Q.

15. Triangles, quadrilaterals, and a hexagon. (Students who are aware of concave polygons may notice more than one hexagon and even some heptagons!)

16.

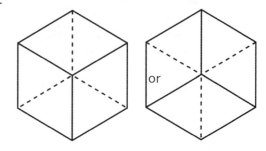

17. 3.

Set II (pages 37–38)

Exercises 19 through 26 are based on the fact that every triangle has three excenters as well as an incenter. Each excenter is the point of concurrence of the lines bisecting exterior angles at two vertices of the triangle (∠DAB and ∠ABE in the figure) and the line that bisects the (interior) angle at the third vertex (∠C in the figure). Point F is one of the excenters of the triangle.

Exercises 27 through 30 are based on the fact that, when the midpoints of the sides of any quadrilateral are connected in order with line segments, the quadrilateral that results is a parallelogram. Furthermore, its perimeter is equal to the sum of the length of the diagonals of the original quadrilateral. These facts are easily proved and will be established later in the course.

•18. Constructions.

19.

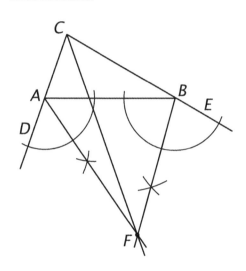

•20. ∠DAB = 110°.

21. ∠ABE = 150°.

22. ∠DCE = 80°.

23. ∠DAF = ∠FAB = 55°.

24. ∠ABF = ∠FBE = 75°.

•25. ∠DCF = ∠FCE = 40°.

26. Line CF seems to bisect ∠DCE.

27.

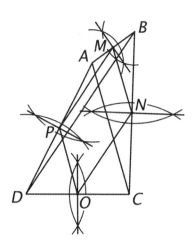

28. The opposite sides of the quadrilateral appear to be equal (and parallel).

29. (Answer depends on student's drawing.)

30. (Answer depends on student's drawing but should be close to the answer to exercise 29.)

Set III (page 38)

The story of Queen Dido and Carthage is told on pages 44–47 of *Mathematics and Optimal Form*, by Stefan Hildebrandt and Anthony Tromba (Scientific American Library, 1985) and on pages 64–71 of the revised and enlarged version of this book titled *The Parsimonious Universe* (Copernicus, 1996).

•1. 1,200 yd. (4 · 300 = 1,200.)

2. 90,000 yd². (300² = 90,000.)

3. 500 yd. (900 − 2 · 200 = 500.)

4. 1,400 yd. (2 · 200 + 2 · 500 = 1,400.)

5. 100,000 yd². (200 · 500 = 100,000.)

6. *(Student drawings.)* (Students who have made several drawings may suspect that the length along the river should be twice the side width to get as much land as possible. In this case, the numbers are 450 yards and 225 yards and the area is 101,250 square yards.)

- 1. Associative for addition.
- 2. Commutative for multiplication.
- 3. Definition of subtraction.
- 4. Identity for addition.
- 5. Definition of division.
- 6. Inverse for multiplication.
- 7. 1,728. ($12 \cdot 12 \cdot 12 = 1,728$.)
- 8. 12,000. ($12 \cdot 10 \cdot 10 \cdot 10 = 12,000$.)
- 9. 20. ($5 + 15 = 20$.)
- 10. -75.
- 11. 225. ($289 - 64 = 225$.)
- 12. 81. ($9^2 = 81$.)
- 13. 49π.
- 14. 14π.
- 15. $5x$.
- 16. x^5.
- 17. $5x$.
- 18. x^5.
- 19. $2x + 3y$.
- 20. x^2y^3.
- 21. $x^2 + y^3$.
- 22. $x^3 + y^2$.
- 23. $x + 2y$.
- 24. xy^2.
- 25. $7x^2$.
- 26. $7x^2$.
- 27. $5x^2$.
- 28. $5x^2$.
- 29. $4x^2 + 3x$.
- 30. $3x^3 + 2x^2 + x$.
- 31. x^{14}.
- 32. x^{14}.
- 33. $6x^5$.
- 34. $6x^5$. ($2x^3 \cdot 3x^2 = 6x^5$.)
- 35. $36x^2$. ($6x \cdot 6x = 36x^2$.)
- 36. $72x^5$. ($8x^3 \cdot 9x^2 = 72x^5$.)
- 37. $4x + 4y$. ($3x + 2y + x + 2y = 4x + 4y$.)
- 38. $4x + 4y$.
- 39. $2x$. ($3x + 2y - x - 2y = 2x$.)
- 40. $2x$.
- 41. 0.
- 42. 1.
- 43. $5x + 10y$.
- 44. $3x + 4y$.
- 45. $5x - 10y$.
- 46. $3x - 4y$.
- 47. $31xy$. ($28xy + 3xy = 31xy$.)
- 48. $25xy$. ($28xy - 3xy = 25xy$.)
- 49. $7 + x^3 + x^5$.
- 50. $10x^8$.

Set I (pages 43–44)

•1. A conditional statement.

2. "*a*" represents the hypothesis and "*b*" represents the conclusion.

3.

A circle labeled *a* inside another circle labeled *b*.

•4. You live in the Ozarks.

5. You live in the United States.

•6. Yes.

7. No.

8. No. If it is cold outside, it isn't necessarily snowing.

9. No.

10. It is cold outside if it is snowing.

11. Yes.

12. Statements 1 and 3.

13. Yes.

•14. If an animal is a koala bear, then it eats only eucalyptus leaves.

15. If the cat is in the birdcage, then it isn't there to sing.

16. If money grew on trees, then Smokey Bear wouldn't have to do commercials for a living.

17. If you are an architect, then you use geometry.

•18. If I don't understand, then I ask questions.

19. If there is a fire, then use the stairs instead of the elevator.

20. If you are an elephant, then you cannot fly. (Or, If you can fly, then you are not an elephant.)

Set II (pages 44–45)

•21. If at first you don't succeed, then try again.

22. Try again if at first you don't succeed.

23. Region 1.

•24. Regions 1 and 2.

25. Regions 2 and 3.

26. Region 3.

27.

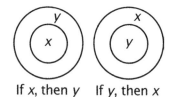

If *x*, then *y*   If *y*, then *x*

•28. The second.

29. The first.

30. Two.

31.

32. Statements a and d.

33. Statements a and d.

34.

•35. Statements c and d.

36. Statements a and b.

Set III (page 45)

The Animal Mind, the source of the information for the Set III exercises, is a fascinating book. This paragraph from the dust jacket describes it well:

"In this engaging volume, James and Carol Gould go in search of the animal mind. Taking a fresh look at the evidence on animal capacities for perception, thought, and language, the Goulds show how scientists attempt to distinguish actions that go beyond the innate or automatically learned. They provide captivating, beautifully-illustrated descriptions of a number of clever and curious animal behaviors—some revealed to be more or less pre-programmed, some seemingly proof of a well-developed mental life."

1. The red monkey shape.

2. In the left-hand column.

3. The red square.

4. If Sarah takes the banana, then Mary will *not* give chocolate to Sarah.

Chapter 2, Lesson 2

Set I (pages 47–48)

Exercises 1 through 4 have an interesting source. In his book titled *Euclid and His Modern Rivals* (1885), Charles L. Dodgson (Lewis Carroll) criticized some of the geometry books of the time. One of them, *Elementary Geometry, following the Syllabus prepared by the Geometrical Association,* by J. M. Wilson (1878) contained on page 17 the following "question":

"State the fact that all 'geese have two legs' in the form of a Theorem." Carroll wrote: "This I would not mind attempting; but, when I read the additional request, to 'write down its converse *theorem,*' it is so powerfully borne in upon me that the writer of the Question is probably himself a biped, that I feel I must, however reluctantly, decline the task."

1. If a creature is a goose, then it has two legs.

2.

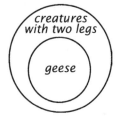

•3. If a creature has two legs, then it is a goose.

4. No.

•5. You are not more than six feet tall.

6. If you are not more than six feet tall, then you are an astronaut.

7. No.

8. No.

•9. If you do not know how to reason deductively.

10. If you cannot comprehend geometry, then you do not know how to reason deductively.

11. You do not know how to reason deductively.

12. If and only if.

•13. They have the same meaning.

14. If and only if.

15. The fear of peanut butter sticking to the roof of your mouth.

16. If you are afraid of peanut butter sticking to the roof of your mouth, then you have arachibutyrophobia.

17. Yes. It must be true because the sentence is a definition.

18. Common, proper, improper.

19. If a fraction is a vulgar fraction, then it is a common fraction that is either proper or improper.

20. If a fraction is a common fraction that is either proper or improper, then it is a vulgar fraction.

•21. Yes. The converse must be true because this is a definition.

22. "New Year's Day" does not mean the same thing as "holiday." Other days are holidays as well, so its converse is not true.

23. Yes. "New Year's Day" means the same thing as "the first day of the year." The converse is true.

Set II (pages 48–49)

The wolf-pack definition in exercises 36 and 37 is an amusing example of recursion. An entire chapter of Clifford Pickover's book titled *Keys to Infinity* (Wiley, 1995) deals with the subject of "recursive worlds."

24. Yes.

25. No.

•26. No.

27. The second sentence. (Only if it is your birthday do you get some presents.)

•28. It is the converse.

29.

30. It is dry ice if and only if it is frozen carbon dioxide.

31. The teacher thought you misspelled "if."

•32. Sentence 2. (If it is a detective story, then it is a whodunit.)

33. Yes.

34. If a car is a convertible, then it has a removable top. If a car has a removable top, then it is a convertible.

•35. It is the converse.

36. Two or more.

37. Because, according to the definition, there is no limit to the number of wolves in a wolf pack.

Set III (page 49)

Since its creation in the early 1970s, Peter Wason's puzzle has appeared in puzzle books in a variety of forms. In *The Math Gene* (Basic Books, 2000), Keith Devlin remarks that most people find the puzzle extremely hard.

Turn over the blue card to see if there is a circle on the other side. Turn over the card with the triangle to see if the other side is blue. (It doesn't matter what is on the other side of the card with the circle. For example, the possibility that the other side is red doesn't contradict the statement

If a card is blue on one side, it has a circle on the other side.

Set I (pages 52–53)

1. If you keep quiet, others will never hear you make a mistake.
   If others never hear you make a mistake, they will think you are wise.

•2. If you keep quiet, others will think you are wise.

3. $a \rightarrow b, b \rightarrow c$. Therefore, $a \rightarrow c$.

•4. All Greeks are humans.
   All humans are mortals.
   Therefore, all Greeks are mortals.

5. All Greeks are statues.
   All statues are mortals.
   Therefore, all Greeks are mortals.

6. Yes.

•7. No.

•8. The hypothesis.

9. The conclusion.

10. $a \rightarrow b, b \rightarrow c, c \rightarrow d$. Therefore, $a \rightarrow d$.

11. The third: If you live where it is cold, you see a lot of penguins.

•12. It also may be false.

13. If Captain Spaulding is in the jungle, he can't play cards.

14. Yes.

15. If NASA launched some cows into space, they would be the herd shot around the world.

•16. A theorem.

17. If you go to Dallas, you will take a plane.
    If you take a plane, you will go to the airport.
    If you go to the airport, you will see all the cabs lined up.
    If you see all the cabs lined up, you will see the yellow rows of taxis.

•18. If you go to Dallas, you will see the yellow rows of taxis.

19. If a duck had sore lips, he would go to a drugstore.
    If he went to a drugstore, a duck would ask for some Chapstick.
    If he asked for some Chapstick, a duck would ask to have it put on his bill.

20. *Second statement:* If they wanted to take along some food, they would try to carry on six dead raccoons.
    Last statement: If the flight attendant objected, they would be told that there is a limit of two carrion per passenger.

21. *First statement:* If a group of chess players checked into a hotel, they would stand in the lobby bragging about their tournament victories.
    Third statement: If the manager asked them to leave, they would ask why.

Set II (pages 53–54)

Dice Proof.

•22. 21. $(1 + 2 + 3 + 4 + 5 + 6 = 21.)$

23. 7. $(\frac{21}{3} = 7.)$

24. 2.

25. 6.

Matchbox Proof.

•26. Red. Because the label is wrong.

27. A red marble and a white marble.

28. Two white marbles.

Tick-tack-toe Proofs.

29.

30. If your opponent marks an X in the upper left, then you will mark an O in the lower right.
    If you mark an O in the lower right, then your opponent will mark an X in the lower center.
    If your opponent marks an X in the lower center, you will mark an O in the upper center.

31. *(Student drawings.)*

32. You will win the game because, no matter where you mark your next O, your opponent will then be forced to allow you to set a winning trap!

Set III (page 54)

Word ladders were invented by Lewis Carroll in 1877 and have been a popular form of word play ever since. An entire chapter in Martin Gardner's *The Universe in a Handkerchief—Lewis Carroll's Mathematical Recreations, Games, Puzzles, and Word Plays* (Copernicus, 1996) deals with word ladders (called "doublets" by Carroll).

Another interesting reference is *Making the Alphabet Dance,* by Ross Eckler (St. Martin's Press, 1996). At the beginning of chapter 4, titled "Transforming One Word into Another", Eckler remarks: "By drawing on ideas from mathematics . . . [word play] gains additional respectability and legitimacy."

1. *Example solution:*
   LESS, LOSS, LOSE, LOVE, MOVE, MORE.

2. *Example solution:*
   HEAD, HEAR, HEIR, HAIR, HAIL, TAIL.

Chapter 2, Lesson 4

Set I (pages 57–58)

It is interesting to note that, like that of Rube Goldberg, Samuel Goldwyn's name has entered the dictionary. According to *The Random House Dictionary of the English Language,* a "Goldwynism" means "a phrase or statement involving a humorous and supposedly unintentional misuse of idiom."

•1. Contradiction.

2. "In–direct."

•3. Suppose that it would not speak foul language.

4. Suppose that he has control over his pupils.

5. Suppose that it does not lead to a contradiction.

•6. No.

7. Yes.

8. No.

9. Indirect.

•10. It is the opposite of the theorem's conclusion.

•11. The earth is not flat.

12. Suppose that the earth is flat.

13. The stars would rise at the same time for everyone, which they do not.

•14. That it is false.

Set II (pages 58–59)

The puzzles about the weights are adapted from ideas in A. K. Dewdney's book titled *200% of Nothing* (Wiley, 1993). The subtitle of the book is *An Eye-Opening Tour through the Twists and Turns of Math Abuse and Innumeracy*. Dewdney uses the idea of equal partition of numbers in a section of the book titled "What Do Mathematicians Do?"

Trading Desks Proof.

•15. *Beginning assumption:*
Suppose that the pupils can obey the teacher.
Contradiction:
There are 13 pupils at the black desks but only 12 brown desks.
Conclusion:
The pupils can't obey the teacher.

Ammonia Molecule Proof.

16. *Beginning assumption:*
Suppose that the atoms of an ammonia molecule are coplanar.
Contradiction:
Each bond angle is 107°.
Conclusion:
The atoms of an ammonia molecule are not coplanar.

Rummy Proof.

17. *Beginning assumption:*
Suppose that someone holds a royal flush.
Contradiction:
You have all the "10" cards.
Conclusion:
In this deal of the cards, no one holds a royal flush.

Balanced Weights Proofs.

18. Puzzle 1. One set is 2, 4, 6, 13; the other set is 7, 18.

19. If a puzzle of this type has a solution, then each set will weigh half the total weight.

20. *Beginning assumption:*
Suppose that there is a solution.
Contradiction:
The sum of all of the weights is odd.
Conclusion:
There is no solution.

Athletes Puzzle.

21. There is only 1 football player and consequently there are 99 basketball players. If there were two or more football players, then, given any two of the athletes, both could be football players. This contradicts the fact that, given any two of the athletes, at least one is a basketball player. Therefore, the assumption that there are two or more football players is false, and so there is only one football player.

Set III (page 59)

A good discussion of David L. Silverman's puzzle about the list of true and false statements appears on page 79 of *Knotted Doughnuts and Other Mathematical Entertainments,* by Martin Gardner (W. H. Freeman and Company, 1986). In its original form, the puzzle contained 1,969 statements (to celebrate the new year of 1969) comparable to those in the list in Set III. The only true statement in the original version was the 1,968th one.

This sort of puzzle suggests other possibilities that students might enjoy exploring. What if the word "false" were changed into "true" in all of the statements? What if the word "exactly" were omitted?

1. No, because any two statements on the list contradict each other.

2. No, because then the list would contain exactly zero false statements.

3. No.

4. No.

5. One, because the preceding conclusions eliminate all of the other possibilities. The true statement must be C.

Chapter 2, Lesson 5

Set I (pages 61–63)

The dog Rin-Tin-Tin was Warner Brothers' first big money-making star. His movie career began in 1922, and his salary at one time was $1,000 a week. According to Clive Hirschhorn's *The Warner Brothers Story*, Rin-Tin-Tin's 19 films kept the studio prosperous during the silent era and saved many theaters from closure. He was voted most popular film performer in 1926!

1. No (because a period has dimensions.)

2. No (because even an atom has dimensions.)

•3. You go around in circles.

4. Star, tail.

5. Because Rin-Tin-Tin was a movie star who had a tail.

•6. Two points determine a line.

7. Because it is taut and hence straight like a line.

•8. Theorems.

9. A statement that is assumed to be true without proof.

10. No.

Three Arrow Table.

11. Three noncollinear points determine a plane.

•12. The points.

13. The plane.

14. The table top might tip over.

15. Noncollinear.

16. If and only if.

17. If points are coplanar, then there is a plane that contains all of them. If there is a plane that contains all of them, then points are coplanar.

•18. Converse.

Intersecting Cubes.

19. True.

•20. True.

21. True.

•22. False.

23. False.

24. True.

25. True.

Set II (pages 63–64)

Japanese Weights.

•26. A picul is 1,600 taels.

27. A picul is 16,000 momme.

28. A momme is one-tenth of a tael.

29. No, because we don't know what any of these units are.

30. Yes. An elephant weighs more than one picul.

31. Because "the load carried by a man" is too vague; some men can carry much heavier loads than others.

Credit Card Rules.

•32. Statements 1 and 3.

33. Transaction finance charge and supercheck.

34. Statements 2, 4, and 5.

35. Statements 2 and 4. (If you go over your credit limit, you will be charged a fee. If you are charged a fee, the fee will be added to your balance.)

36. If you go over your credit limit, a fee will be added to your balance.

Crooked Lines.

•37. Two points determine a line.

38. Yes.

39. We did not define the word "line" (nor did we define the word "straight").

40. A theorem is a statement that is proved.

41. *Beginning assumption:*
Suppose two lines can intersect in more than one point.
Contradiction:
Two points determine a line.
Conclusion:
Two lines cannot intersect in more than one point.

Set III (page 64)

1. 3 lines.

2. 6 lines.

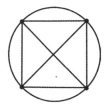

3. 10 lines.

4. 15 lines.

5. 20 points: 190 lines.

Some students who answer this correctly will probably do it by observing the pattern of increasing differences.

Number of points	2	3	4	5	6
Number of lines	1	3	6	10	15
		2	3	4	5

Each additional point results in one more line than the point before it.

Other students may deduce a formula for the number of lines, *l*, in terms of the number of points, *p*:

$$l = \frac{p(p-1)}{2}$$

(This is the formula for the binomial coefficient: $\binom{p}{2}$.)

Set I (pages 67–68)

Sherman Stein makes some scathing remarks in his book titled *Strength in Numbers* (Wiley, 1996) about the "tear off the corners" of a triangle procedure illustrated in exercises 8 through 10.

He writes:

"Here is how one of the seventh-grade texts has the pupils 'discover' that the sum of the three angles of a triangle is always 180°. Under the heading 'Work Together,' it has these directions:

Begin with a paper triangle that is a different shape from those of the other members of your group. Number the angles of the triangle and tear them off the triangle. Place the three angles side-by-side so that pairs of angles are adjacent and no angles overlap.

What seems to be true? Compare your results with those of your group. Make a conjecture.

On the very next page, facing these directions, we find, 'In the Work Together activity you discovered that the following statement is true.' In boldface follows, 'The sum of the measures of the angles of a triangle is 180°.'

What kind of discovery or constructing knowledge out of experience is that? From what I know about pupils, they would glance at the bold type on the next page and stop experimenting."

•1. The length of the hypotenuse of a right triangle.

2. The speed of light.

3. The circumference of a circle.

4. The length of one of the other sides of a right triangle.

•5. The area of a circle.

6. The Pythagorean Theorem.

7. The squares on the sides contain 16, 9, and 25 small squares; 16 + 9 = 25.

8. The sum of the angles of a triangle is 180°.

•9. No.

10. No.

Target Circles.

11. 2, 4, 6, 8, and 10 units.

•12. 2π, 4π, 6π, 8π, and 10π units.

•13. 9π square units. ($\pi 3^2 = 9\pi$.)

14. 9π square units. ($\pi 5^2 - \pi 4^2 = 9\pi$.)

15. It is because the purple region appears to many people to be much larger than the yellow region.

16. 100. 36 + 64 = 100 by the Pythagorean Theorem.

17. 144. 169 – 25 = 144 by the Pythagorean Theorem.

•18. 30°. (The Triangle Angle Sum Theorem.)

19. 45°.

20. $(90 - n)°$.

•21. 70°. ($180 - 40 = 140$, $\dfrac{140}{2} = 70$.)

22. 65°. ($180 - 50 = 130$, $\dfrac{130}{2} = 65$.)

23. $\left(\dfrac{180 - n}{2}\right)°$.

•24. No.

25. This is the converse of the Pythagorean Theorem, and the converse of a true statement isn't necessarily true. (We will see later that this particular converse is true, but to show why requires a separate argument.)

Set II (pages 69–70)

David Gale, Professor Emeritus of Mathematics at the University of California at Berkeley recalls the rotating pencil demonstration in exercises 38 through 40 as his first encounter with a mathematical proof. Another student showed it to him when he was in the fifth grade. (*Tracking The Automatic Ant,* by David Gale, Springer, 1998).

Eye Pupil.

•26. Approximately 50 mm². ($\pi 4^2 = 16\pi \approx 50$.)

27. Approximately 3 mm². ($\pi 1^2 = \pi \approx 3.14$.)

28. Approximately 16 times as much.

Triangle Angle Sum Theorem.

29. 90°.

30. 60°.

31. No, because then the third angle would have a measure of 0°.

•32. Yes, because the sum of the other two angles could be 1°.

33. Yes, because the sums of the angles in the two triangles are the same.

The Stretched Cord.

•34. a^2.

35. $2a^2$.

36. $2a^2$.

37. By the Pythagorean Theorem.

Pencil Experiment.

•38. To the left.

39. That the pencil has turned through 180°.

40. The sum of the angles of a triangle is 180°.

Aryabhata on the Circle.

41. Yes. $a = \dfrac{1}{2}c\dfrac{1}{2}d = \dfrac{1}{2}(2\pi r)\dfrac{1}{2}(2r) = \pi r^2$.

42. 3.1416. If the circumference is $8(4 + 100) + 62{,}000 = 62{,}832$ and the diameter is 20,000, then, because $c = \pi d$,

$62{,}832 = \pi(20{,}000)$; so $\pi = \dfrac{62{,}832}{20{,}000} = 3.1416$.

43. Dilcue's answer is a good one because not only is PIE an old way to spell PI but his answer reflected in a vertical mirror is 3.14.

Set III (page 70)

Greg N. Frederickson, the author of *Dissections, Plane and Fancy* (Cambridge University Press, 1997), is a professor of computer science at Purdue University. His book is the definitive work on the amazing topic of geometric dissections and should be on the bookshelf of every geometry classroom.

When the three pieces are arranged as in the second figure, they seem to form a square whose side is the hypotenuse of one of the right triangles. When the pieces are arranged as in the fourth figure, seem to form two squares whose sides are equal to the other two sides of the right triangle. The total area of the three pieces is the same no matter how they are arranged; so the two arrangements illustrate the Pythagorean Theorem.

Chapter 2, Review

Set I (pages 71–73)

A. Bartlett Giamatti was president of Yale University and, later, of the National Baseball League. He also served briefly as Commissioner of Major League Baseball until his death in 1989. *Take Time for Paradise* is a nice little book on sports and play.

The Morton Salt Company originally considered the slogan "Even in rainy weather, it flows freely" to promote the message that its salt would pour easily in damp weather. They then settled on a variation of the old proverb: "It never rains but it pours" for their slogan and have been using it on their packages since 1914.

The SAT problem was a multiple-choice question in its original format, and the problem asked only for *x*.

The judge in the trial excerpt was Judge Ito, and the conversation took place in Los Angeles on July 26, 1995, in the case of The People of the State of California vs. Orenthal James Simpson.

•1. If something is a limerick, then it has five lines.

2. If I perfect my perpetual motion machine, then I will make a fortune.

3. If something is a toadstool, then it is not edible.

4. $a \rightarrow b$, $b \rightarrow c$; therefore, $a \rightarrow c$.

5. If we have known freedom, then we fear its loss.

6. One or both of the premises being false.

7. If you are a daredevil, then you are recklessly bold.
   If you are recklessly bold, then you are a daredevil.

8. Each is the converse of the other.

•9. Collinear.

10. Two.

•11. A postulate.

12. An Euler diagram.

•13. If it rains, then it pours.

14. It pours if it rains.

15. If it pours, then it rains.

16. No.

17. 100°. (180 – 48 – 32 = 100.)

•18. 50°. ($2x = 100$, $x = 50$.)

19. 98°. (180 – 50 – 32 = 98.)

•20. Three noncollinear points determine a plane.

21. If the feet of the four legs were not coplanar, the tripod might tip back and forth.

Geometry in Spanish.

22. In every right triangle, the square of the hypotenuse is equal to the sum of the squares of the other two sides.

•23. Square.

24. Equal.

25. They are defined by other words, which results in going around in circles.

26. Point, line, plane.

27. Because we must have a starting point to avoid going around in circles. These words are especially simple and so serve as good starting points. (Note: The answer is *not* that we cannot define them.)

Courtroom Questions.

28. The witness thought π might be 2.12 or 2.17. The witness also said that pi times 2.5 squared is 19 but, rounded to the nearest whole number, the answer would be 20.

29. No. The judge guessed that pi was 3.1214, perhaps influenced by the witness's guess of 2.12, and 3.1214 is too small.

30. No.

Set II (pages 73–74)

In his book titled *Science Awakening* (Oxford University Press, 1961), B. L. van der Waerden said of the Egyptians' method for finding the area of a circle: "It is a great accomplishment of the Egyptians to have obtained such a good approximation. The Babylonians, who had reached a much higher stage of mathematical development, always used π = 3."

Although it is unlikely that your students will be surprised that the circumference/perimeter ratio of the circle to its circumscribed square is equal to the area/area ratio, this relation is remarkable. It holds true even if the circle and square are stretched into an ellipse and rectangle. John Pottage wrote a nice dialogue between Galileo's characters Simplicio and Segredo about this topic in the first chapter of his *Geometrical Investigations—Illustrating the Art of Discovery in the Mathematical Field* (Addison-Wesley, 1983).

The planet so small that someone can walk around it in 20 minutes is included in Lewis Carroll's *Sylvie and Bruno Concluded* (1893). More about Carroll's mathematical recreations can be found in Martin Gardner's *The Universe in a Handerchief* (Copernicus, 1996). Carroll wrote of the planet: "There had been a great battle . . . which had ended rather oddly: the vanquished army ran away at full speed, and in a very few minutes found themselves face-to-face with the victorious army, who were marching home again, and who were so frightened at finding themselves between *two* armies, that they surrendered at once!"

31. Postulates.

•32. Theorems.

33. Direct and indirect.

•34. 144 square units. ($12^2 = 144$.)

35. 256 square units. ($16^2 = 256$.)

36. 400 square units. ($144 + 256 = 400$.)

•37. 20 ft. ($x^2 = 16^2 + 12^2 = 400$, $x = \sqrt{400} = 20$.)

38. $(\frac{16}{9}r)^2 = (\frac{256}{81})r^2$; so π would be equal to $\frac{256}{81}$, or approximately 3.16.

39. They would be too large.

Panda Proof.

40. *Second statement:* If he had a sandwich, he would take out a gun.
    Fourth statement: If he shot the waiter, he would leave without paying.

•41. Direct.

Clue Proof.

42. *Beginning assumption:* Suppose that Colonel Mustard didn't do it.
    Missing statement: If Miss Scarlet did it, then it was done in the dining room.
    Contradiction: It happened after 4 P.M.
    Conclusion: Colonel Mustard did it.

43. Indirect.

Circle in the Square.

•44. $2\pi r$.

45. $8r$.

46. $\frac{2\pi r}{8r} = \frac{\pi}{4}$. ($\frac{\pi}{4}$ is approximately 0.785.)

47. πr^2.

•48. $4r^2$.

49. $\frac{\pi r^2}{4r^2} = \frac{\pi}{4}$.

50. Approximately 1,680 feet. If someone can walk three miles in an hour, they can walk one mile in 20 minutes. Because $c = \pi d$,

$$d = \frac{c}{\pi} = \frac{5,280}{\pi} \approx 1,680 \text{ feet.}$$

Chapter 2, Algebra Review (page 76)

•1. $10x$.

•2. $16x^2$.

•3. $6x + 5$.

•4. $4x - 5$.

•5. $6x^2$.

•6. $-6x^2$.

•7. $9x + 4y$.

•8. $3x + 2y$.

•9. $5x + 15$.

•10. $44 - 11x$.

•11. $18x + 6$.

•12. $40 - 56x$.

•13. $9x^2 - 18x$.

•14. $20x + 6x^2$.

•15. $x^2y + xy^2$.

•16. $2x^2 - 4x + 8$.

•17. $5x - 3 = 47$,
$5x = 50$,
$x = 10$.

•18. $9 + 2x = 25$,
$2x = 16$,
$x = 8$.

•19. $4x + 7x = 33$,
$11x = 33$,
$x = 3$.

•20. $10x = x + 54$,
$9x = 54$,
$x = 6$.

•21. $6x - 1 = 5x + 12$,
$x - 1 = 12$,
$x = 13$.

•22. $2x + 9 = 7x - 36$,
$-5x + 9 = -36$,
$-5x = -45$,
$x = 9$.

•23. $8 - x = x + 22$,
$8 - 2x = 22$,
$-2x = 14$,
$x = -7$.

•24. $3x - 5 = 10x + 30$,
$-7x - 5 = 30$,
$-7x = 35$,
$x = -5$.

•25. $4(x - 11) = 3x + 16$,
$4x - 44 = 3x + 16$,
$x - 44 = 16$,
$x = 60$.

•26. $x(x + 7) = x^2$,
$x^2 + 7x = x^2$,
$7x = 0$,
$x = 0$.

•27. $5(x + 2) = 2(x - 13)$,
$5x + 10 = 2x - 26$,
$3x + 10 = -26$,
$3x = -36$,
$x = -12$.

•28. $6(4x - 1) = 7(15 + 3x)$,
$24x - 6 = 105 + 21x$,
$3x - 6 = 105$,
$3x = 111$,
$x = 37$.

•29. $8 + 2(x + 3) = 10$,
$8 + 2x + 6 = 10$,
$2x + 14 = 10$,
$2x = -4$,
$x = -2$.

•30. $x + 7(x - 5) = 2(5 - x)$,
$x + 7x - 35 = 10 - 2x$,
$8x - 35 = 10 - 2x$,
$10x - 35 = 10$,
$10x = 45$,
$x = 4.5$.

•31. $4(x + 9) + x(x - 1) = x(6 + x)$,
$4x + 36 + x^2 - x = 6x + x^2$,
$3x + 36 = 6x$,
$36 = 3x$,
$x = 12$.

•32. $2x(x + 3) + x(x + 4) = 5x(x + 2) - 2x^2$,
$2x^2 + 6x + x^2 + 4x = 5x^2 + 10x - 2x^2$,
$3x^2 + 10x = 3x^2 + 10x$,
$0 = 0$;
so x can be any number.

Set I (pages 80–81)

Zero among all numbers is surely the cause of the most confusion. Ask anyone other than someone knowledgeable in mathematics what number one divided by zero is or what zero divided by zero is, and you will probably not get a meaningful answer. The problem, of course, is that there is no number equal to $\frac{n}{0}$ if $n \neq 0$, whereas quantities that approach $\frac{0}{0}$ may yield any number in the limit, a fact of which beginning calculus students become aware when they first encounter l'Hôpital's Rule. That is why $\frac{0}{0}$ itself must remain undefined as well.

Charles Seife remarks in his book titled *Zero— The Biography of a Dangerous Idea* (Viking, 2000): "If you wantonly divide by zero, you can destroy the entire foundation of logic and mathematics. Dividing by zero once—just one time—allows you to prove, mathematically, anything at all in the universe."

- •1. Subtraction.

- •2. Substitution.

- 3. Multiplication.

- 4. Substitution.

- 5. Division.

- 6. $a + b + c + d = b + e + f + g$.

- 7. Subtraction.

- 8. Subtraction.

- 9. Substitution.

- •10. Addition.

- 11. Substitution.

- 12. Division (or multiplication by $\frac{1}{2}$).

- 13. 2.

- •14. No number (because there is no number when multiplied by 0 that gives 6.)

- 15. 0.

- 16. Any number!

17. Any number can replace ?, although this fact makes the first equation seem strange for numbers other than 0 or 1.

18. Substitution.

- •19. If $a = b$ and $c \neq 0$, then $\frac{a}{c} = \frac{b}{c}$.

  If $c = d$, then $\frac{a}{c} = \frac{b}{d}$ by substitution.

Correcting Mistakes.

20. In the first step, 7 was subtracted from the left side of the equation but added to the right side. The equation should be $2x = 12$.

- •21. Substitution.

22. Indirect.

Set II (pages 81–82)

We will return to the "sum of the area of the crescents equal to the area of the triangle" problem later in the course.

The figure from the Chinese text illustrating the Pythagorean Theorem is, of course, the one from Euclid's *Elements*. Although the Chinese text dates back only about 400 years, the Pythagorean Theorem may have been known in China long before Pythagoras was born. The theorem appears in the oldest Chinese mathematics book known, the *Chou Pei*, possibly written as long ago as 1100 B.C.

Triangle and Crescents.

23. They are equal.

24. By subtracting II + IV from both sides of the equation I + II + IV + V = II + III + IV, we get I + V = III.

Binomial Square.

- •25. a^2 and b^2.

- •26. $2ab$.

27. $a + b$.

- •28. $(a + b)^2$.

29. $a^2 + 2ab + b^2$.

30. $(a + b)^2 = a^2 + 2ab + b^2$.

Chinese Proof.

31. The Pythagorean Theorem.

•32. Substitution.

33. Substitution.

34. Addition.

35. The square of the hypotenuse of a right triangle is equal to the sum of the squares of the other two sides.

Circle Formula.

36. $a = \pi r^2$ and $r = \dfrac{d}{2}$; so $a = \pi(\dfrac{d}{2})^2 = \dfrac{\pi}{4}d^2$.

Quadratic Formula.

37. $ax^2 + bx + c = 0$.

•38. Division.

39. Subtraction.

40. Addition.

•41. A direct proof.

42. Substitution.

43. $x = \dfrac{-5 \pm \sqrt{25 - 24}}{4}$, $x = \dfrac{-5 \pm 1}{4}$;

$x = \dfrac{-5 + 1}{4} = \dfrac{-4}{4} = -1$;

$x = \dfrac{-5 - 1}{4} = \dfrac{-6}{4} = -1.5$.

Dilcue's Pie.

44. The area and circumference of the pie were evidently the same number.

45. If $\pi r^2 = 2\pi r$, then $\dfrac{\pi r^2}{\pi r} = \dfrac{2\pi r}{\pi r}$, so $r = 2$. The radius of the pie must have been 2 feet. The circumference must have been $2\pi(2) = 4\pi$ (or approximately 12.6) feet.

46. 4π (or approximately 12.6) *square* feet.

Set III (page 83)

There are quite a number of fallacious algebraic "proofs" based on dividing by zero in a disguised form. Carroll's version, accompanied by his charming whimsy, is undoubtedly the best—especially because it results in such an unexpected answer to one of the first multiplication problems that every child encounters.

Because $x = y$, $x - y = 0$. In dividing each side of the equation by $(x - y)$, we are dividing by 0, so the rest does not follow. This is especially obvious if we substitute 0 for $(x - y)$ in the equation $2(x + y)(x - y) = 5(x - y)$. It is certainly true that $2 \times 2 \times 0 = 5 \times 0$, but not that $2 \times 2 = 5$.

Chapter 3, Lesson 2

Set I (pages 86–88)

Although the triple jump was included in the first modern Olympics in 1896, it was probably not part of the ancient Greek Olympics. The triple jump is now thought to have originated as a result of a misunderstanding. The Greeks added the lengths of the three best jumps in the long-jump competition, which may have led to the idea that they were performing a "triple" jump. The world record for the triple jump is currently a little more than 18 meters.

The closest footwear that we currently have to "seven league boots" was reported in the July 15, 2000, issue of *New Scientist* magazine. A team of Russian engineers has developed fuel-powered boots that enable the wearer to jump over 6-foot walls, take 12-foot steps, and walk at speeds of 35 miles an hour!

Hotels and apartment buildings commonly have floor numbers that skip from 12 to 14, owing mainly to the possible superstition of potential residents. Even room numbers ending in 13 are often omitted.

1. The points on a line can be numbered so that positive number differences measure distances.

2. A point is between two other points on the same line iff its coordinate is between their coordinates.

3. If A-B-C, then AB + BC = AC.

•4. The definition of betweenness of points.

5. The Betweenness of Points Theorem.

Triple Jump.

6. 18 meters. (58 − 40 = 18.)

•7. 52. (47 + 5 = 52.)

8. 6 meters. (58 − 52 = 6.)

•9. BC + CD = BD, or 5 + 6 = 11.

10. AB + BD = AD, or 7 + 11 = 18.

11. AC + CD = AD, or 12 + 6 = 18.

Eclipse Problems.

12. Collinear.

13. Figure B.

14. 93,240,000 mi. (SE + EM = SM, 93,000,000 + 240,000 = 93,240,000.)

•15. 92,760,000 mi. (SM + ME = SE, SM + 240,000 = 93,000,000, SM = 92,760,000.)

Seven League Boots.

16. B, 11; C, 18; D, 25; E, 32; F, 39.

•17. 18.

18. 39.

19. How far (in leagues) the person is from home.

Skyscraper.

•20. 9 − 2 = 7.

21. 15 − 5 = 10.

22. No. The second answer is wrong because the ruler omits 13. (The skyscraper does not have a "13th" floor.)

Football Field.

23. A, 25; B, 35.

24. AB = 35 − 25 = 10.

25. The lines are not numbered like a ruler because to most numbers there correspond two points, not one.

•26. A, −25; B, 15.

27. AB = 15 − (−25) = 40.

28. Yes.

SAT Problem.

29.

•30. B.

31. 6 units. [5 − (−1) = 6.]

Set II (pages 88–90)

When we observe stars at night, any three stars that seem to be collinear almost certainly are not. Such stars in space generally determine the vertices of a triangle that we happen to be seeing "on edge."

At first glance, the arrangement of the numbers on the yardstick in exercise 48 does not seem especially remarkable. The story behind them, however, is rather intriguing. Martin Gardner in his book titled *The Magic Numbers of Dr. Matrix* (Prometheus Books, 1985) noted: "Henry E. Dudeney, in problem 180, *Modern Puzzles* (1926), asked for the placing of eight marks on a 33-inch ruler and gave 16 solutions. Dudeney undoubtedly believed that at least nine marks were required for any ruler longer than 33 inches." Gardner reported that the yardstick problem of exercise 46 was solved by John Leech in 1956 and first published in a paper titled "On the Representation of 1, 2, . . . , *n* Differences" in the *Journal of the London Mathematical Society*. It has since been proved that Leech's solution for marking the yardstick with only 8 marks is the only one possible.

Pole Vaulter.

32. A point is between two other points on the same line iff its coordinate is between their coordinates.

33. The pole is bent; so the points are not collinear.

34. If A-B-C, then AB + BC = AC.

•35. AB + BC is greater than AC.

36. No.

Ladder Problem 1.

37. (*Student answer.*) ("Yes" seems reasonable.)

•38. AC = AB + BC.

39. BD = BC + CD.

40. Substitution.

41. Subtraction.

Ladder Problem 2.

42. B, 23; C, 35; D, 47; E, 59; F, 71; G, 83; H, 95; I, 107; Y, 118.

43. 107 − 11 = 96 and 8 × 12 = 96.

44. $2 \times 11 + 8 \times 12 = 22 + 96 = 118$.

Stars Proof.

45. Proof.
    Suppose that star Y is between star X and star Z.
    If X-Y-Z, then XY + YZ = XZ.
    If XY + YZ = XZ, then 7.2 + 9.8 = 16.6.
    This contradicts the fact that 17.0 ≠ 16.6.
    Therefore, our assumption is false and
    *   star Y is not between star X and star Z.*

Signal Problem.

46. If you don't change your speed, you will
    travel $3 \times 44 = 132$ feet in 3 seconds. Because
    A-B-C, AB + BC = AC; so $x + 60 = 132$ and
    $x = 72$. You can be 72 feet from the intersec-
    tion.

47.
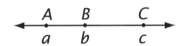

    Proof.
    (1) A-B-C. (The hypothesis.)
    (2) $a > b > c$. (Definition of betweenness of
        points and this case.)
    (3) $AB = a - b$ and $BC = b - c$. (Ruler
        Postulate.)
    (4) $AB + BC = (a - b) + (b - c) = a - c$.
        (Addition.)
    (5) $AC = a - c$. (Ruler Postulate.)
    (6) $AB + BC = AC$. (Substitution.)

Yardstick Distances.

48.

1	AB, IJ	19	BF
2	BC	20	AF
3	AC, CD	21	DG
4	GH, HI	22	EI
5	BD, HJ	23	EJ
6	AD	24	CG
7	DE, EF, FG	25	DH
8	GI	26	BG
9	GJ	27	AG
10	CE	28	CH
11	FH	29	DI
12	BE	30	BH, DJ
13	AE	31	AH
14	DF, EG	32	CI
15	FI	33	CJ
16	FJ	34	BI
17	CF	35	AI, BJ
18	EH	36	AJ

Set III (page 90)

It is always fun to see good mathematics
problems posed and solved in unexpected places.
The oar problem is adapted from an example in
the entry on "Oars" in John Lord's fascinating
book titled *Sizes—The Illustrated Encyclopedia*
(HarperCollins, 1995).

1. Half of CB is 21 inches and half of the
   overlap is 2 inches; so AB and CD should
   each be $21 + 2 = 23$ inches.

2. Because $AB = 7x = 23$, $x = \dfrac{23}{7}$. Because
   $AF = 25x$, $AF = 25(\dfrac{23}{7}) \approx 82$ inches. (Notice
   that the length of the paddle is not needed.)

3. Because 82 inches ≈ 6.8 feet, you should
   probably order 7-foot oars.

Chapter 3, Lesson 3

Set I (pages 93–95)

More examples of rotations in sports include
somersaults in gymnastics and spins in figure
skating. Single, double, and triple axels in skating
are named for their inventor, Norwegian skater
Axel Paulsen.

1. The rays in a half-rotation can be numbered
   from 0 to 180 so that positive number
   differences measure angles.

2. A ray is between two others in the same
   half-rotation iff its coordinate is between
   their coordinates.

3. If OA-OB-OC, then
   $\angle AOB + \angle BOC = \angle AOC$.

4. Ray HE.

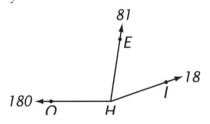

5. $\angle OHE = 99°$, $\angle EHI = 63°$, and $\angle OHI = 162°$.

•6. The definition of betweenness of rays.

7. The Protractor Postulate.

8. Rays.

9. Its sides.

10. ∠A, ∠CAB, ∠PAC.

Danger Area.

•11. The coordinates.

12. 75°.

•13. Acute.

14. One revolution. (The player jumps, spins around once, and releases the ball before landing.)

15. The diver turns one and a half rotations, or 540°, during the dive.

•16. 15. (One each at A, E, and G, and three each at B, C, D, and F.)

17. All 15 of them.

18. ∠CBG, ∠CDE, ∠CFE, ∠CFG.

19. ∠ABG, ∠BCF, ∠DCF, ∠ADE, ∠E, ∠G.

20. No. We do not know from what perspective the photograph was taken.

Bubble Angles.

21. Right.

•22. Obtuse.

23. 90°. ($\frac{360}{4}$ = 90.)

24. 120°. ($\frac{360}{3}$ = 120.)

Cactus Spokes.

•25. 14.4°. (There are 25 spokes and hence 25 such angles; $\frac{360}{25}$ = 14.4.)

•26. Yes.

27. Its coordinate is between their coordinates. (CS-CP-CK because 43.2 < 57.6 < 86.4.)

28. 14.4°. (57.6 – 43.2 = 14.4.)

•29. 28.8°. (86.4 – 57.6 = 28.8.)

30. 43.2°. (86.4 – 43.2 = 43.2.)

•31. ∠SCP + ∠PCK = ∠SCK.

32. The Betweenness of Rays Theorem.

33. If OA-OB-OC, then ∠AOB + ∠BOC = ∠AOC.

Set II (pages 95–97)

Runway Numbers.

34. 18.

•35. 27.

36. They are $\frac{1}{10}$th of the numbers on the protractor.

37. 23. (50° + 180° = 230°.)

•38. They are the same.

39. 110°. (360° – 250°.)

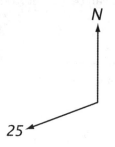

40. The left runway and the right runway.

Pool Ball Angles.

•41. Betweenness of Rays Theorem. (If OA-OB-OC, then ∠AOB + ∠BOC = ∠AOC.)

42. Substitution.

43. Subtraction.

NBC Peacock.

•44. 20°. ($\frac{180}{9}$ = 20.)

45. 18°. ($\frac{180}{10}$ = 18.)

46. 15°. ($\frac{180}{12}$ = 15.)

47. 16.363636 . . .° (or $16\frac{4}{11}$°). ($\frac{180}{11} = 16\frac{4}{11}$.)

Minutes and Seconds.

48. An hour.

49. 3,600. (1° = 60' = 60 × 60" = 3,600".)

•50. 360.

•51. 15. ($\frac{360°}{24 \text{ hours}}$.)

52. 15.

53. 15.

54. *Proof.*

    (1) OA-OB-OC. (The hypothesis.)

    (2) $a < b < c$. (Definition of betweenness of rays and this case.)

    (3) $\angle AOB = b - a$ and $\angle BOC = c - b$. (Protractor Postulate.)

    (4) $\angle AOB + \angle BOC = (b - a) + (c - b) = c - a$. (Addition.)

    (5) $\angle AOC = c - a$. (Protractor Postulate.)

    (6) $\angle AOB + \angle BOC = \angle AOC$. (Substitution.)

Set III (page 97)

This exercise is another example of the need for deductive reasoning and proof. No matter how accurately the figure may be drawn, measurements of $\angle EFB$ can neither give its exact value nor reveal why it should be 20°. J. L. Heilbron calls it "The Tantalus problem" after Tantalus, the character in Greek mythology from which the word "tantalize" is derived. The name is fitting because the prospective solver of the problem is "tormented with something desired but out of reach."

A nice discussion of the problem and a similar but easier one are included on pages 292–295 of Heilbron's *Geometry Civilized* (Clarendon Press, 1998). More interesting information about the Tantalus problem is included in *Tracking the Automatic Ant,* by David Gale (Springer, 1998). See pages 123–125 on "Configurations with Rational Angles" and the research done on the problem by Armando Machado on pages 227–232.

1. (*Student drawing of figure.*)

2. $\angle AGB = 50°$, $\angle AGF = 130°$, $\angle BGF = 130°$, $\angle C = 20°$.

3. On an accurate drawing, it should appear to be about 20°.

Chapter 3, Lesson 4

Set I (pages 100–101)

The arrow figure in exercises 12 through 14 is a variation of the "Muller-Lyer illusion." It was discovered in 1899 by C. H. Judd, who mentioned it in an article on geometrical illusions published in *Psychological Review.* Other variations of this illusion are included in *Can You Believe Your Eyes?* by J. R. Block and Harold E. Yuker (Brunner/Mazel, 1992).

Origami is the subject of one of Martin Gardner's early "Mathematical Games" columns in *Scientific American* (July 1959), reprinted in *The 2nd Scientific American Book of Mathematical Puzzles and Diversions* (Simon & Schuster, 1961). As a child, Peter Engel was inspired by this column and has written a wonderful book on the subject, *Folding the Universe* (Vintage Books, 1989). Beyond including detailed procedures for folding a large variety of figures from simple to extremely complex, Engel relates the history of origami and its connections with many other fields. For example, in a section of the book titled "The Psychology of Invention," Engel includes a summary of French mathematician Jacques Hadamard's remarks on the subject of invention in mathematics. *Folding the Universe* is without doubt the definitive book on origami for the mathematics classroom.

1. Points that are on the same line.

2. Lines that contain the same point.

•3. Figures that can be made to coincide (or fit exactly together).

4. A drawing made with a straightedge and compass.

5. A statement formed by interchanging the hypothesis and conclusion of another statement.

6. A theorem that is easily proved as a consequence of another statement.

•7. Then it divides it into two equal segments.

8. If a point divides a line segment into two equal segments, then it is the midpoint of it.

•9. Each is the converse of the other.

10. A line segment has exactly one midpoint.

11. An angle has exactly one ray that bisects it.

Arrow Illusion.

12. (Most people would say M.)

13. No, because a line segment has exactly one midpoint.

14. AN = NB.

Origami Duck.

•15. No.

16. Yes.

17. ∠BAD.

18. ∠BCD, ∠ACD, and ∠BCA.

•19. 45°.

20. 22.5°.

21. 45°.

•22. 67.5°.

23. 67.5° (because in triangle DFC, ∠D = 90° and ∠FCD = 22.5°).

Bisection Constructions.

24.

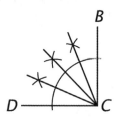

25.

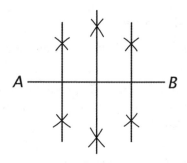

Clock Puzzle.

•26. The midpoint of a line segment divides it into two equal segments.

27. 1.5.

•28. 1.5.

29. D, 4.5; E, 6; F, 7.5.

30. 7.5 seconds.

Set II (pages 102–104)

Students taking chemistry might enjoy identifying the atoms in the acetylene molecule: two carbon atoms (points B and C in the figure) and two hydrogen atoms (points A and D). According to Linus Pauling, the H–C bond length is 1.1 angstroms and the C–C bond length is 1.2 angstroms.

Euclid defined an angle as "the inclination to one another of two lines in a plane which meet one another and do not lie in a straight line"; so, for Euclid, straight angles are not considered.

David W. Henderson, in his valuable book titled *Experiencing Geometry* (Prentice Hall, 2001), offers some important insights with regard to the question, "What is an angle?" He writes:

"I looked in all the plane geometry books in the university library and found their definitions for 'angle.' I found nine different definitions! Each expressed a different meaning or aspect of 'angle' . . . .

There are at least three different perspectives from which one can define 'angle,' as follows:

a *dynamic* notion of angle—angle as *movement*;

angle as *measure*; and,

angle as *geometric shape* . . . .

Each of these perspectives carries with it methods for checking angle congruency. You can check the congruency of two dynamic angles by verifying that the actions involved in creating or replicating them are the same. If you feel that an angle is a measure, then you must verify that both angles have the same measure. If you describe angles as geometric shapes, then one angle should be made to coincide with the other using isometries in order to prove angle congruence."

Yen Measurement.

•31. 2 cm.

•32. 2π (or about 6.28). (c = πd = 2π.)

33. π (or about 3.14).

Acetylene Proofs.

34. Addition.
    Betweenness of Points Theorem.
    Substitution.

35. Suppose that B is the midpoint of AC.
    The midpoint of a line segment divides it into two equal segments.
    Substitution.
    The fact that AC > 2AB.
    B is not the midpoint of AC.

Miter Joint Proof.

36. (1) PA-PC-PB. (By hypothesis.)
   (2) ∠1 + ∠2 = ∠APB. (Betweenness of Rays Theorem.)
   (3) PB bisects ∠APB. (By hypothesis.)
   (4) ∠1 = ∠2. (If a line bisects an angle, it divides the angle into two equal angles.)
   (5) ∠1 + ∠1 = ∠APB; so 2∠1 = ∠APB. (Substitution.)
   (6) $\angle 1 = \dfrac{\angle APB}{2}$. (Division.)

Angle Definition.

37. No. The second figure looks like a line.

38. Unless it is marked on the angle, we don't know where it is.

39. It is hard to tell because we don't know on which side of the line to draw them.

40. It seems as if a straight angle has two bisectors, one on either side of the line.

Passion Flower.

41.

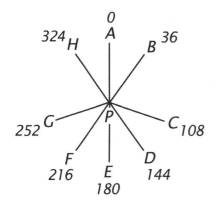

•42. 180.

43. 72°. ($\dfrac{360}{5} = 72$.)

•44. They are each 36°. ($\dfrac{72}{2} = 36$.)

45. PB, 36; PC, 108; PD, 144; PF, 216; PG, 252; PH, 324.

46. No, because the rays do not lie in a "half-rotation."

47. Yes, because it divides ∠GPC into two equal angles.

In the chapter on Mascheroni constructions in *Mathematical Circus* (Knopf, 1979), Martin Gardner wrote:

"It is often said that the ancient Greek geometers, following a tradition allegedly started by Plato, constructed all plane figures with a compass and a straightedge (an unmarked ruler). This is not true. The Greeks used many other geometric instruments, including devices that trisected angles. They did believe, however, that compass-and-straightedge constructions were more elegant than those done with other instruments. . . .

In later centuries geometers amused themselves by imposing even more severe restrictions used in construction problems. The first systematic effort of this kind is a work ascribed to the 10th-century Persian mathematician Abul Wefa, in which he described constructions possible with the straightedge and a 'fixed compass,' later dubbed the 'rusty compass.'"

That it is the 120° angle that can be most easily bisected with a rusty compass will be confirmed in a later lesson.

Rusty Compass Constructions.

1.

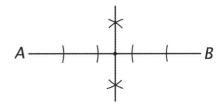

2.

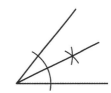

3.

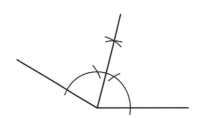

4. An angle of 120°.

Set I (pages 107–108)

• 1. 90°.

2. 180°.

3. A *compliment*.

Safe Angle.

• 4. 87°.

5. They are complementary.

6. 177°.

7. They are supplementary.

Hour Angle Measure.

• 8. 15°. ($\frac{360}{24} = 15$.)

• 9. 165°.   (180 − 15 = 165.)

10. 75°.

11. The earth turns through one "hour angle" in one hour.

12.

13. 15°. ($\frac{90}{6} = 15$.)

14. They are multiples of 15°.

Skier Forces.

15.

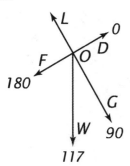

• 16. ∠DOW + ∠WOF = ∠DOF.

17. ∠DOW = 117 − 0 = 117°;
∠WOF = 180 − 117 = 63°;
∠DOF = 180 − 0 = 180°.

18. They are supplementary.

• 19. 27°. (∠WOG = 117 − 90 = 27°.)

20. They are complementary.

21. ∠WOG, ∠WOF.

22. ∠LOW, ∠DOW.

Triangle in the Woods.

23. The word supplementary is defined in reference to two angles, not three.

24. No. If two of its angles were supplementary, their sum would be 180°; there would be no degrees left for the third angle.

25. Yes. Two angles could have a sum of 90°; the third angle would consequently be a right angle.

Hammer Throw.

• 26. Betweenness of Rays Theorem.

27. A right angle is equal to 90°.

• 28. Substitution.

29. If a line bisects an angle, it divides the angle into two equal angles.

30. Substitution.

31. Division.

Set II (pages 108–109)

The Arctic Circle problems are based on information in Ernest Zebrowski's *History of the Circle—Mathematical Reasoning and the Physical Universe* (Rutgers University Press, 1999). Zebrowski writes:

"Shown on every globe and drawn (in projection) on every world map, the Arctic Circle connects points around Earth at 66.54° north latitude, and the Antarctic Circle is at 66.54° south latitude. But why would anyone attach special significance to such an unusual number? Clearly, it must not be the number itself that is important, but rather some geographical concept that gives rise to this numerical value. In fact, the latitude of the Arctic Circle is but an artifact of a geometrical relationship, and it is this underlying geometry that is significant.

We've already seen that the Tropic of Capricorn is a circle of latitude where the sun is directly overhead at noon around the date of December 22, and that this happens at a latitude of 23.46° south of the equator. On that day of the year, the sun's path is quite low in the sky in the Northern Hemisphere, and the farther north we go, the lower the path of the sun. At points 90° north of the Tropic of Capricorn, we reach a circle of latitude where the sun barely peeks over the horizon at noon at the winter solstice. This is the Arctic Circle. North of this circle, the sun does not rise at all during the first day of winter. The Arctic Circle has a latitude of 23.46° south latitude, and Earth's spherical geometry requires that there be a 90° angle between the two.

Meanwhile . . . we find a circle of latitude around the South Pole that basks in sunlight all twenty-four hours of the day when locations within the Arctic Circle are in twenty-four hour darkness. Six months later, of course, the roles of the Arctic and Antarctic Circles are reversed."

Clock Hands.

32. 8:30.

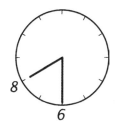

33. 60°.

34. If the big hand were on the 6, the small hand would be halfway from 8 to 9.

•35. 75°.

Complement and Supplement.

36.

∠A	∠C	∠S
10°	80°	170°
20°	70°	160°
30°	60°	150°

•37. ∠S seems to be 90° larger than ∠C.

38. $\angle C = (90 - x)°$; $\angle S = (180 - x)°$.

39. $\angle S - \angle C = (180 - x)° - (90 - x)° = 90°$, or $\angle S = \angle C + 90°$.

Earth Angles.

40. They are complementary.

41. On the horizon.

•42. They are supplementary.

43. The sun never rises.

44. OB bisects ∠AOC. ∠1 = ∠2 because complements of the same angle are equal.

45. Yes. $\angle 1 + \angle AOC = 90°$ and $\angle AOC = \angle 1 + \angle 2$; so $\angle 1 + \angle 1 + \angle 2 = 90°$. Also $\angle 1 = \angle 2$; so $3\angle 1 = 90°$ and so $\angle 1 = 30°$. Therefore $\angle 2 = 30°$ and $\angle AOC = 60°$.

46. *Proof.*
   (1) ∠1 and ∠2 are supplements of ∠3. (Given.)
   (2) $\angle 1 + \angle 3 = 180°$; $\angle 2 + \angle 3 = 180°$. (If two angles are supplementary, their sum is 180°.)
   (3) $\angle 1 + \angle 3 = \angle 2 + \angle 3$. (Substitution.)
   (4) $\angle 1 = \angle 2$. (Subtraction.)

Set III (page 109)

The angles of the squares are each 90°. Three squares and three star points surround a point just below the center of the figure.

Because the sum of the three angles of the squares is 270°, the sum of the three star points is $360° - 270° = 90°$. The angles at the points of the stars are evidently each 30°.

Chapter 3, Lesson 6

Set I (pages 112–114)

The perception example of exercises 3 through 6 is derived from the work of computer scientist A. Guzmán as reported in *Human Information Processing—An Introduction to Psychology*, by Peter H. Lindsay and Donald A. Norman (Harcourt Brace Jovanovich, 1977). Guzmán analyzed pictures of overlapping figures and discovered that, wherever three lines meet to form a linear pair in a figure, it usually indicates that one object is in front of another. This rule is just one of a number of geometric rules that we automatically use in interpreting line drawings.

A rule of thumb followed by civil engineers in applying linear pairs to the construction of road intersections is that the smaller of the two angles in which two streets meet at a "T intersection" should not be less than 80°.

•1. Because they are vertical angles.

2. Because they form linear pairs.

3. As a four-sided figure in front of a triangle.

4. A, B, D, E; C, B, I, H; D, F, G; I, F, E.

5. ∠CBA and ∠ABI (ABH);
   ∠HIA and ∠AIB (AIC); ∠DFE and ∠EFG.

6. No.

•7. They are a linear pair and they are supplementary.

8. 125°.

9. $(180 - x)°$.

10. ∠1.

Sun Directions.

11.

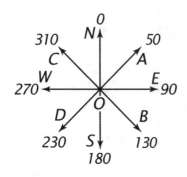

•12. 80°. $(130 - 50 = 80.)$

13. 80°.

•14. They are vertical angles.

15. 100°.

•16. No.

17. If two angles are equal, then they are vertical angles.

18. No.

19. No.

20. Yes.

Ship Position.

21. Concurrent.

22. (*Figure similar to the one below. Measurements of angles will vary.*)

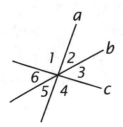

•23. Two.

24. 180°.

25. Yes. Because ∠1 + ∠2 + ∠3 = 180° and ∠5 = ∠2, ∠1 + ∠3 + ∠5 = 180° by substitution.

•26. 180°.

27. The sum of the angles of a triangle is 180°.

28. 180°.

29. These angles are vertical angles with the angles of the triangle and so equal to them.

•30. These angles are linear pairs; so they are supplementary.

31. Addition.

32. 360°.

33. Because ∠2 + ∠3 + ∠5 + ∠6 + ∠8 + ∠9 = 540° and ∠3 + ∠6 + ∠9 = 180°,
    ∠2 + ∠5 + ∠8 = 360° by subtraction.

Set II (pages 114–115)

The bending of light in a gravitational field certainly complicates our ability to perceive the physical universe. In the part titled "The Rise of Geometrical Analysis" in their book *Mathematics in Civilization* (Dover, 1984), H. L. Resnikoff and R. O. Wells, Jr., write:

"All terrestrial experiments appear to show that light traveling in a vacuum follows a straight-line path, and it was natural to assume that this would be true of light passing through any region in which there is a gravitational force . . . .

The fundamental idea underlying Einstein's model, that the force of gravity acts by creating curvature in space . . . suggests that the paths of light-rays should be deformed from straight lines wherever the geodesics due to gravity are not straight. In particular, this model suggests that the path of a light particle should bend toward the more curved regions (in physical terms, toward the *sources* of gravitational force). For a light-ray just grazing the limb of the sun the Einstein model predicts that the direction of the ray will be bent toward the sun by an angle of 1.75". This prediction has been verified . . . ."

Steep Slope.

34.

∠1	∠2	∠3	∠4
25°	65°	115°	65°
33°	57°	123°	57°
48°	42°	138°	42°

35. ∠4 = ∠2.

•36. Vertical angles are equal.

•37. ∠3 = ∠1 + 90°, or ∠3 − ∠1 = 90°.

38. ∠2 = $(90 − x)°$;
∠3 = $[180 − (90 − x)]° = (90 + x)°$;
∠4 = $(90 − x)°$.

39. $(90 + x)° − x° = 90°$.

Hedge Shears.

40. ∠ESR and ∠HSA.

41. SA is always between SE and SR;
SR is always between SA and SH.

42. ∠ESA and ∠HSR.

43. *Proof.*
(1) ∠ESR = ∠HSA. (Given.)
(2) ∠ESA + ∠ASR = ∠ESR;
∠ASR + ∠HSR = ∠HSA. (Betweenness of Rays Theorem.)
(3) ∠ESA + ∠ASR = ∠ASR + ∠HSR. (Substitution.)
(4) ∠ESA = ∠HSR. (Subtraction.)

Bent Light Ray.

44. There are 60 minutes in 1 degree and 60 seconds in 1 minute.

•45. 179° 59' 58.25".

Pinhole Camera.

46. 45°. (∠1 = ∠2 and ∠1 + ∠2 = 90°.)

47. *Proof.*
(1) ∠1 and ∠2 are vertical angles. (Given.)
(2) ∠1 = ∠2. (Vertical angles are equal.)
(3) ∠1 and ∠2 are complementary. (Given.)
(4) ∠1 + ∠2 = 90°. (If two angles are complementary, their sum is 90°.)
(5) ∠1 + ∠1 = 90° so 2∠1 = 90°. (Substitution.)
(6) ∠1 = 45°. (Division.)

SAT Problem.

48. The angles labeled $(5z)°$ and $(6z + 4)°$ form a linear pair; so $(5z) + (6z + 4) = 180$, $11z = 176$, and $z = 16$. The angles labeled $(5z)°$ and $(8w)°$ also form a linear pair; so $5z + 8w = 180$. Because $z = 16$, $5(16) + 8w = 180$, $8w = 100$, and so $w = 12.5$.

Set III (page 116)

The Chinese proof of the vertical angle theorem is identical with that given by Euclid (Book I, Proposition 15).

Euclid's equivalent of our linear pair theorem is his Proposition 13, which states: "If a straight line set up on a straight line makes angles, it will make either two right angles or angles equal to two right angles." Hence the appearance of "2∠R" in the Chinese proof rather than reference to either "supplements" or "180°".

1. AB and CD intersect at O.

2. ∠AOD and ∠BOC are vertical angles, ∠AOC and ∠BOD are vertical angles.

3. To prove.

4. Proof.

5. The measures of two right angles, or 180°.

6. The angles in a linear pair are supplementary.

7. Substitution.

8. Subtraction.

Chapter 3, Lesson 7

Set I (pages 119–120)

It is interesting to note that the scuba diver figure comes from a section titled "Geometrical Optics" in a physics textbook (Maschuri L. Warren's *Introductory Physics*, W. H. Freeman and Company, 1979). The author points out that Isaac Newton used five postulates about the geometry of optics as well as geometric constructions in his analysis of the behavior of light. Newton's book, *Opticks*, was published in 1704 and was immediately accepted as the definitive description of light. Newton modeled *Opticks*, as well as his most famous work, the *Principia*, after Euclid's *Elements* by starting with definitions and postulates and then proceeding to develop the subject matter through a logical sequence of theorems.

In working the exercises on the Necker cube, your students may discover it "flipping back and forth" and consequently be ready for the ambiguity of exercise 16. Donald D. Hoffman includes a long discussion of this illusion in *Visual Intelligence— How We Create What We See* (Norton, 1998). Hoffman remarks that, although the picture "feels flat," it "looks like a cube." The viewer "constructs the cube" by following certain "rules" critical to vision. He says:

"No one teaches you these rules. Instead, you acquire them early in life in a genetically predetermined sequence that requires, for its unfolding, visual experience."

The difficulty with the Necker cube is that these rules produce two conflicting but equally coherent interpretations of it as a figure in three dimensions.

•1. Two angles are complementary if their sum is 90°.

2. Two lines are perpendicular if they form a right angle.

3. All right angles are equal.

•4. Supplements of the same angle are equal.

5. A line bisects an angle if it divides it into two equal angles.

•6. If the angles in a linear pair are equal, then their sides are perpendicular.

7. Two lines are parallel iff they lie in the same plane and do not intersect.

8. They may not lie in the same plane. They may lie in lines that intersect.

9. They are concurrent.

•10. Vertical angles are equal.

11. The angles in a linear pair are supplementary.

12. Five.

Cube Illusion.

13. DC, HG, and EF.

•14. DH, AE, and, if perspective is taken into account, HG, EF.

15. (There are four: DA, HE, AB, and EF.)

16. Either one, depending on how you look at the cube.

Mud Cracks.

•17. They are a linear pair.

18. They are supplementary.

19. 90°.

20. A right angle.

•21. They are perpendicular.

22. If the angles in a linear pair are equal, then their sides are perpendicular.

Set II (pages 120–121)

The waterfall exercise comes, of course, from the fact that, if a line is perpendicular to a plane, it is perpendicular to every line in the plane that passes through the point of intersection. This definition will be considered formally in the chapter on geometric solids.

A Deceptive Figure.

23. Yes. AO ⊥ OD because ∠1 is a right angle.

•24. ∠1 and ∠3 look like vertical angles, yet they are not equal.

25. ∠1 and ∠4 look like a linear pair as do ∠1 and ∠2, yet they are not supplementary.

26. Except for AO and OD, the lines that look perpendicular do not form right angles.

27. No. If they were vertical angles, the figure would contain two pairs of opposite rays and all four angles would be right angles.

Bisecting a Linear Pair.

28. 130°.

29.

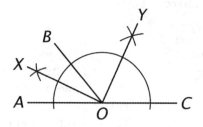

•30. ∠BOX, 25°; ∠BOY, 65°.

31. 90°.

32. The angles in a linear pair are supplementary.

•33. 180°.

34. 90°.

35. A right angle.

36. Because it has a measure of 90°.

•37. They are perpendicular.

38. Because they form a right angle.

39. $\angle 1 + \angle 2 + \angle 3 + \angle 4 = 360°$ so
$90 + n + 2n + 3n = 360$, $6n = 270$, and $n = 45$.

40. $\angle 1 = 90°$, $\angle 2 = 45°$, $\angle 3 = 90°$,
$\angle 4 = 135°$.

41. They are equal.

42. They are supplementary.

43. *Proof.*
(1) WA ⊥ AB and WA ⊥ AC. (Given.)
(2) ∠WAB and ∠WAC are right angles.
(Perpendicular lines form right angles.)
(3) ∠WAB = ∠WAC. (All right angles are equal.)

44. *Proof.*
(1) ∠1 and ∠2 are complementary. (Given.)
(2) $\angle 1 + \angle 2 = 90°$. (The sum of two complementary angles is 90°.)
(3) PB-PC-PD. (Given.)
(4) ∠BPD = ∠1 + ∠2. (Betweenness of Rays Theorem.)
(5) $\angle BPD = 90°$. (Substitution.)
(6) ∠BPD is a right angle. (A 90° angle is a right angle.)
(7) AB ⊥ PD. (Two lines that form a right angle are perpendicular.)

Set III (page 122)

The FD Puzzle was sold by the Perry and Elliott Company of Massachusetts and is a variation of the well-known "T-Puzzle," which will be explored in the lesson on trapezoids.

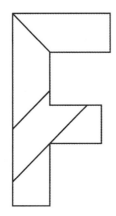

Chapter 3, Review

Set I (pages 124–125)

•1. Yes. E-F-H because $6.1 < 7.7 < 10.8$.

2. Point D is closer because $CB = 1.7$ and $CD = 1.6$.

3. No. $AE = 6.1$ and $EI = 5.9$.

Sliding Bevel.

4. They are always supplementary.

•5. $(180 - x)°$.

6. ∠BEV is obtuse and ∠VEL is acute.

7. 90°.

•8. The handle is perpendicular to the blade.

9. If the angles in a linear pair are equal, their sides are perpendicular.

Old Quadrant.

10. 47°. ($65 - 18 = 47$.)

•11. CE-CS-CD.

12. 23.5°.

•13. 41.5.

14. $65 - 41.5 = 23.5$.

Ollie's Logic.

15. Yes. It is true that either two lines intersect or they do not.

16. If two lines intersect, they must form a right angle to be perpendicular. If two lines do not intersect, they must lie in the same plane to be parallel.

SAT Problem.

17.

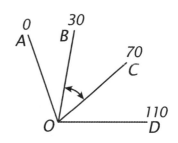

•18. 110.

19. 30.

20. 40°.

Shoe Laces.

21. They are vertical angles and they are equal.

22. 45°.

•23. They would be perpendicular.

24. They would be equal.

25. Supplements of the same angle are equal.

Axioms and Corollaries.

26. An axiom is like a postulate. It is a statement that is assumed to be true without proof.

27. A corollary is a theorem that can be easily proved as a consequence of another statement that is already accepted.

•28. Any number is equal to itself.

29. A line segment has *exactly one midpoint.*

30. An angle has exactly one ray *that bisects it.*

31. All right angles *are equal.*

Set II (pages 126–128)

The chair-design exercises are from the field of biotechnology. According to *Human Factors in Engineering and Design,* by Mark S. Sanders and Ernest J. McCormick (McGraw-Hill, 1992),

"The field of human factors—referred to as *ergonomics* in Europe and elsewhere—deals with the consideration of human characteristics, expectations, and behaviors in the design of the things people use in their work and everyday lives and of the environments in which they work and live."

Your students may be surprised by the extensive use of mathematics, including geometry and statistics, in this field.

Clock Problems.

32. 150° at 5 o'clock; 60° at 10 o'clock.

33.

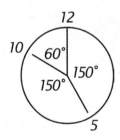

$60° = 360° - 2(150°).$

Stilt Distances.

•34. $(6 + x)$ ft.

35. x ft.

36. *Proof.*
(1) S-F-T and F-T-H. (Given.)
(2) SF + FT = ST and FT + TH = FH. (Betweenness of Points Theorem.)
(3) ST = FH. (Given.)
(4) SF + FT = FT + TH. (Substitution.)
(5) SF = TH. (Subtraction.)

Chair Design.

37. $\angle 2 = (90 - x)°$.

•38. $\angle AOB = (90 - x + y)°$.

39. *Proof.*
Suppose that ∠AOB is a right angle.
∠AOB = 90°. *(A right angle equals 90°.)*
$90 - x + y = 90$. *(Substitution.)*
$-x + y = 0$. *(Subtraction.)*
$y = x$. *(Addition.)*
This contradicts the fact that if x is between 5 and 8 degrees and y is between 20 and 25 degrees, $y \neq x$.
Therefore, our assumption is false and *∠AOB cannot be a right angle.*

40. No. [∠AOB = $(90 - x + y)°$. To find the smallest possible value of ∠AOB, let $x = 8$ and $y = 20$. ∠AOB = $90 - 8 + 20 = 102°$. ∠AOB is at least 102°.]

Another SAT Problem.

41.

B · 6 · D · 4 · A · 2 · C

42. 4 units.

Graphite and Diamond.

•43. Hexagons.

44. 720°.
$(\angle 1 = \dfrac{360°}{3} = 120°; 6\angle 1 = 6(120°) = 720°.)$

45. *Proof.*
(1) $\angle 1 + \angle 2 + \angle 3 = 360°$ and all of the angles in the figure are equal. (Given.)
(2) $\angle 1 + \angle 1 + \angle 1 = 360°$; so $3\angle 1 = 360°$. (Substitution.)
(3) $6\angle 1 = 720°$. (Multiplication.)

•46. Six.

47. ∠AOB, ∠BOC, ∠COD, ∠AOD, ∠AOC, ∠BOD.

48. *Proof.*
Suppose that ∠AOB and ∠BOC are a linear pair.
∠AOB and ∠BOC are supplementary. *(The angles in a linear pair are supplementary.)*
∠AOB + ∠BOC = 180°. *(The sum of two supplementary angles is 180°.)*
This contradicts the fact that
∠AOB + ∠BOC = 109.5° + 109.5° = 219°.
Therefore, our assumption is false and *∠AOB and ∠BOC cannot be a linear pair.*

49. No. Any three points are always coplanar. (If they are collinear, they lie in many different planes. If they are noncollinear, they determine a plane.)

March Times.

•50. 150.

51. 3,000. ($100 \cdot 30 = 3,000$.)

52. $30n$.

•53. 120 steps per minute.

54. 3,600 inches per minute. ($120 \cdot 30 = 3,600$.)

55. Approximately 3.4 miles per hour. (3,600 inches per minute = 300 feet per minute = 18,000 feet per hour $\approx$ 3.4 miles per hour.)

•56. 6,480 inches per minute. ($180 \cdot 36 = 6,480$.)

57. No. Twice 3,600 is 7,200, not 6,480.

Chapter 3, Algebra Review (page 130)

•1. $6x + 1$.

•2. $2x - 4y$.

•3. $4x^2 + 8$.

•4. $3x^2 + 6x - 5$.

•5. $12x - 3$.

•6. $x^3 + 4x^2 - x - 8$.

•7. $14x - 5y$.

•8. $7x + 4$.

•9. $6x$.

•10. $-x^2 - 11x$.

•11. $2x^2 + 10x - 1$.

•12. $2x^3 + y^3$.

•13. $x^2 + 14x + 24$.

•14. $6x^2 - 7x - 20$.

•15. $16x^2 - 1$.

•16. $x^3 - 10x^2 + 19x + 6$.

•17. $4y^3 - 8y^2 + 13y - 5$.

•18. $x^3 + y^3$.

•19. $5x^3 + 3x^2 - 20x - 12$.

•20. $(x^2 - x - 6)(x + 6) = x^3 + 5x^2 - 12x - 36$.

•21. $10x - 35 - 10x - 14 = -49$.

•22. $(10x - 1)(10x - 1) = 100x^2 - 20x + 1$.

•23. $(x + 4)(x + 4)(x + 4) = (x^2 + 8x + 16)(x + 4) = x^3 + 12x^2 + 48x + 64$.

•24. $x^2 - (x^2 + 4x - 5) = -4x + 5$.

Set I (pages 134–136)

David Wells tells about Nicholas Saunderson in *The Penguin Book of Curious and Interesting Mathematics* (Penguin Books, 1997). Although blind from the age of 12, Saunderson became a noted mathematician and was appointed to the professorship first held by Sir Isaac Newton at Cambridge University. His pin board, now known as the geoboard, is popular even today.

Pin-Board Geometry.

1. The axes.

2. The origin.

•3. A and E.

4. D.

•5. B.

6. G.

•7. A (4, 3), C (–3, –4).

8. D (6, 4), E (4, –2), F (–2, 0), G (0, 6).

•9. On the y-axis.

10. On the x-axis.

Grids.

11. The x-axis.

•12. Two.

13. To remind them to read (list) the x-coordinate before the y-coordinate.

14. (154, 263).

15. Both are divided into sections by parallel and perpendicular lines.

Birds and Bats.

•16. The fourth (IV).

17. The second and third (II and III).

18. The parrots.

•19. The hummingbirds.

20. The storks.

21. The albatrosses.

22. Approximately (–2.5, 0) and (–35, 0).

An Early Graph.

•23. Point A.

24. Their x-coordinates are negative numbers and their y-coordinates are zero. (Also, the x-coordinate of D is less than the x-coordinate of P.)

25. Both are negative. (Also, the x-coordinate of T is equal to the x-coordinate of P.)

Set II (pages 136–138)

Although we can show at this point that the current arrows have equal lengths, we have not yet learned how to prove them parallel. Some of your students may know that such arrows represent a vector, a quantity with magnitude and direction.

The Japanese home problems provide a preview of the translation transformation, a topic to be explored later.

It is amusing to realize that computer users are unknowingly selecting coordinates every time they click the mouse. Without a coordinate system, the mouse couldn't operate as it does.

It will not be long until we can prove the conjecture suggested in exercise 49—that it is sufficient to know that the sides of two triangles are equal in order to know that they are congruent.

Students intrigued by the "distances in different dimensions" problems will be interested in Thomas Banchoff's book titled *Beyond the Third Dimension,* cited in the footnote in the text. Banchoff, Professor of Mathematics at Brown University, is a specialist in the geometry of higher dimensions and has created award-winning films on the subject.

Tidal Current.

26. They seem to be parallel and to be equal in length.

•27. A (4, 1), B (6, 6).
$$AB = \sqrt{(6-4)^2+(6-1)^2} = \sqrt{4+25} = \sqrt{29} \approx 5.39.$$

28. C (–4, –2), D (–2, 3).
$$CD = \sqrt{[-2-(-4)]^2+[3-(-2)]^2} = \sqrt{4+25} = \sqrt{29}.$$

29. E (0, –6), F (2, –1).
$$EF = \sqrt{(2-0)^2+[-1-(-6)]^2} = \sqrt{4+25} = \sqrt{29}.$$

30. Yes (assuming that all of the arithmetic has been done correctly!)

Japanese Floor Plan.

• 31. A_1 (5, 2), A_2 (15, 2).

32. B_1 (7, 6), B_2 (17, 6).

33. The x-coordinates are 10 units larger; the y-coordinates are the same.

34. The floor plan would appear (10 units) above the post pattern.

Computer Screen.

• 35. (900, 200).

36. (100, 800).

• 37. 250 mm. ($AC = \sqrt{(900-100)^2+(800-200)^2} = \sqrt{640,000+360,000} = \sqrt{1,000,000} = $ 1,000 pixels. 1,000 × 0.25 mm = 250 mm.)

38. 250 mm. ($BD = \sqrt{(100-900)^2+(800-200)^2}$, etc. The same distance as AC.)

39. The eBay icon.

Distance Formulas.

40. $\sqrt{(x_2-x_1)^2+(y_2-y_1)^2}$.

• 41. $\sqrt{3^2+3^2} = \sqrt{18}$, or $3\sqrt{2}$.

42. Yes. ($\sqrt{(x_2-x_1)^2} = |x_2-x_1|$), the positive difference of the coordinates.)

43. $\sqrt{(4-1)^2} = \sqrt{3^2} = 3$.

44. $\sqrt{3^2+3^2+3^2} = \sqrt{27}$, or $3\sqrt{3}$.

45. $\sqrt{(x_2-x_1)^2+(y_2-y_1)^2+(z_2-z_1)^2+(w_2-w_1)^2}$.

46. $\sqrt{3^2+3^2+3^2+3^2} = \sqrt{36} = 6$.

Triangle Coordinates.

47. $AB = \sqrt{20^2+15^2} = \sqrt{400+225} = \sqrt{625} = 25$; $BC = 20$; $AC = 15$.

48. $CD = \sqrt{16^2+12^2} = \sqrt{256+144} = \sqrt{400} = 20$; $OC = \sqrt{9^2+12^2} = \sqrt{81+144} = \sqrt{225} = 15$; $OD = 25$.

49. It seems as if the tracing would fit because the sides of the triangles are equal.

Set III (page 138)

A Circle Challenge.

To find out, we need to find the distances of the six points from the center of the circle.

$OA = \sqrt{63^2+16^2} = \sqrt{4,225} = 65$.

$OB = \sqrt{25^2+60^2} = \sqrt{4,225}$.

$OC = \sqrt{(-36)^2+54^2} = \sqrt{4,212} \approx 64.9$.

$OD = \sqrt{(-52)^2+39^2} = \sqrt{4,225}$.

$OE = \sqrt{(-33)^2+(-56)^2} = \sqrt{4,225}$.

$OF = \sqrt{20^2+(-62)^2} = \sqrt{4,244} \approx 65.1$.

The radius of the circle is 65 so point C is inside the circle and point F is outside.

Chapter 4, Lesson 2

Set I (pages 141–143)

Chapter 11 of *Mathematical Gems II*, by Ross Honsberger (M.A.A., 1976) deals with the art-gallery problem. Honsberger reports:

"At a conference in Stanford in August, 1973, Victor Klee asked the gifted young Czech mathematician Václav Chvátal (University of Montreal) whether he had considered a certain problem of guarding the paintings in an art gallery. The way the rooms in museums and galleries snake around with all kinds of alcoves and corners, it is not an easy job to determine the minimum number of guards that are necessary to survey the entire building. The guards are to remain at fixed posts, but they are able to turn around on the spot. The walls of the gallery are assumed to be straight. In a short time, Chvátal had the whole matter figured out. He showed that if the gallery has n walls (i.e., if the floor plan is an n-gon), then, whatever its zig-zag shape, the minimum number of guards needed is never more than $\left[\dfrac{n}{3}\right]$, the integer part of $\dfrac{n}{3}$."

The symmetry of the art gallery in exercises 20 through 24 makes four guards sufficient. (Note that one guard standing at one of the corners of the central room can see all of it as well as one entire wing.)

To realize that the "magic magazine" polygon of exercise 25 can be divided into *any* number of congruent polygons is surprising at first but obvious when you see how!

Connect the Dots.

1. The sides.

•2. The vertices.

3. 32.

•4. A 32-gon.

5. No. The first and last line segments would intersect only one other line segment.

•6. That they are congruent.

7. That they are equal.

Paper-Clip Patent.

•8. They are right angles (and equal).

9. 45°.

10. They are vertical angles.

•11. Line segments AB and EF intersect two others but not at their endpoints.

Origami Duck.

•12. ΔAEG.

13. ΔACB.

14. ΔCGF, ΔCGE, ΔCBE.

15. ΔACF.

16. ΔCEF.

•17. CBEG.

18. DCEF.

19. ABCF.

Art Gallery.

20. Two.

21. 16.

22. A 16-gon.

23. 16.

24. Four (a guard could stand in each corner of the central square.)

Puzzle Figure.

•25. An octagon (an 8-gon).

26. Hexagons (6-gons).

•27. D.

28. B.

29. H.

30. ABHJGD ↔ HBCFIE.

•31. H.

32. C.

33. F.

•34. HBCFIE ↔ MNIEHK.

35. HBCFIE ↔ GLMKHJ.

36. MNIEHK ↔ GLMKHJ.

Set II (pages 143–144)

In *The Penguin Book of Curious and Interesting Geometry* (Penguin Books, 1991), David Wells says of the tessellation pattern of hexagons and stars:

"The regular polygons which tessellate individually can be transformed into tessellations of polygons and stars by moving them slightly apart, and then dividing the space between the tiles. This can be interpreted as a *hinged tessellation*. Each side of a star which is not a side of a hexagon is a strap, hinged at both ends, which holds two hexagons together. As the hexagons separate further, the star becomes fatter, then momentarily appears as a large equilateral triangle, and finally as a hexagon, identical to the original hexagon."

Tessellation Pattern.

37. Both have six sides (or both are hexagons). (Also, both are equilateral, symmetric, etc.)

38. No (as the tessellation shows).

39. Concave.

•40. Convex.

41. (*Drawings of convex and concave pentagons.*)

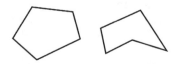

42. (Drawings of convex and concave quadrilaterals.)

43. No.

Polygons on a Grid.

44.

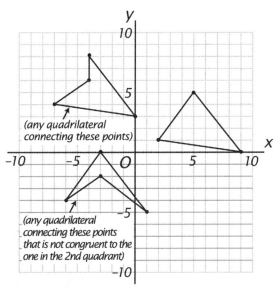

(any quadrilateral connecting these points)

(any quadrilateral connecting these points that is not congruent to the one in the 2nd quadrant)

•45. No.

46. Yes, if they connected them in a different order.

47. Three.

Dividing a Rectangle.

•48. 21 square units.

49. 3.5 square units. ($\frac{21}{6}$).

50. That congruent polygons have equal areas.

51. Yes. Each piece contains 3.5 square units.

52. 45°, 90° (two angles), 135°.

•53. No.

54. No.

Proof Without a Figure.

55. (1) Given.
    (2) If two triangles are congruent, their corresponding sides and angles are equal.
    (3) Substitution.
    (4) If their corresponding sides and angles are equal, two triangles are congruent.

Set III (pages 144–145)

In chapter 3, titled "Mandelbrot's Fractals," of *Penrose Tiles to Trapdoor Ciphers* (W. H. Freeman and Company, 1989), Martin Gardner explains that the polygons considered in the Set III exercises are among the "monster curves" discovered in the second half of the nineteenth century. David Hilbert created open-ended versions of these figures, and Waclaw Sierpinski conceived of the polygon equivalents.

1. 16.

2. 64.

3. 256 and 1,024. (Each one has four times the number of sides of the preceding one.)

4. 4,194,304.

5. The nth one would have 4^{n+1} sides. (The first one has 4^2 sides, the second has 4^3 sides, the third has 4^4 sides, the fourth has 4^5 sides, and the tenth has 4^{11} sides.)

Chapter 4, Lesson 3

Set I (pages 148–149)

According to the U.S. Coast Guard and U.S. Navy, the Bermuda, or Devil's, Triangle is "an imaginary area located off the southeastern Atlantic coast of the United States which is noted for a high incidence of unexplained losses of ships, small boats, and aircraft." The vertices of the triangle are generally taken as Bermuda, Miami, Florida, and San Juan, Puerto Rico. A sensationalized book on the Triangle was a bestseller in 1974 (*The Bermuda Triangle,* by Charles Berlitz.)

The exercises concerning intersections of a line with a triangle were inspired by the following postulate of Pasch: "If a line in the plane of a triangle intersects one side, it also intersects another side." Benno Artmann tells of Pasch in *Euclid—The Creation of Mathematics* (Springer, 1999):

"Euclid had employed the notions of ordering and betweenness of points on a line intuitively . . . . The subject of ordering was taken up by Moritz Pasch in a book *Lectures on Recent Geometry* in 1882. His axioms about betweenness essentially completed the axiomatization of plane geometry. (Pasch had such an acute sense for fine points in logic that in his later days he was the foremost expert on the bylaws of his university in Giessen. His son-in-law C. Thaer produced the standard

German translation of Euclid."

Bermuda Triangle.

- •1. BF and BP.

  2. FB.

- •3. ∠B.

  4. ∠F and ∠B.

French Postulate.

- •5. SAS.

  6. If two sides and the included angle of one triangle are equal to two sides and the included angle of another triangle, the triangles are congruent.

  7. CAC.

  8. ACA.

  9. If two angles and the included side of one triangle are equal to two angles and the included side of another triangle, the triangles are congruent.

Congruence Correspondences.

- •10. TR = AN.

- •11. ∠I = ∠G.

  12. IT = GA.

  13. LS = AG.

  14. ∠E = ∠N.

- •15. Two triangles congruent to a third triangle are congruent to each other.

Triangle Names.

  16. All the sides are equal. (Some students may guess: All the angles are equal.)

- •17. All the angles are acute.

  18. Two sides are equal. (The labels imply that equilateral triangles are not isosceles, although this is not generally considered to be the case.)

  19. One of the angles is obtuse.

- •20. None of the sides are equal.

  21. One of the angles is a right angle.

Intersecting a Triangle.

- •22. All three.

  23. Two.

  24. Two.

  25. No.

  26. Yes.

Set II (pages 149–150)

In his book titled *Triangles—Getting Ready for Trigonometry* (Crowell, 1962), Henry M. Neely wrote about the eye-focusing problem:

"Every waking moment, while your eyes are open and you are looking at something, you are solving triangles . . . . Did you ever see a little baby reach out and try to grasp its mother's finger? It doesn't do very well. The baby will reach out in the general direction of the finger and then will swing its little hand around until its sense of touch finishes the job. The trouble is that the baby is working with a calculating machine that has not yet been completely hooked up.

The baby's eyes can measure the angles fairly well and so it is able to detect the general direction of the finger. But its eyes cannot focus to the correct distance because the brain's memory circuit has not yet stored the most important bit of information—the length of the base line.

Two angles are not enough to solve a triangle. You must also have the length of at least one side."

The astronomy exercise shows how the base line AB is measured in preparation for solving a second triangle with AB as a side and a point on the moon as its third vertex. Points A and B mark the locations of two widely separated observatories on the same meridian: Stockholm (60° N) at A and the Cape of Good Hope (34° S) at B.

Eye Focus Triangle.

  27.

  28. 5 inches (measured from the drawing.)

- •29. 35°.

30. Yes, because 80° + 65° + 35° = 180°.

•31. They get larger.

32. It gets smaller.

Astronomy Problem.

33.

•34. About 5.9 cm.

35. We used 1 cm to represent 1,000 miles; so 5.9 cm represents 5,900 miles.

36. ∠A ≈ 43°, ∠B ≈ 43°.

37. They seem to be equal.

Molecule Path.

•38. Perpendicular lines form right angles.

39. All right angles are equal.

40. The reflexive property.

•41. ASA.

42. They are equal.

Pond Problem.

43. They mean that point O is between points X and A and is also between points Y and B.

•44. Both are pairs of opposite rays.

45. Vertical angles are equal.

46. SAS.

47. The distance across the pond, XY, is equal to the distance from B to A. The distance BA can be measured or paced off.

Set III (page 150)

Match Puzzles.

1. One way is shown below.

2. One way is shown below.

3. One way is shown below.

Chapter 4, Lesson 4

Set I (page 153)

Roof Plan.

1.

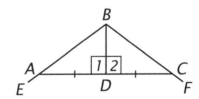

•2. The midpoint of a line segment divides it into two equal segments.

3. Perpendicular lines form right angles.

4. All right angles are equal.

5. Reflexive.

6. SAS.

•7. Corresponding parts of congruent triangles are equal.

8.

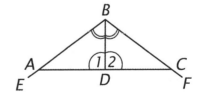

9. Reflexive.

•10. ASA.

11. Corresponding parts of congruent triangles are equal.

12.

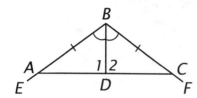

13. If a line bisects an angle, it divides it into two equal angles.

14. Reflexive.

15. SAS.

16. Corresponding parts of congruent triangles are equal.

•17. If the angles in a linear pair are equal, then their sides are perpendicular.

Repeating Patterns.

18. Squares.

•19. Hexagons.

•20. △ABX ≅ △CBX.

21. ASA.

22. △CBX ≅ △ECY.

23. SAS.

24. Two triangles congruent to a third triangle are congruent to each other.

25. Corresponding parts of congruent triangles are equal.

Set II (pages 154–155)

The Muddy Boggy River is in Oklahoma.

Strange Results.

•26. No. △BDC is larger than △ADB.

27. ASA does not apply, because, in △ADB, DB is not included by ∠A and ∠ADB.

28. AB is less than AC.

29. SAS does not apply, because the equal angles in the two triangles are not included by the equal sides.

Congruence Proofs.

30. *Proof.*
    (1) HN = NR. (Given.)
    (2) NO bisects ∠HNR. (Given.)

(3) ∠HNO = ∠RNO. (If an angle is bisected, it is divided into two equal angles.)
(4) NO = NO. (Reflexive.)
(5) △HNO ≅ △RNO. (SAS.)

31. *Proof.*
    (1) ∠BGU and ∠EGL are vertical angles. (Given.)
    (2) ∠BGU = ∠EGL. (Vertical angles are equal.)
    (3) BG = GE and UG = GL. (Given.)
    (4) △BUG ≅ △ELG. (SAS.)
    (5) BU = LE. (Corresponding parts of congruent triangles are equal.)

32. *Proof.*
    (1) ∠C = ∠O. (Given.)
    (2) ∠R and ∠N are supplements of ∠1. (Given.)
    (3) ∠R = ∠N. (Supplements of the same angle are equal.)
    (4) CR = ON. (Given.)
    (5) △CRE ≅ △ONT. (ASA.)

33. *Proof.*
    (1) ∠T and ∠2 are complements of ∠1. (Given.)
    (2) ∠T = ∠2. (Complements of the same angle are equal.)
    (3) TA = AU and TU = UB. (Given.)
    (4) △TAU ≅ △UAB. (SAS.)
    (5) AU = AB. (Corresponding parts of congruent triangles are equal.)

Grid Exercise.

34.

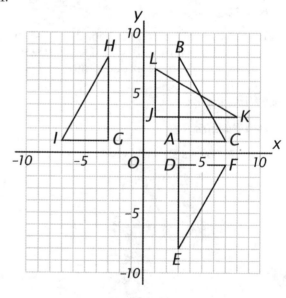

•35. $AB = 7$, $AC = 4$, $BC = \sqrt{65} \approx 8.1$.

36. The fourth (IV).

•37. ΔABC and ΔDEF would coincide.

•38. The x-coordinates are equal;
the y-coordinates are opposites.

39. $DE = 7$, $DF = 4$, $EF = \sqrt{65}$.

•40. SAS.

41. The second (II).

42. G(–3, 1); H(–3, 8); I(–7, 1).

43. The x-coordinates are opposites; the
y-coordinates are equal.

•44. Yes.

45. The x- and y-coordinates are reversed.

46. Yes. Fold the paper so that the x-axis
coincides with the y-axis.

River Problem.

47.

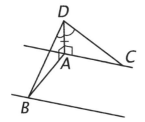

48. Alice adjusted the rim of her hat to a point
directly across the river on the opposite
bank. She then turned and looked to see
where the corresponding point would be
along the bank on her side of the river. The
triangles are congruent by ASA, so
$AC = AB$.

Set III (page 156)

The Handcuffed Prisoners problem is adapted
from Kirkman's schoolgirls problem. Both versions
are discussed at length by Martin Gardner in
"Dinner Guests, Schoolgirls, and Handcuffed
Prisoners" in his book titled *The Last Recreations*
(Copernicus, 1997). The following information is
taken from Gardner's chapter.

The Reverend Thomas Kirkman, a nineteenth-
century amateur British mathematician, first
published his schoolgirls problem in 1847. He
stated the problem in the following way: "Every
day of the week a teacher takes 15 schoolgirls on
a walk. During the walk the girls are grouped in
triplets. Can the teacher construct the triplets so
that after the seven walks each pair of girls has
walked in the same triplet once and only once?"

Any solution to the problem is a "Steiner
triple system," named after nineteenth-century
Swiss geometer Jacob Steiner, who investigated it
and other problems like it belonging to the area
of combinatorics called "block-design theory."
The solution for the "nine prisoners in four days"
version of the problem is unique.

Dudeney's actual puzzle of the nine prisoners
is a more complicated version of the schoolgirls
problem in that only pairings of prisoners handcuffed *to each other* were not to be repeated.

Gardner cites quite a number of articles in
various mathematics journals about these problems, with titles such as "Kirkman's Schoolgirls in
Modern Dress" (*The Mathematical Gazette*),
"The Nine Prisoners Problem" (*Bulletin of the
Institute of Mathematics and Its Applications*),
and "Generalized Handcuffed Designs" (*Journal
of Combinatorial Theory*).

The Problem of the Handcuffed Prisoners.

The four arrangements are:

123, 456, 789;
167, 258, 349;
148, 269, 357;
159, 247, 368.

Chapter 4, Lesson 5

Set I (pages 159–160)

The figure from Euclid's *Elements* illustrates his
proof of the isosceles triangle theorem (Book I,
Proposition 5). Amazingly, this theorem was as
far as the study of geometry went in some of the
universities of the Middle Ages. The name
commonly associated with it, *pons asinorum*,
Latin for "bridge of asses" or "bridge of fools" may
derive from the appearance of Euclid's diagram.
In his *History of Mathematics—An Introduction*
(Allyn & Bacon, 1985), David M. Burton describes
it as resembling "a trestle-bridge so steep that a
horse could not climb the ramp, though a sure-
footed animal such as an ass could. Perhaps only
the sure-footed student could proceed beyond
this stage in geometry."

The proof by Pappus (c. 300 A.D.) is much
simpler than that by Euclid and is the one

presented in the text. One of the first computer programs developed to prove theorems of geometry in 1959 "rediscovered" Pappus's proof.

1.

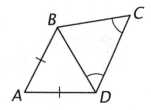

2. ∠ADB.

•3. ∠ABD and ∠ADB.

•4. BD.

5. BD and BC.

•6. No.

7. Isosceles.

•8. If two sides of a triangle are equal, the angles opposite them are equal.

9. ∠A.

10. SAS.

11. Corresponding parts of congruent triangles are equal.

"Isosceles."

•12. That it has at least two equal sides.

13. Yes. A triangle all of whose sides are equal has at least two equal sides.

14. They are opposite the legs.

15. *Example answer:*

16. Not possible.

17. *Example answer:*

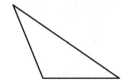

Alcoa Building.

18.

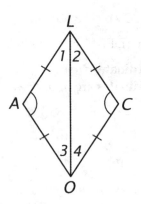

19. Isosceles.

20. If two sides of a triangle are equal, the angles opposite them are equal.

•21. Addition.

•22. Betweenness of Rays Theorem.

•23. Substitution.

24. SAS.

25. Corresponding parts of congruent triangles are equal.

26. It bisects them.

Shuffleboard Court.

27. It appears to be perpendicular to them and to bisect them.

•28. They appear to be parallel.

29. Isosceles.

•30. ∠S = ∠B = 72°. ($\frac{180-36}{2} = \frac{144}{2} = 72$.)

31. Approximately 72°.

32. Approximately 108°. (180 − 72 = 108.)

Set II (pages 160–162)

In *The Architecture of Molecules* (W. H. Freeman and Company, 1964), Linus Pauling discusses the bond angle in the water molecule and similar molecules. The sixth column of elements in the periodic table begins with oxygen, followed by sulfur, selenium, and tellurium. The bond angle for water (H_2O) is 104.5°. For hydrogen sulfide (H_2S), it is 92.2°; for hydrogen selenide (H_2Se), it is 91°; and, for hydrogen telluride (H_2Te), it is 90°. Because the bond angle of hydrogen sulfide is fairly close to 90°, assuming that the triangle is a right triangle in calculating the hydrogen–hydrogen distance gives a fairly good result: 1.90 angstroms. The Pythagorean Theorem is, of course, a special case of the Law of Cosines. Applying the Law of Cosines to the obtuse triangle gives a distance of 1.93 angstroms.

33. *Theorem 10.* If two angles of a triangle are equal, the sides opposite them are equal.
    Given: In $\triangle$EFG, $\angle$E = $\angle$F.
    Prove: GE = GF.
    Proof.
    (1) In $\triangle$EFG, $\angle$E = $\angle$F. (Given.)
    (2) EF = FE. (Reflexive.)
    (3) $\angle$F = $\angle$E. (Given.)
    (4) $\triangle$EFG $\cong$ $\triangle$FEG. (ASA)
    (5) GE = GF. (Corresponding parts of congruent triangles are equal.)

34. *Corollary.* An equilateral triangle is equiangular.
    Given: $\triangle$ABC is equilateral.
    Prove: $\triangle$ABC is equiangular.
    Proof.
    (1) $\triangle$ABC is equilateral. (Given.)
    (2) BC = AC and AC = AB. (All of the sides of an equilateral triangle are equal.)
    (3) $\angle$A = $\angle$B and $\angle$B = $\angle$C. (If two sides of a triangle are equal, the angles opposite them are equal.)
    (4) $\angle$A = $\angle$C. (Substitution.)
    (5) $\triangle$ABC is equiangular. (A triangle is equiangular if all of its angles are equal.)

35. *Corollary.* An equiangular triangle is equilateral.
    Given: $\triangle$ABC is equiangular.
    Prove: $\triangle$ABC is equilateral.
    Proof.
    (1) $\triangle$ABC is equiangular. (Given.)
    (2) $\angle$A = $\angle$B and $\angle$B = $\angle$C. (All of the angles of an equiangular triangle are equal.)
    (3) BC = AC and AC = AB. (If two angles of a triangle are equal, the sides opposite them are equal.)
    (4) BC = AB. (Substitution.)
    (5) $\triangle$ABC is equilateral. (A triangle is equilateral if all of its sides are equal.)

Hydrogen Sulfide.

36. Isosceles.

37. 43.9°.

•38. 92.2°. $(180 - 2 \cdot 43.9 = 92.2.)$

•39. Obtuse.

40. The Pythagorean Theorem applies only to right triangles, and this triangle is obtuse. (However, it is close to being a right triangle; so the Pythagorean Theorem gives a good approximation: $H_1H_2 \approx 1.9$ angstroms.)

41.
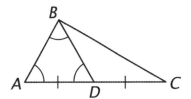

42. $\triangle$BDC is isosceles. Because $\triangle$ABD is equiangular, it is also equilateral; so BD = AD. D is the midpoint of AC; so AD = DC. It follows that BD = DC by substitution and that $\angle$DBC = $\angle$DCB because the sides opposite them are equal. (It is also possible to find the measures of the angles by using the Triangle Angle Sum Theorem.)

Star Puzzles.

•43. A pentagon.

•44. Because its sides intersect in points other than their endpoints.

45. Yes. The triangles ABC, BCD, CDE, DEA, and EAB are congruent by SAS. The line segments that form the red star are corresponding parts of these triangles.

46. No. They don't look equal and the figure is said to be accurately drawn.

47. Yes. The triangles ACD, BDE, CEA, DAB, and EBC would be congruent by SAS. The sides of the blue polygon are corresponding parts of these triangles.

SAT Problem.

48. $y = 20°$. $\triangle$PQT and $\triangle$TRS are isosceles; so $\angle$PTQ = $\angle$P = 40° and $\angle$RTS = $\angle$S = 40°. So $\angle$PTS = 40° + $y°$ + 40°. Also $\angle$PTS + $\angle$P + $\angle$S = 180°; so $\angle$PTS = 100°. Therefore, 100 = 80 + y; so y = 20.

Drawing Exercise.

49.

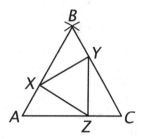

• 50. They seem to be congruent.

51. Because ΔABC is equilateral, it is also
equiangular; so ∠A = ∠B = ∠C;
AX = BY = CZ = 1 inch and
XB = YC = ZA = 2 inches; so the triangles
are congruent by SAS.

• 52. It seems to be equilateral (and
equiangular).

53. Because ΔAXZ ≅ ΔBYX ≅ ΔCZY,
XY = YZ = XZ (corresponding parts of
congruent triangles are equal).

Set III (page 162)

Escher's reptile design can be extended to fill the
plane. Among the many geometric patterns
underlying it is the fact that the points at which
the heads of three lizards touch are the vertices
of equilateral triangles.

1. Three.

2. 120°.

3. It is the mirror image of the lizards in
Escher's drawing.

Chapter 4, Lesson 6

Set I (pages 165–166)

Richard Pawley wrote an article in *The Mathematics Teacher* (May 1967) concerning the
conditions under which two triangles can have
five parts of one of them equal to five parts of
the other and yet not be congruent. The simplest
pair of such triangles whose sides are integers
have sides of lengths 8-12-18 and 12-18-27. (The
pair used in exercises 9 and 10 have sides of
lengths 343-392-448 and 392-448-512.) An analysis
of such triangles requires a knowledge of the
triangle inequality theorem as well as of similarity.
The golden ratio even enters into the picture!

Force Triangles.

• 1. SSS.

• 2. Corresponding parts of congruent triangles
are equal.

Plumb Level.

3. SSS.

4. Corresponding parts of congruent triangles
are equal.

• 5. They are a linear pair.

6. If the angles in a linear pair are equal, their
sides are perpendicular.

7. Line BC is horizontal as long as the weight
hangs over the midpoint of segment BC.

8. Yes. They would be congruent by SAS.

Equal Parts.

9. No. One triangle is larger than the other.

• 10. Five.

• 11. No.

12. No.

13. No.

14. Yes. If they have six pairs of equal parts,
they must be congruent by SSS.

Angle Bisection.

15.

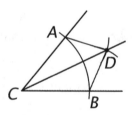

• 16. CA = CB and AD = BD.

17. Reflexive.

• 18. SSS.

19. Corresponding parts of congruent triangles
are equal.

Segment Bisection.

20.

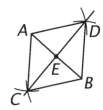

21. AD = BD = AC = BC.

22. SSS.

23. Corresponding parts of congruent triangles are equal.

•24. SAS. (AC = BC, ∠ACD = ∠BCD, CE = CE.)

25. Corresponding parts of congruent triangles are equal.

Set II (pages 166–167)

In *Geometry and the Imagination*, David Hilbert refers to the study of linkages as "a part of kinematics that is intimately connected with elementary metric geometry." He continues:

"The term 'plane linkage' applies to every plane system of rigid rods which are interconnected, or connected with fixed points of the plane about which they are free to turn, in such a way that the system is capable of certain motions in the plane. The simplest mechanism of this kind consists of a single rod with one end attached to a point of the plane; in effect, this is a compass. Just as the free end of a compass traces out a circle, so the points of the rods in all other plane linkages move on algebraic curves, i.e. curves satisfying an algebraic equation in a system of Cartesian coordinates. Conversely, it is possible to find a system of jointed rods that is suitable for the construction (at least piecemeal) of any given algebraic curve no matter how complicated."

The linkage considered in exercises 36 through 43 is Peaucellor's inversor, a device that links circular and linear motion.

Linkage Problem 1.

26.

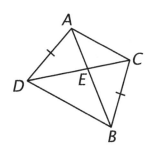

27. They seem to be congruent.

28. Yes. ΔABC ≅ ΔCDA by SSS.

•29. ∠ADC = ∠CBA because they are corresponding parts of congruent triangles.

30. No. E does not appear to be the midpoint of either line segment.

31. Yes. ∠AEC = ∠DEB because they are vertical angles.

32. ΔDAB ≅ ΔBCD (by SSS).

33. ∠DAB = ∠BCD.

34. ΔAED ≅ ΔCEB by ASA. ∠ADC = ∠CBA was shown in exercise 29, AD = CB was given, and ∠DAB = ∠BCD was shown in exercise 33.

35. They are isosceles. Each has a pair of equal angles: in ΔAEC, ∠CAB = ∠ACD because ΔABC ≅ ΔCDA, shown in exercise 28; in ΔDEB, ∠ABD = ∠CDB, because ΔDAB ≅ ΔBCD, shown in exercise 32.

Linkage Problem 2.

36.

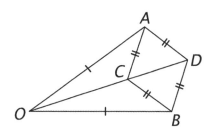

37. They seem to be collinear.

•38. SSS.

39. They are corresponding parts of ΔAOC and ΔBOC.

•40. OC bisects ∠AOB.

41. OD bisects ∠AOB because ΔAOD ≅ ΔBOD (SSS).

•42. Because an angle has exactly one ray that bisects it (the corollary to the Protractor Postulate).

43. They are collinear.

44.

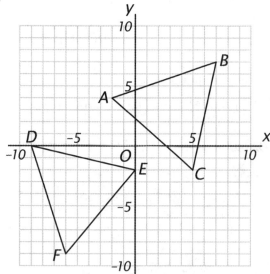

45. It looks as if it might be equilateral (and equiangular).

46. AB = $\sqrt{81+9}$ = $\sqrt{90}$.

   BC = $\sqrt{4+81}$ = $\sqrt{85}$.

   AC = $\sqrt{49+36}$ = $\sqrt{85}$.

47. (Student answer.) (△ABC is isosceles.)

48. DE = $\sqrt{81+4}$ = $\sqrt{85}$.

   EF = $\sqrt{36+49}$ = $\sqrt{85}$.

   DF = $\sqrt{9+81}$ = $\sqrt{90}$.

49. △ABC and △DEF are congruent by SSS.

50. ∠A = ∠B = ∠D = ∠F.

51. ∠E = ∠C.

Set III (page 168)

The puzzle of bracing the square was devised by Raphael M. Robinson of the University of California at Berkeley. Robinson's solution for bracing the square in the plane required 31 rods. When Martin Gardner included it in his "Mathematical Games" column, 57 readers sent in a solution using 25 rods, and about 20 other readers discovered the solution illustrated in the Set III exercise. Gardner remarked that solutions sent in with fewer than 23 rods turned out to be either nonrigid or geometrically inexact. His *Sixth Book of Mathematical Games from Scientific American,*

cited in the text, includes several illustrations of alternative solutions proposed for the puzzle, some correct and some incorrect.

Bracing a square in space turns out to be surprisingly simple: eight rods can be attached to it, four "above" and four "below" to form a regular octahedron.

Bracing the Square.

1.

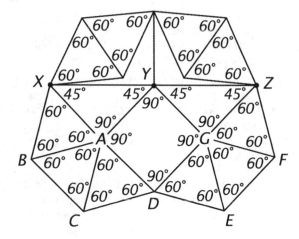

2. Because ∠XYA = 45°, ∠AYG = 90°, and ∠GYZ = 45°, ∠XYZ = 180°. X, Y, and Z must be collinear because ∠XYZ is a straight angle.

Chapter 4, Lesson 7

Set I (pages 172–173)

In the law of reflection, the angles of incidence and reflection are considered to be the angles formed by the ray of light with the normal to the mirror. Because your students are unlikely to know this, they are as likely to copy ∠1 as ∠2; the end result, of course, is the same.

- •1. A compass.

- •2. Distances between pairs of points.

- 3. Yes.

- 4. No.

- •5. A line segment is bisected.

- 6. An angle is copied.

- 7. An angle is bisected.

- 8. A triangle is copied.

Light Reflection.

9.

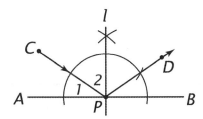

10. Line *l* is perpendicular to line AB.

•11. ∠2 is complementary to ∠1.

•12. Copying an angle.

Bridge Girder.

13.

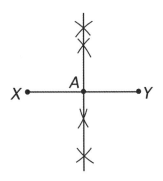

14. SSS.

15. An equilateral triangle is equiangular, and corresponding parts of congruent triangles are equal.

16. The two lines appear to be parallel.

Equidistant Points.

17.

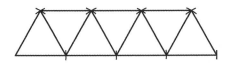

•18. It means equally distant.

19. One.

20. Infinitely many.

21. On the line constructed in exercise 17 to find the midpoint of segment XY (the perpendicular bisector of XY).

Car Jack.

22.

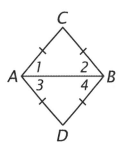

23. Isosceles.

24. If two sides of a triangle are equal, the angles opposite them are equal.

•25. Addition.

•26. Substitution.

27. SSS.

28. Corresponding parts of congruent triangles are equal.

29. It bisects them.

Set II (pages 174–175)

The surprise in the equilateral triangle construction is the illusion of a curve; specifically, a parabola. The line segments form its envelope.

The triangle suggested by Vitruvius for use in designing stair steps is the 3-4-5 right triangle. We can't prove that such a triangle is a right triangle at this point, however, because we have not yet established the converse of the Pythagorean Theorem.

An amazing story goes with the guitar fret construction exercise. In *Another Fine Math You've Got Me Into* . . . (W. H. Freeman and Company, 1992), Ian Stewart reports that the inventor of the construction, Daniel Strähle, explained it in an article in the *Proceedings of the Swedish Academy* in 1743. One of the members of the academy, a noted mathematician, checked the construction and claimed that it was far more inaccurate than a musician would find acceptable. More than two centuries later (1957), J. M. Barbour of Michigan State University discovered that this noted mathematician had made a mistake. In fact, his mistake was so large that it could have easily been discovered by measuring the drawing! (He computed the measure of ∠AIU as 49°14', whereas it is actually 33°32'.) Strähle, the man with no mathematical training, was vindicated and the expert discovered to be wrong!

A Construction with an Unexpected Result.

30.

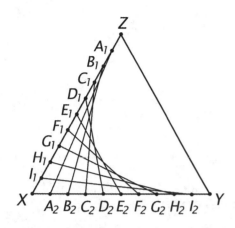

31. A curve.

32. $\triangle XI_1I_2$.

33. SAS.

Stair Design.

34.

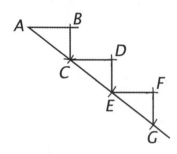

35. Right triangles.

•36. SSS.

•37. Height = 7.5 in; width = 10 in.

(10 ft = 120 in; $\dfrac{120 \text{ in}}{16 \text{ steps}}$ = 7.5 in per step;

3 units = 7.5 in; so 1 unit = 2.5 in and
4 units = 10 in.)

38. 13 ft, 4 in. (16 × 10 in = 160 in = 13 ft, 4 in.)

Guitar Frets.

39.

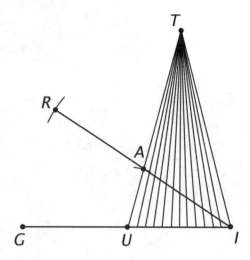

40. They are not evenly spaced (they are spaced
more closely together toward the body of
the guitar).

Set III (page 175)

1.

2. 6.

3. 14.

4. 12.

Chapter 4, Review

Set I (pages 176–178)

Approximate angles of repose of various materials
can be found in handbooks for architects, civil
engineers, and contractors. Such angles are
especially important in estimating the stability of
excavations. Round gravel, for example, has an
angle of repose of about 30°, whereas "angular"
gravel has an angle of repose of about 38°.

Polar-Bear Pattern.

1. It is not a polygon, because two of its sides are not line segments; they are curved.

•2. Because they have the same size and shape (they can be made to coincide).

3. The bears fill the plane.

Turkish Geometry.

•4. They are all equal.

5. Two of them are equal.

6. None of them are equal.

From Euclid.

7. Statement 3.

•8. Statement 4.

9. A postulate is a statement that is assumed to be true without proof; a theorem is a statement that is proved.

10. Substitution.

11. Statement 2.

Angles of Repose.

12.

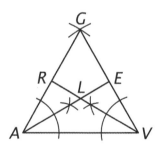

•13. An equilateral triangle is equiangular.

14. Division.

•15. Substitution.

16. If two angles of a triangle are equal, the sides opposite them are equal.

•17. 60°. (By the Triangle Angle Sum Theorem.)

18. 30°.

•19. 120°.

20. 90°.

•21. 30°.

22. 90°.

23. 45°.

24. Yes, because we can construct a 90° angle and we can bisect it to construct a 45° angle.

Congruence in 3-D.

•25. SSS.

26. Corresponding parts of congruent triangles are equal.

27. ∠ABD and ∠ACD. (They are the other angles opposite sides marked with three tick marks.)

28. ∠BAD, ∠BCD, and ∠ADC. (They are the other angles opposite sides marked with two tick marks.)

Set II (pages 178–180)

One of the elementary figures of projective geometry is the "complete quadrilateral," the subject of exercises 35 and 36. Four lines in a plane, no two of which are parallel and no three of which are concurrent, intersect in six points. The six points determine three line segments that are called the "diagonals" of the quadrilateral. That the midpoints of these diagonals are always collinear is a theorem of projective geometry.

SAT Problem.

•29. 90°. [$x + y + (x - y) = 180, 2x = 180, x = 90.$]

30. 90°.

31. 90°.

32. The value of x does not depend on the value of y.

•33. 45°.

34. The triangle cannot be equilateral, because $y = 90°$.

Lines and Midpoints.

35.

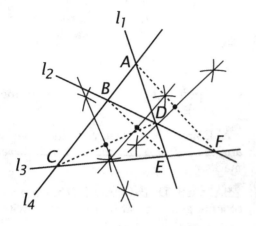

36. They appear to be collinear.

Ollie's Mistake.

37.

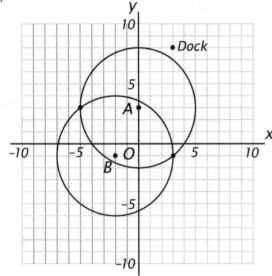

•38. $\sqrt{4+16} = \sqrt{20}$ (or $2\sqrt{5} \approx 4.5$).

39. Two.

•40. At $(-5, 3)$.

41. At $(3, -1)$.

42. The distance between the dock at $(3, 8)$ and $(-5, 3)$ is $\sqrt{64+25} = \sqrt{89}$. The distance between the dock at $(3, 8)$ and $(3, -1)$ is 9; so Ollie's money is at $(3, -1)$.

Rigid Structures.

43. *Example answer:*

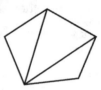

44. *Example answer:*

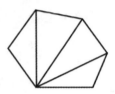

45. *Example answer:*

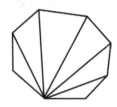

•46. Seven.

47. $n - 3$.

Pool Proof.

48. (1) Given.
 (2) Betweenness of Points Theorem.
 (3) Given.
 (4) Perpendicular lines form right angles.
 (5) All right angles are equal.
 (6) Reflexive.
 (7) SAS.
 (8) Corresponding parts of congruent triangles are equal.
 (9) Substitution (steps 2 and 8).

Carpenter's Square and Compass.

49. First use the scale on the square to mark two points X and Y on the sides of the angle at the same distance from the vertex (so that XB = YB). Then position the square so that the distances from its corner to these two points are equal. $\triangle BXP \cong \triangle BYP$ by SSS; so $\angle ABP = \angle CBP$.

50. Because BP bisects the angle, ∠ABP = ∠CBP. Also BX = BY and BZ = BZ; so $\triangle$BXZ $\cong$ $\triangle$BYZ by SAS. ∠1 and ∠2 are a linear pair and ∠1 = ∠2 (they are corresponding parts of the congruent triangles); so XY ⊥ BP because, if the angles in a linear pair are equal, their sides are perpendicular.

51. *Proof.*
    (1) ∠A = ∠1 and ∠2 = ∠C. (Given.)
    (2) BD = AB and CD = BD. (If two angles of a triangle are equal, the sides opposite them are equal.)
    (3) AB = CD. (Substitution.)

52. *Proof.*
    (1) O is the midpoint of XY. (Given.)
    (2) OX = OY. (The midpoint of a line segment divides it into two equal segments.)
    (3) ∠1 = ∠2 and OA = OB. (Given.)
    (4) $\triangle$AOX = $\triangle$BOY. (SAS)
    (5) ∠A = ∠B. (Corresponding parts of congruent triangles are equal.)

Chapter 4, Algebra Review (page 182)

•1. $2(3x - 1)$.

•2. $3(2x - 1)$.

•3. $2(3x - 2)$.

•4. $4(x + 3y)$.

•5. $x(x + 10)$.

•6. $x^2(3x - 2)$.

•7. $5(3\pi - 8)$.

•8. $2x(4 + y)$.

•9. $x(a + b - c)$.

•10. $2\pi r(h + r)$.

•11. $(x + 4)(x + 4)$ or $(x + 4)^2$.

•12. $(x + 4)(x - 4)$.

•13. $(x + 3)(x + 10)$.

•14. $(x - 2)(x + 15)$.

•15. $(x + 7)(x - 6)$.

•16. $(2x - 1)(x - 7)$.

•17. $(3x - 5)(x + 2)$.

•18. $x(x - 4y)$.

•19. $(x - 2y)(x - 2y)$ or $(x - 2y)^2$.

•20. $(x + 2y)(x - 2y)$.

•21. $\pi(a^2 - r^2) = \pi(a + r)(a - r)$.

•22. $5x(x^2 - 4) = 5x(x + 2)(x - 2)$.

•23. $x(x^2 + x - 12) = x(x + 4)(x - 3)$.

•24. $z(2x - 3)$.

•25. $(x + 2)(2x - 3)$.

Set I (pages 186–187)

A chapter of Lawrence Wright's *Perspective in Perspective* (Routledge & Kegan Paul, 1983) is titled "Grand Illusions" and concerns various tricks played by architects with perspective. Plato mentions the use of such tricks in the design of Greek temples such as the Parthenon. Giotto's campanile in Florence is unusual in its use of "counter-perspective." The tower widens measurably upward, and the upper levels become successively taller toward the top. From the ground, they can look almost equal.

Hat Illusion.

•1. Either $h > w$, $h = w$, or $h < w$.

2. The "three possibilities" property.

3. *(Student answer.)* (For most people, the hat appears to be taller than it is wide, and so $h > w$.)

4. The hat is 1.7 cm high and its brim is 1.8 cm wide.

Pecking Order.

5. Hen A will always peck hen C.

•6. The transitive property.

Equalities and Inequalities.

7. The "three possibilities" property.

•8. Multiplication.

9. Subtraction.

10. Substitution.

•11. Addition.

12. Division.

Deceiving Appearances.

13. Isosceles. (Also acute.)

•14. AB > BC.

•15. Substitution.

16. Scalene. (Also acute.)

17. DF < DE.

18. The transitive property.

19. Because one has two equal sides and the other does not. (One is isosceles, the other scalene.)

Perspective Inequalities.

•20. ∠APB > ∠HPI.

21. ∠BPC > ∠CPD.

•22. ∠PIH < ∠PBA.

23. ∠PHG < ∠PGF.

24. ∠PHI > ∠PGH and ∠PGH > ∠PFG; so ∠PHI > ∠PFG.

25. So that the upper levels would appear to be the same height when viewed from the ground.

Set II (pages 187–189)

SAT Problem.

•26. Transitive.

27. Addition theorem of inequality.

28. Substitution. (Substituting AC for AB + BC and CE for CD + DE in exercise 27.)

29. Could be false. (To see why, imagine that DE is much longer than it is.)

•30. The "whole greater than part" theorem.

31. Transitive. (AB < BC is given and BC < BD in exercise 30.)

Scalene Triangle.

32.

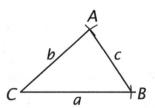

33.

•34. ∠A ≈ 83°, ∠B ≈ 56°, ∠C ≈ 41°.

35. ∠A > ∠B > ∠C.

36. Yes. The order corresponds to the order of the lengths of the opposite sides.

Rectangle Inequalities.

•37. 9 units. $(2 \cdot 4 + 2 \cdot 0.5 = 9.)$

38. Yes. The "whole greater than part" theorem.

39. 2 square units. $(4 \cdot 0.5 = 2.)$

40. No. This area problem shows why: $4 > 0$ and $0.5 > 0$, but $4 \times 0.5 = 2 < 4.$

Vertical Lines and Angles.

•41. That the line is "upright" (contains the center of the earth).

42. The sides of one angle must be opposite rays to the sides of the other angle. (It must also be true that two angles are equal if they are vertical angles.)

43. Yes: $\angle 1$ and $\angle 2$ are not vertical angles. If they were vertical angles, they would be equal. Because $AB = AC$, $\angle 2 = \angle 3$ (if two sides of a triangle are equal, the angles opposite them are equal). But $\angle 1 < \angle 3$; so $\angle 1 < \angle 2$ (substitution).

44. No. Because $\angle 2 = \angle 3$ ($AB = AC$) and $\angle 2 = \angle 4$, $\angle 3 = \angle 4$. The fact that two angles are equal does not mean, however, that they are necessarily vertical angles.

45. (1) Given.
    (2) Addition.
    (3) Given.
    (4) Addition.
    (5) Transitive (steps 2 and 4).

46. (1) Given.
    (2) Addition.
    (3) Given.
    (4) Substitution.

Optics Figure.

47. (1) Given.
    (2) Betweenness of Rays Theorem.
    (3) Protractor Postulate.
    (4) The "whole greater than part" theorem.

48. *Proof.*
    (1) A-B-C. (Given.)
    (2) $AB + BC = AC.$ (Betweenness of Points Theorem.)
    (3) $AC > AB.$ ("Whole greater than part.")
    (4) $\angle ADB = \angle DAB.$ (Given.)

(5) $AB = DB.$ (If two angles of a triangle are equal, the sides opposite them are equal.)

(6) $AC > DB.$ (Substitution.)

Set III (page 189)

In *Human Information Processing*, Peter H. Lindsay and Donald A. Norman deal with the logic of choice. Concerning the three statements of the Set III exercises, they write:

"Almost without exception, people agree that their own decision processes ought to be logical. Moreover, formal theories of decision making assume logical consistency: Preferences among objects ought to be consistent with one another. If A is preferred to B and B to C, then logically A should be preferred to C. Transitivity of this sort is a basic property that just ought to hold anytime different objects are compared to one another. . . .

These three rules constitute a sensible set of postulates about decision behavior. Indeed, if the rules are described properly, so that the mathematical framework is removed, they simply sound like common sense. Those who study decision making, however, have discovered that there is a difference between the rules that people believe they follow and the rules that they actually do follow. Thus, although these three assumptions form the core of most decision theories, it is also realized that they do not always apply. There are some clear examples of instances in which each rule is violated."

(The authors then give anecdotal examples of violations for each of the three rules.)

Martin Gardner treated the subject of non-transitive paradoxes in one of his "Mathematical Games" columns. It is included in his *Time Travel and Other Mathematical Bewilderments* (W. H. Freeman and Company, 1988).

1. If you prefer an apple to a banana and a banana to a cookie, then you prefer an apple to a cookie.

2. The transitive property.

3. If you have no preference between an apple and a banana or between a banana and a cookie, then you have no preference between an apple and a cookie.

4. Substitution.

5. If you have no preference between an apple and a banana and you prefer a cookie to nothing at all, then you would prefer having an apple and a cookie to having just a banana.

6. Yes.
   If C > 0, then A + C > A. (Addition.)
   But A = B, so A + C > B. (Substitution.)

Chapter 5, Lesson 2

Set I (pages 192–193)

Aristotle included a section on the causes of the rainbow in *De Meteorologia*. According to Carl Boyer, Aristotle's work is "the first truly systematic theory of the rainbow that has come down to us." Boyer comments on the soundness of some of Aristotle's geometrical arguments and says: "Had his successors continued his work at the same high level, the story of the rainbow might not have been such a tale of frustration as it was destined to be."

Garage Door.

•1. ∠BCY, ∠BAX, ∠CAD(∠YAD).

2. It gets larger.

•3. It gets smaller.

4. It gets smaller.

5. If two sides of a triangle are equal, the angles opposite them are equal.

•6. An exterior angle of a triangle is greater than either remote interior angle.

7. Vertical angles are equal.

8. They are always equal. Because ∠B = ∠BAY and ∠BAY = ∠DAX, ∠B = ∠DAX by substitution.

Exterior Angles.

•9. ∠2, ∠5, ∠8.

•10. Two.

11. Six.

12. Yes. If the triangle is equiangular, its exterior angles must all be equal because they are supplements of equal angles.

13. No. The exterior angles at each vertex are vertical angles and vertical angles are equal.

14. *Example figure:*

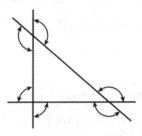

15. *Example figure:*

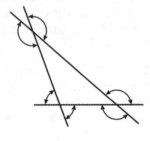

•16. Two.

17. Two.

18. Six.

Rainbow.

•19. ΔROS.

•20. ∠ROA > ∠SRO and ∠ROA > ∠S.

21. ΔROC.

22. ∠ROS > ∠ORC and ∠ROS > ∠RCO(∠RCS).

23. ΔRCS and ΔRCO.

24. ∠RCA > ∠SRC, ∠RCA > ∠S, ∠RCA > ∠ORC, ∠RCA > ∠ROC.

•25. No. It does not form a linear pair with an angle of the triangle.

Lines and Angles.

•26. 360°.

27. 1,080°. (3 · 360 = 1,080.)

•28. 180°.

29. 180°.

30. 720°. (1,080 – 180 – 180 = 720.)

31. It indicates that the sum is 720°.

We are indebted to Proclus for his commentary on the first book of Euclid's *Elements*. It and the work of Pappus are our two main sources of information on the history of Greek geometry. Proclus defended Euclid from the charge that he proved things that had no need of proof. (More on this is included in Lesson 4 of this chapter.)

Exterior Angle Theorem.

•32. A line segment has exactly one midpoint.

33. The Ruler Postulate (or by Construction 1, to copy a line segment).

34. SAS.

35. Corresponding parts of congruent triangles are equal.

36. Vertical angles are equal.

•37. The "whole greater than part" theorem.

38. Substitution.

Angle Sum.

•39. Two points determine a line.

40. An angle is an exterior angle of a triangle if it forms a linear pair with an angle of the triangle.

•41. The angles in a linear pair are supplementary.

42. If two angles are supplementary, their sum is 180°.

•43. An exterior angle of a triangle is greater than either remote interior angle.

44. Addition.
45. Substitution.

Angle in a Triangle.

46.

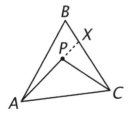

47. ∠APC > ∠AXC and ∠AXC > ∠B because an exterior angle of a triangle is greater than either remote interior angle. So ∠APC > ∠B by the transitive property.

Proclus's Proof.

48. If two sides of a triangle are equal, the angles opposite them are equal.

49. ∠1 = ∠C and ∠2 = ∠A.

50. The Exterior Angle Theorem (∠1 is an exterior angle of ∆PBC and ∠2 is an exterior angle of ∆PBA.)

51. Indirect.

Set III (page 194)

In his *History of Mathematics*, David Eugene Smith includes a paragraph on "drumhead trigonometry." He wrote: "The continual warfare of the Renaissance period shows itself in many ways in the history of mathematics . . . . One of them is related to the subject now under consideration. Several writers of the 16th century give illustrations of the use of the drumhead as a simple means of measuring angles of elevation in computing distances to a castle or in finding the height of a tower." The illustration is from S. Belli's *Libro del Misvrar con la Vista (Book of Measuring with Eyesight)*, published in Venice in 1569.

Drumhead Geometry.

1.

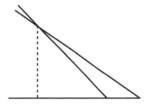

2. (About 4 in; more precisely, about $4\frac{1}{8}$ in.)

3. About 80 ft.

Chapter 5, Lesson 3

Set I (pages 197–198)

There are now many more "anamorphic artists" painting streets than pictures. The idea in using anamorphic figures as traffic markers is clearly not to make them difficult to recognize but rather just the opposite. The driver of a moving car who is looking far ahead sees the markers from a sharp angle and hence sees them regain their "normal" shape. An entire "Mathematical Games" column focuses on various types of anamorphic art (*Scientific American*, January 1975). It is included in Martin Gardner's *Time Travel and Other Mathematical Entertainments* (W. H. Freeman and Company, 1988).

"Triangles in perspective" is an interesting topic. Suppose we have some wire-frame models of triangles of various shapes. What kinds of shadows can they cast? Can the shadow of an equilateral triangle look scalene? Can the shadow of an isosceles triangle look equilateral? Can the shadow of an obtuse triangle look acute? Does the shadow of a triangle always look like a triangle? What determines the smallest shadow that a triangle can cast?

1. Bicycle lane.

2. So that it can be seen more easily to a driver viewing it "on edge."

•3. Each is the converse of the other.

•4. BC < AC.

5. The "three possibilities" property.

6. ∠A = ∠B.

7. If two sides of a triangle are equal, the angles opposite them are equal.

•8. ∠A > ∠B.

9. ∠A < ∠B.

•10. If two sides of a triangle are unequal, the angles opposite them are unequal in the same order.

11. ∠A > ∠B.

12. BC > AC.

13. Indirect.

Triangle Drawing 1.

•14. ∠B.

15. ∠A.

16.

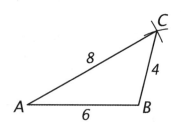

17. Obtuse. (Also, scalene.)

•18. ∠A ≈ 29°, ∠B ≈ 104°, ∠C ≈ 47°.

Triangle Drawing 2.

•19. DE.

20. DF.

21.

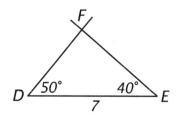

•22. DF ≈ 4.5 cm, EF ≈ 5.4 cm.

Triangle in Perspective.

•23. ∠X > ∠Y.

24. If two sides of a triangle are unequal, the angles opposite them are unequal in the same order.

25. ∠Z > ∠Y.

26. ∠Z > ∠X (given) and ∠X > ∠Y (exercise 23); so ∠Z > ∠Y by the transitive property.

27. XY > XZ.

28. If two angles of a triangle are unequal, the sides opposite them are unequal in the same order.

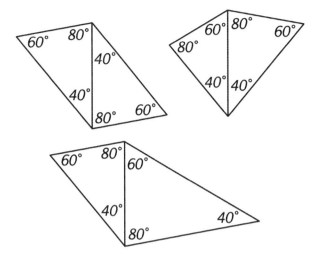

The exercises on the pairs of "not quite equilateral" triangles are more challenging than they first appear. Some comparable drawings (above) in which the angles are 40°, 60°, and 80° instead are revealing. (Although the triangles in the second and third pairs are not congruent, they are obviously similar. We will return to these figures in the chapter on similarity.)

Folding Experiment.

29. *(Triangle cut out and folded).*

•30. ∠BDE.

31. ∠BDE is an exterior angle of △DEC.

32. ∠BDE > ∠C.

33. If two sides of a triangle are unequal, the angles opposite them are unequal in the same order.

•34. BE bisects ∠ABC.

35. △ABE ≅ △DBE (SAS); so ∠BDE = ∠A (corresponding parts of congruent triangles are equal).

Not Quite Equilateral.

•36. Yes. They are congruent by ASA.

37. AC.

38. BD.

39. They are equal because corresponding parts of congruent triangles are equal.

40. FG.

41. GH.

•42. FG < GH because both are sides of △FGH in which GH is the longest side.

43. No. They can't be congruent because FG and GH are corresponding parts of the triangles but they are not equal.

44. Yes. They are not congruent, because the shortest side of △IJK is IJ and the shortest side of △JKL is JK. IJ < JK because they are both sides of △IJK in which IJ is the shortest side. As with the preceding pair of triangles, these triangles cannot be congruent, because IJ and JK are corresponding parts of the triangles but they are not equal.

•45. An equilateral triangle is equiangular.

46. Betweenness of Rays Theorem.

•47. The "whole greater than part" theorem.

48. Substitution.

49. If two angles of a triangle are unequal, the sides opposite them are unequal in the same order.

50. Because ∠PXA is an exterior angle of △PXY, ∠PXA > ∠PYX. Because PX ⊥ AB, ∠PXA and ∠PXY are right angles; so ∠PXA = ∠PXY = 90°. Therefore, in △PXY, 90° > ∠PYX (substitution) and ∠PXY = 90°; so ∠PXY > ∠PYX (substitution again). It follows that PY > PX because, if two angles of a triangle are unequal, the sides opposite them are unequal in the same order.

Set III (page 199)

The message says "HELLO." (As Martin Gardner explains, "hold the page horizontally [with the bottom of the page] near the tip of your nose, close one eye, and read the message on a sharp slant.")

Chapter 5, Lesson 4

Set I (pages 202–203)

In *Euclid—The Creation of Mathematics* (Springer, 1999), Benno Artmann quotes Proclus on the Triangle Inequality Theorem:

"The Epicureans are want to ridicule this theorem, say it is evident even to an ass and needs no proof

. . .they make [this] out from the observation that, if hay is placed at one extremity of the sides, an ass in quest of provender will make his way along the one side and not by way of the other two sides."

Artmann adds some pertinent remarks of his own:

"(The Epicureans of today might as well add that one could see the proof on every campus where people completely ignorant of mathematics traverse the lawn in the manner of the ass.) Proclus replies rightly that a mere perception of the truth of a theorem is different from a scientific proof of it, which moreover gives reason why it is true. In the case of Euclid's geometry, the triangle inequality can indeed be derived from the other (equally plausible) axioms. On the other hand, the Epicureans win in the modern theory of metric spaces, where the triangle inequality is the fundamental axiom of the whole edifice."

The transits of Venus (the occasions when Venus is between the sun and Earth) are few and far between. In fact, there have been only six of them since the invention of the telescope (1631, 1639, 1761, 1769, 1874, 1882). The next transits will be on Dec. 11, 2117 and Dec. 8, 2125. In the past, these alignments were used to determine the distance between Earth and the sun by timing the beginning and ending of a transit from widely separated geographical locations, but such methods are now obsolete. More information on the transits of Venus can be found in *June 8, 2004—Venus in Transit*, by Eli Maor (Princeton University Press, 2000).

Donkey Sense.

1. The sum of any two sides of a triangle is greater than the third side.

•2. $DP + PH > DH$.

3. $DH + HP > DP$; $PD + DH > PH$.

4. A postulate.

5. No.

Earth, Sun, and Venus.

•6. They are collinear.

•7. Yes; 160 million miles. ($93 + 67 = 160$.)

8. Yes; 26 million miles. ($93 - 67 = 26$.)

9. Not possible.

Spotter Problem.

•10. $PA + PB > 12$, $PA + PC > 12$, and $PB + PC > 12$.

11. $(PA + PB) + (PA + PC) + (PB + PC) > 36$.

12. $2PA + 2PB + 2PC > 36$.

13. $PA + PB + PC > 18$.

14. It is more than 18 km.

SAT Problem.

15. No. The sides of the triangle cannot be 2, 7, and 3, because $2 + 3 < 7$.

16. One triangle. The only integer that will work for x is 6.

Distance and Collinearity.

•17. A, B, and C are not collinear.

18. In $\triangle ABC$, $AB + BC > AC$. The sum of any two sides of a triangle is greater than the third side.

•19. $AB + BC = AC$.

20. A, B, and C must be collinear.

The Third Side.

21. Yes. Because the triangle is isosceles, the length of the third side must be either 4 or 9. It can't be 4, because $4 + 4 < 9$; so it must be 9.

22. Yes. From the Triangle Inequality Theorem we know that, if x is the length of the third side, then $5 + 7 > x$, and $5 + x > 7$. So $2 < x < 12$.

23. If the length of the third side is x, then either $x^2 = 6^2 + 8^2$ or $6^2 + x^2 = 8^2$. So either $x^2 = 36 + 64 = 100$ and $x = 10$ or $36 + x^2 = 64$ so that $x^2 = 28$ and $x = \sqrt{28} \approx 5.3$.

Set II (pages 203–204)

Heron's Proof.

•24. An angle has exactly one line that bisects it.

25. If an angle is bisected, it is divided into two equal angles.

26. An exterior angle of a triangle is greater than either remote interior angle.

•27. Substitution.

•28. If two angles of a triangle are unequal, the sides opposite them are unequal in the same order.

29. Addition.

30. Betweenness of Points Theorem.

31. Substitution.

Quadrilateral Inequality.

32.

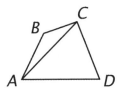

Proof.
(1) ABCD is a quadrilateral. (Given.)
(2) Draw AC. (Two points determine a line.)
(3) AB + BC > AC and AC + CD > AD. (The sum of any two sides of a triangle is greater than the third side.)
(4) AB + BC + CD > AC + CD. (Addition.)
(5) AB + BC + CD > AD. (Transitive.)

Light Path.

•33. The sum of any two sides of a triangle is greater than the third side.

34. Betweenness of Points Theorem.

•35. Substitution.

36. SAS.

37. Corresponding parts of congruent triangles are equal.

38. Substitution.

39. Vertical angles are equal.

40. Corresponding parts of congruent triangles are equal.

41. Substitution.

42. AY + YB > AX + XB (the statement in exercise 38).

43. ∠1 = ∠3 (the statement in exercise 41).

Work Triangle.

•44. 5.5 m. ($7 - 1.5 = 5.5$.)

45. $(5.5 - x)$ m.

46.

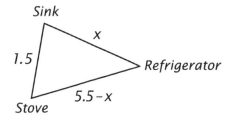

47. (1) $x + 1.5 > 5.5 - x$; so $2x > 4$, and so $x > 2$.
   (2) $x + (5.5 - x) > 1.5$; so $5.5 > 1.5$ (which doesn't tell us anything about x).
   (3) $1.5 + (5.5 - x) > x$; so $7 - x > x$; so $7 > 2x$, and so $x < 3.5$.

48. It should be more than 2 m but less than 3.5 m.

Set III (page 205)

We are, of course, assuming that our plane geometry is a reasonably good approximation for what is actually a problem in spherical geometry. The distances in this problem are "air distances" as reported in the *World Almanac*. The incorrect number is the distance between Paris and Rome; it is actually 690 miles.

1. *(London, Paris, and Cairo)*

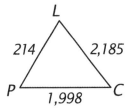

2. *(London, Rome, and Cairo)*

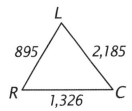

3. (Paris, Rome, and Cairo)

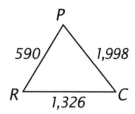

4. The London-Paris-Rome triangle, because
214 + 590 = 804 < 895.
The Paris-Rome-Cairo triangle, because
590 + 1,326 = 1,916 < 1,998.

5. The length that the two impossible triangles
have in common is 590, which suggests that
the distance between Paris and Rome is
wrong.

Chapter 5, Review

Set I (pages 206–208)

The Gateway Arch in St. Louis, at 630 ft, is
approximately twice as tall as the Statue of
Liberty (305 ft) and half as tall as the Empire State
Building (1,250 ft). Designed by Eero Saarinen in
the shape of an inverted catenary, the arch was
completed in 1965. The cross sections of its legs
are equilateral triangles with sides 54 ft long at
ground level, tapering to 17 ft at the top. Like the
top-hat illusion in Lesson 1, the arch gives the
impression of being taller than it is wide; the two
dimensions are actually the same.

1. In the tree trunk.

2. From the right edge at a sharp angle.

3. Diamond will scratch glass.

4. The transitive property.

Gateway Arch.

• 5. The "three possibilities" property.

6. (*Student answer.*) (Most people see the arch
as looking taller than it is wide.)

• 7. Both dimensions are 2.5 in.

8. Each is 630 ft. (2.5 × 252 = 630.)

Portuguese Theorem.

9. The Exterior Angle Theorem.

10. An exterior angle of a triangle is greater
than either remote interior angle.

11. An exterior angle of a triangle is an angle
that forms a linear pair with an angle of the
triangle.

• 12. One angle has a measure of 179° and the
other two angles are each less than 1°.

13. One angle is a right angle and the other two
angles are acute.

14. *Every* triangle has several exterior angles
that are obtuse; so this doesn't tell us
anything.

Soccer Angle.

15.

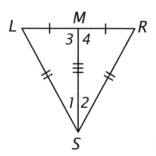

• 16. The midpoint of a line segment divides it
into two equal segments.

• 17. Reflexive.

18. SSS.

19. Corresponding parts of congruent triangles
are equal.

• 20. If the angles in a linear pair are equal, their
sides are perpendicular.

21.

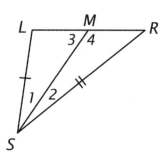

• 22. If two sides of a triangle are unequal, the
angles opposite them are unequal in the
same order.

23. An exterior angle of a triangle is greater
than either remote interior angle.

•24. Transitive.

25. If two angles of a triangle are unequal, the sides opposite them are unequal in the same order.

Roman Column.

•26. The small angles each appear to be equal to 10°.

27. They increase as you look upward.

28. So that the sections would look equal to someone standing at the base of the column.

•29. An exterior angle of a triangle is greater than either remote interior angle.

30. Transitive.

Set II (pages 208–209)

Integers and Triangles.

31. $1 + 3 < 5$ and $2 + 4 = 6$. For a triangle to be possible, both of these sums would have to be *greater* than the third length.

32. They are collinear.

33. *Example figure:*

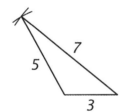

•34. $n > 2$. $(2n + 2 > n + 4, n > 2.)$

35. Yes. We've just shown that a triangle is possible as long as $n > 2$.

36. An equilateral triangle.

Different Definition.

•37. One.

38. Our definition says that a triangle is isosceles if it has *at least* two equal sides; so the third side could also be equal.

•39. Yes.

40. (1) If the triangle is equilateral, all three sides can be 10.
    (2) If the triangle has exactly two equal sides and the side of length 10 is one of them, then the second side is 10 and the third side is less than 20.
    (3) If the triangle has exactly two equal sides and the side of length 10 is not one of them, then each of the other sides must be more than 5.

Screen Display.

•41. SSS.

42. Corresponding parts of congruent triangles are equal.

•43. Betweenness of Rays Theorem.

44. The "whole greater than part" theorem.

45. Substitution.

46. If two angles of a triangle are unequal, the sides opposite them are unequal in the same order.

SAT Problem.

47. $x + (x - 2) > 7 - x$.
    $x + (7 - x) > x - 2$.
    $(x - 2) + (7 - x) > x$.

48. $2x - 2 > 7 - x$, so $3x > 9$; so $x > 3$.
    $7 > x - 2$; so $9 > x$, and so $x < 9$.
    $5 > x$; so $x < 5$.
    Therefore, $3 < x < 5$.

49. *Proof.*
    (1) AB > AC. (Given.)
    (2) $\angle C > \angle B$. (If two sides of a triangle are unequal, the angles opposite them are unequal in the same order.)
    (3) $\angle A$ and $\angle B$ are complementary. (Given.)
    (4) $\angle A + \angle B = 90°$. (The sum of two complementary angles is 90°.)
    (5) $\angle A + \angle C > \angle A + \angle B$. (Addition.)
    (6) $\angle A + \angle C > 90°$. (Substitution.)

50. *Proof.*
   (1) XB = XC. (Given.)
   (2) ∠C = ∠XBC. (If two sides of a triangle are equal, the angles opposite them are equal.)
   (3) ∠ABC = ∠ABX + ∠XBC. (Betweenness of Rays Theorem.)
   (4) ∠ABC > ∠XBC. (The "whole greater than part" theorem.)
   (5) ∠ABC > ∠C. (Substitution.)
   (6) AC > AB. (If two angles of a triangle are unequal, the sides opposite them are unequal in the same order.)

Chapter 5, Algebra Review (page 210)

•1. $\dfrac{32}{40} = \dfrac{4(8)}{5(8)} = \dfrac{4}{5}$.

•2. $\dfrac{6x}{16} = \dfrac{2(3x)}{2(8)} = \dfrac{3x}{8}$.

•3. $\dfrac{x}{x^2} = \dfrac{1(x)}{x(x)} = \dfrac{1}{x}$.

•4. $\dfrac{5(x+1)}{10} = \dfrac{5(x+1)}{5(2)} = \dfrac{x+1}{2}$.

•5. $\dfrac{4x+4}{7x+7} = \dfrac{4(x+1)}{7(x+1)} = \dfrac{4}{7}$.

•6. $\dfrac{x-7}{3x-21} = \dfrac{x-7}{3(x-7)} = \dfrac{1}{3}$.

•7. $\dfrac{x+5}{x^2-25} = \dfrac{x+5}{(x-5)(x+5)} = \dfrac{1}{x-5}$.

•8. $\dfrac{x^2-xy}{xy-y^2} = \dfrac{x(x-y)}{y(x-y)} = \dfrac{x}{y}$.

•9. $\dfrac{x^2+5x+6}{10x+20} = \dfrac{(x+2)(x+3)}{10(x+2)} = \dfrac{x+3}{10}$.

•10. $\dfrac{x^2+x-20}{x^2-x-30} = \dfrac{(x+5)(x-4)}{(x+5)(x-6)} = \dfrac{x-4}{x-6}$.

•11. $\dfrac{1}{x} = \dfrac{4(1)}{4(x)} = \dfrac{4}{4x}$.

•12. $\dfrac{1}{x} = \dfrac{(x^3)(1)}{(x^3)(x)} = \dfrac{x^3}{x^4}$.

•13. $\dfrac{3}{2-x} = \dfrac{(-1)(3)}{(-1)(2-x)} = \dfrac{-3}{x-2}$.

•14. $\dfrac{4(2)}{4(3)}$ and $\dfrac{3(3)}{3(4)}$, $\dfrac{8}{12}$ and $\dfrac{9}{12}$.

•15. $\dfrac{7(4)}{7(x)}$ and $\dfrac{2}{7x}$, $\dfrac{28}{7x}$ and $\dfrac{2}{7x}$.

•16. $\dfrac{y(1)}{y(x)}$ and $\dfrac{x(1)}{x(y)}$, $\dfrac{y}{xy}$ and $\dfrac{x}{xy}$.

•17. $\dfrac{3(x)}{3(x+2)}$ and $\dfrac{2}{3x+6}$, $\dfrac{3x}{3x+6}$ and $\dfrac{2}{3x+6}$.

•18. $\dfrac{5(x+1)}{(x-1)(x+1)}$ or $\dfrac{5x+5}{x^2-1}$ and $\dfrac{x}{(x-1)(x+1)}$ or $\dfrac{x}{x^2-1}$.

•19. $\dfrac{x(x)}{x(yz)}$ and $\dfrac{y(y)}{y(xz)}$, $\dfrac{x^2}{xyz}$ and $\dfrac{y^2}{xyz}$.

Set I (pages 214–215)

The exercises on folding a paper chain suggest some possible exploration on the student's part. Cutting and manipulating a paper chain of the type illustrated reveals examples of how the strip could have been folded any number of times from three (a "binary" folding approach) to seven (an "accordion" folding approach). In general, 2^n penguins require at least n folds but could be produced by as many as $2^n - 1$ folds. Even exploring the physical limit on the possible value of n for the binary-folding approach yields an answer that surprises most students.

Rope Trick.

1.

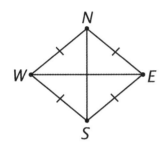

2. Two points each equidistant from the endpoints of a line segment determine the perpendicular bisector of the line segment.

3. Yes.

•4. Yes, because W and E are equidistant from the endpoints of NS.

Lines of Symmetry.

5.

6.

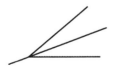

7.

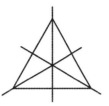

8.

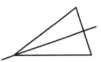

9.

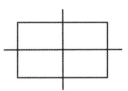

10.

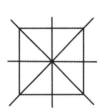

Baseball Diamond.

11.

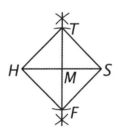

•12. 6 cm.

•13. 90 ft. ($6 \cdot 15 = 90$.)

14. 127.5 ft. ($8.5 \cdot 15 = 127.5$.)

•15. In the west.

16. So that the sun doesn't get in the batter's eyes.

17. Because, as he faces the batter, his left arm is on the south.

Paper Chain.

18. *(Student answer.)* (Any number from three to seven is correct.)

•19. Seven.

20. Four times. ($2^4 = 16$. Unlike exercise 18, this exercise asks for the *minimum* rather than for simply the possible number of folds.)

Stealth Bomber.

• 21. Line *l* is the perpendicular bisector of ST.

• 22. If a line segment is bisected, it is divided into two equal segments.

23. Perpendicular lines form right angles.

24. All right angles are equal.

25. Reflexive.

26. SAS.

27. Corresponding parts of congruent triangles are equal.

Set II (pages 215–217)

The "Star Geometry" problem is another version of the surfer's puzzle but with a regular pentagon instead of an equilateral triangle. An explanation of why the sum of the five distances is independent of the choice of the point is based on area. For this figure, the "exact" sum is $\frac{55}{8 \tan 36°}$, or approximately 9.46 inches. We will return to the surfer's puzzle later.

In his book titled *Inversions* (Byte Books, 1981), Scott Kim wrote of his amazing PROBLEM/SOLUTION figure: "Reflecting on this PROBLEM yields a surprising SOLUTION. Notice that reflection about a diagonal axis preserves a pen angle of 45 degrees, so that calligraphic conventions are maintained."

The figure drawn by Kepler was for Albrecht von Wallenstein, one of the most famous military commanders of his time. He was Emperor Ferdinand II's major general during Europe's Thirty Year War. Charles Blitzer in his book titled *Age of Kings* (Time-Life Books, 1967) wrote that, although hundreds of books have been written about Wallenstein, "the most illuminating analysis of this complex man" is presented in Kepler's diagram.

Race to the Fence.

28.

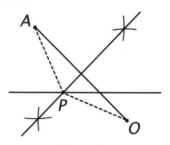

29. It would have been the point midway between them.

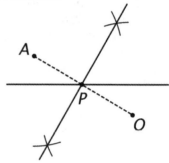

Star Geometry.

30. *Example answer:*

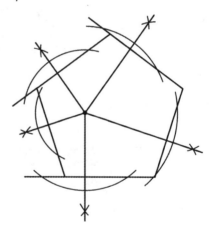

• 31. 9.5 in.

32. The result doesn't depend on the choice of the point.

Linkage Problem.

33. Because points A and C are equidistant from the endpoints of BD (two points each equidistant from the endpoints of a line segment determine the perpendicular bisector of the line segment.)

•34. C and E; D and F.

35.

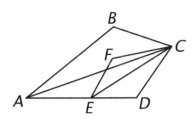

•36. Because they are corresponding parts of congruent triangles (ΔABC ≅ ΔADC by SSS).

37. Yes. ∠B = ∠D (exercise 36) and ∠F = ∠D because ΔEFC ≅ ΔEDC; so ∠B = ∠F by substitution.

One Problem, Two Solutions.

38. At a 45° angle from the lower left to the upper right.

•39. N.

40. T.

41. The word SOLUTION is readable on both sides of the symmetry line.

Kepler's Diagram.

42. *(There are various ways to construct this figure.)*

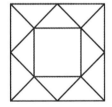

•43. Three.

44. Isosceles right triangles.

45. 16.

Set III (page 218)

In the reference cited in the footnote, Martin Gardner reports that Hallard Croft, a mathematician at Cambridge University, "asked if there existed a finite set of points on the plane such that the perpendicular bisector of the line segment joining any two points would always pass through at least two other points of the set." The figure in the Set III exercise is the solution using only eight points discovered by Leroy Kelly of Michigan State University. (The four central

points are the vertices of a square, and the four outer points are the vertices of four equilateral triangles erected on the sides of the square.) Although there are 28 possible pairings of the eight points, the symmetries of the figure are such that only six of them are distinct. These six possibilities are illustrated in the figures below.

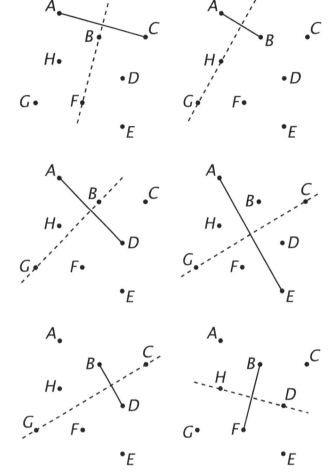

An Unusual Set of Points.

Every perpendicular bisector passes through two other points of the figure.

Chapter 6, Lesson 2

Set I (pages 221–222)

The figure for exercises 17 through 23 illustrates a theorem of projective geometry. The theorem states that, if the vertices of a [self-intersecting] hexagon lie alternately on two lines and if two pairs of opposite sides are respectively parallel, then the third pair of opposite sides also are parallel. If the two lines are parallel, this theorem is easily proved. In contrast, the fact that it is true even if the lines are not parallel seems surprising from the perspective of Euclidean geometry, and how it might be proved is not obvious.

- •1. (1) $\angle 1 = \angle 2$.
  (2) $\angle 2 = \angle 3$.
  (3) $\angle 1 = \angle 3$.
  (4) $a \parallel b$.

  2. (1) $\angle 1$ and $\angle 2$ are supplementary.
  (2) $\angle 2$ and $\angle 3$ are supplementary.
  (3) $\angle 1 = \angle 3$.
  (4) $a \parallel b$.

  3. (1) $a \perp c$ and $b \perp c$.
  (2) $\angle 1$ and $\angle 2$ are right angles.
  (3) $\angle 1 = \angle 2$.
  (4) $a \parallel b$.

Folded Paper.

- •4. Alternate interior angles.

  5. Equal alternate interior angles mean that lines are parallel.

- •6. In a plane, two lines perpendicular to a third line are parallel.

Snake Track.

- •7. Corresponding angles.

  8. They are parallel.

- •9. Equal corresponding angles mean that lines are parallel.

  10. Interior angles on the same sides of the transversal.

  11. They are supplementary.

  12. They are parallel.

  13. Supplementary interior angles on the same side of a transversal mean that lines are parallel.

I-Beam.

- •14. No.

  15. Yes.

  16. Yes.

Parallel Lines.

17.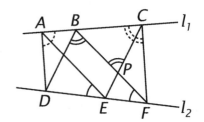

- •18. AE $\parallel$ BF.

  19. Equal corresponding angles mean that lines are parallel.

  20. DB $\parallel$ EC(PC).

  21. Equal alternate interior angles mean that lines are parallel.

  22. AD $\parallel$ CF.

  23. Supplementary interior angles on the same side of a transversal mean that lines are parallel.

Set II (pages 223–224)

Drafting Triangles.

24. $\triangle ABC \cong \triangle FED$ because the drafting triangles are identical; so $\angle A = \angle F$. AB $\parallel$ EF because equal alternate interior angles mean that lines are parallel.

25. $\angle BCA = \angle EDF$; so BC $\parallel$ DE because equal alternate interior angles mean that lines are parallel.

Optical Illusion.

26.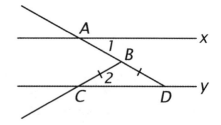

- •27. No. These angles are not formed by a transversal. They could be equal and yet lines x and y could intersect either to the left or to the right.

28. Yes. If BD = BC, $\angle 2 = \angle BDC$ (if two sides of a triangle are equal, the angles opposite them are equal). Because $\angle 1 = \angle 2$, $\angle 1 = \angle BDC$ (substitution). So $x \parallel y$ (equal alternate interior angles mean that lines are parallel).

Different Proofs.

29. (2) Two lines are perpendicular if they form a right angle.
    (3) In a plane, two lines perpendicular to a third line are parallel.

30. (2) All right angles are equal.
    (3) Equal alternate interior angles mean that lines are parallel.

Construction Exercise.

31.

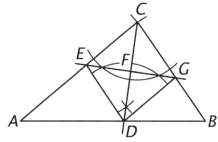

•32. Four.

33. ASA (∠ECF = ∠GCF, CF = CF, ∠CFE = ∠CFG.)

34. Corresponding parts of congruent triangles are equal.

•35. SAS. [CF = FD and CF ⊥ FD (construction), EF = FG (exercise 34).]

36. Corresponding parts of congruent triangles are equal.

37. Equal alternate interior angles mean that lines are parallel.

•38. No.

39. *Proof.*
    (1) AE = AD and ∠E = ∠BCE. (Given.)
    (2) ∠E = ∠ADE. (If two sides of a triangle are equal, the angles opposite them are equal.)
    (3) ∠ADE = ∠BCE. (Substitution.)
    (4) AD ∥ BC. (Equal corresponding angles mean that lines are parallel.)

40. *Proof.*
    (1) AB = CD and AD = BC. (Given.)
    (2) BD = BD. (Reflexive.)
    (3) ΔABD ≅ ΔCDB. (SSS).
    (4) ∠ABD = ∠CDB. (Corresponding parts of congruent triangles are equal.)
    (5) AB ∥ DC. (Equal alternate interior angles mean that lines are parallel.)

The cross-bar figures are included in a section titled "The Importance of Rules" in the chapter on human perception in Lindsay and Norman's *Human Information Processing*. They write:

"When a rectilinear object is viewed straight on, the intersecting lines formed by its contours form right angles. As the object is tilted and rotated in space, the angles in the retinal image diverge from right angles. The degree of divergence depends on the orientation of the object in space. By interpreting that divergence from right angles as a result of depth, distance information can be extracted."

What Do You See?

1. *(Student answer.)* (The top bar.)

2. *(Student answer.)* (The bottom bar.)

3. The first figure is viewed "in two dimensions," seen as lying in the plane of the paper. The second figure is viewed "in three dimensions," seen as in perspective on the side of a box.

4. Yes. A third possibility is that the two cross bars do not lie in a common plane, in which case both could be perpendicular to the vertical bar.

5. It doesn't follow that they are parallel. Our theorem says that *in a plane*, two lines perpendicular to a third line are parallel.

Chapter 6, Lesson 3

Set I (pages 226–228)

Euclid's assumption that "if two lines form interior angles on the same side of a transversal whose sum is less than 180°, then the two lines meet on that side of the transversal" is his fifth postulate (his version of the Parallel Postulate).

In his commentary on Euclid's *Elements*, Sir Thomas Heath wrote: "When we consider the countless successive attempts made through more than twenty centuries to prove the Postulate, many of them by geometers of ability, we cannot but admire the genius of the man who concluded that such a hypothesis, which he found necessary to the validity of his whole system of geometry, was really indemonstrable.

From the very beginning, as we know from Proclus, the Postulate was attacked as such, and attempts were made to prove it as a theorem or to get rid of it by adopting some other definition of parallels; while in modern times the literature of the subject is enormous."

An entire chapter of David Henderson's valuable book titled *Experiencing Geometry* (Prentice Hall, 2001) is on the subject of parallel postulates. In his historical notes, Henderson remarks:

"Playfair's Parallel Postulate got its current name from the Scottish mathematician, John Playfair (1748–1819), who brought out successful editions of Euclid's *Elements* in the years following 1795. After 1800 many commentators referred to Playfair's [parallel] postulate (PPP) as the best statement of Euclid's [fifth] postulate (EFP), so it became a tradition in many geometry books to use PPP instead of EFP."

Henderson goes on to explain that in *absolute geometry*, the part of geometry that can be developed without using a parallel postulate, Playfair's postulate is equivalent to Euclid's. He then points out:

"Because absolute geometry was the focus of many investigations the statement 'PPP is equivalent to EFP' was made and is still being repeated in many textbooks and expository writings about geometry even when the context is not absolute geometry . . . . EFP is true on spheres and so cannot be equivalent to PPP, which is clearly false."

Optical Illusion.

- •1. 12.

- •2. If they form a right angle (or if they form four equal angles).

- 3. If they lie in the same plane and do not intersect.

- 4. *(Student answer.)* (Possibly white circles or squares at the "breaks" in the grid. Also broad white diagonal strips.)

Windshield Wipers.

- •5. ∠3.

- •6. Equal corresponding angles mean that lines are parallel.

- 7. Interior angles on the same side of the transversal.

- 8. Supplementary interior angles on the same side of a transversal mean that lines are parallel.

9. ∠2 and ∠4.

10. Yes. ∠1 and ∠2 (or ∠3 and ∠4) are a linear pair; so they are supplementary. If ∠1 and ∠4 are supplementary, then ∠2 = ∠4 (or ∠1 = ∠3) because supplements of the same angle are equal. It follows that the wipers are parallel because they form equal corresponding angles with the transversal.

Exactly One.

11. That a cat has at least one flea.

- •12. At least, no more than, exactly.

13. At least, no more than, exactly.

- •14. No more than.

15. No more than.

16. At least.

17. At least, no more than, exactly.

18. At least, no more than, exactly.

Parallelogram.

19.

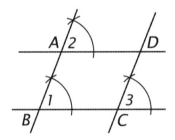

- •20. They form equal corresponding angles with a transversal.

21. They seem to be equal.

- •22. They seem to be supplementary.

23. They seem to be equal.

24. No. It can't be folded so that the two halves coincide.

Euclid's Assumption.

- •25. All four of them.

26.

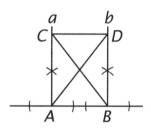

•27. Down, below the figure.

28. To the left.

Set II (pages 228–229)

The restriction to the plane in Theorem 18 is, of course, unnecessary, because the theorem is also true in space. (The proof in solid geometry is based on theorems on perpendicular lines and planes.)

Gateway.

29.

a   *b*
C   D

A   B

•30. One.

31. One.

32. They are parallel.

•33. In a plane, two lines perpendicular to a third line are parallel.

34. They seem to be parallel.

35. One.

•36. Through a point not on a line, there is exactly one line parallel to the line.

•37. Yes. (SAS; AC = BD, ∠CAB = ∠DBA, AB = AB.)

38. Yes. (Corresponding parts of congruent triangles are equal.)

39. Yes. (SSS; AC = BD, AD = BC, CD = CD.)

40. Yes. (Corresponding parts of congruent triangles are equal.)

41. No. (It will be possible to prove these triangles congruent by methods learned in future lessons, however.)

Theorem 18.

•42. *a* and *b* are not parallel.

•43. Given.

44. Through a point not on a line, there is exactly one line parallel to the line.

45. *a* ∥ *b*.

Proclus's Claim.

46. Yes.

47. If *c* doesn't intersect *b*, then *c* is parallel to *b*. This would mean that, through P, lines *c* and *a* are both parallel to *b*, which contradicts the Parallel Postulate.

Set III (page 229)

The "dragon curve" was discovered by NASA physicist John Heighway. Martin Gardner explains in the chapter titled "The Dragon Curve and Other Problems" in his book titled *Mathematical Magic Show* (Knopf, 1977) that the curve gets its name from the fact that, when it is drawn with each right angle rounded, it "vaguely resembles a sea dragon paddling to the right with clawed feet, his curved snout and coiled tail just above an imaginary waterline."

The curve can be generated in various ways including paper folding, by geometric construction, and from a sequence of binary digits. The version in the text is called an order-4 dragon because it is produced by four folds. The order-6 dragon shown below with its right angles rounded makes the origin of the name more evident.

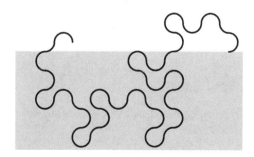

The length of the strip of paper suggested for folding the dragon was chosen so that, if a student uses graph paper of 4 units per inch, the unfolded dragon will fit nicely on the graph. Each time the paper is unfolded, the half produced is a 90° rotation. The original fold point has been placed at the origin so that someone studying the graph may discover the connection between the 90° clockwise turn and the coordinate transformation: $(a, b) \rightarrow (b, -a)$.

A Dragon Curve.

1. (The folded paper should look like the dragon curve shown in the text.)

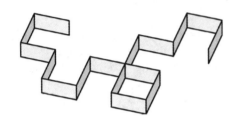

2.

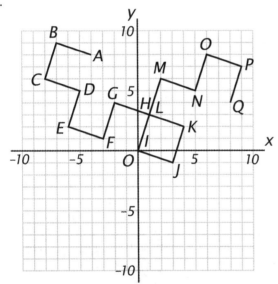

3. It looks like the edge of the folded strip.

4. Point I.

5. Points E and M.

6. It has the same shape but is rotated 90° clockwise.

7. For each point with coordinates (a, b) in the first half, there is a corresponding point with coordinates $(b, -a)$ in the second half.

Set I (pages 232–234)

The Poggendorff Illusion

The fishpole puzzle is based on an illusion created by J. C. Poggendorff in 1860. The original illusion is illustrated above. Although the oblique segments are aligned, the upper one appears raised in relation to the lower one. The Poggendorff illusion and the theories developed to explain it are discussed in depth in *Perception,* by Irvin Rock (Scientific American Library, 1984).

Which Fishpole?

•1. They are equal.

•2. Parallel lines form equal corresponding angles.

3. Fishpole A.

4. It doesn't look like fishpole A.

Corollary Proofs.

•5. (1) Given.
 (2) Parallel lines form equal corresponding angles.
 (3) Vertical angles are equal.
 (4) Substitution.

6. (1) Given.
 (2) Parallel lines form equal corresponding angles.
 (3) The angles in a linear pair are supplementary.
 (4) The sum of two supplementary angles is 180°.
 (5) Substitution.
 (6) Two angles are supplementary if their sum is 180°.

7. (1) Given.
   (2) Perpendicular lines form right angles.
   (3) A right angle is 90°.
   (4) Given.
   (5) Parallel lines form equal corresponding angles.
   (6) Substitution.
   (7) A 90° angle is a right angle.
   (8) Lines that form a right angle are perpendicular.

Bent Pencil.

8. Parallel lines form equal corresponding angles.

•9. Betweenness of Rays Theorem.

10. Substitution.

11. Subtraction.

•12. Equal corresponding angles mean that lines are parallel.

Find the Angles.

•13. ∠1 = 115°, ∠2 = 65°, ∠3 = 115°.

14. ∠1 = 90°, ∠2 = 90°, ∠3 = 90°.

15. ∠1 = 70°, ∠2 = 55°.

16. ∠1 = 110°, ∠2 = 70°, ∠3 = 110°.

City Streets.

•17. They must be perpendicular because, in a plane, a line perpendicular to one of two parallel lines is also perpendicular to the other.

18. They must be parallel because, in a plane, two lines perpendicular to a third line are parallel.

19. 113°, 67°, and 113°.

•20. 67°, 113°, 67°, and 113°.

21. Parallel lines form equal corresponding angles.

T Puzzle.

•22. ∠1 and ∠2 are supplementary because parallel lines form supplementary interior angles on the same side of a transversal.

23. ∠2 = ∠3 because parallel lines form equal corresponding angles.

•24. ∠3 = ∠4 because parallel lines form equal alternate interior angles.

25. ∠2 = ∠4 by substitution.

26. ∠4 and ∠5 are complementary because they together with the right angle add up to 180°.

27. ∠5 and ∠6 are supplementary because parallel lines form supplementary interior angles on the same side of a transversal.

Set II (pages 234–235)

In his book titled *The Flying Circus of Physics* (Wiley, 1977), Jearl Walker reports: "Looking down on desert sand dunes from a high altitude airplane, one sees curious long, narrow dune belts running across the desert, roughly from north to south, in almost straight lines, as if one were viewing well-designed parallel streets. The dune belts are characteristic of every major desert in the world." Walker explains that the belts are caused by air currents and that the dominant winds in every desert of the world are north or south.

Parallel Construction.

28.

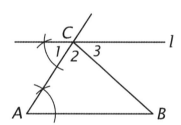

29. Equal alternate interior angles mean that lines are parallel.

•30. Through a point not on a line, there is exactly one line parallel to the line (the Parallel Postulate).

•31. Parallel lines form equal alternate interior angles.

32. 180°.

33. 180°.

Overlapping Angles.

34. Yes. ∠B = ∠DGC and ∠DGC = ∠E because parallel lines form equal corresponding angles. So ∠B = ∠E by substitution.

35. No. If one angle were turned with respect to the other, their sides would obviously not be parallel.

Sand Dune Lines.

36.

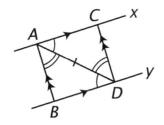

•37. Two points determine a line.

38. Parallel lines form equal alternate interior angles.

•39. In a plane, two lines perpendicular to a third line are parallel.

40. Parallel lines form equal alternate interior angles.

41. Reflexive.

42. ASA. ($\angle$CAD = $\angle$BDA, AD = AD, $\angle$BAC = $\angle$CDA.)

43. Corresponding parts of congruent triangles are equal.

•44. The length of a perpendicular segment between them.

SAT Problem.

45. $x + (2x - 30) = 180$.

•46. $3x = 210$; so $x = 70$. The acute angle is 70°.

47. $2x - 30 = 2(70) - 30 = 110°$. The obtuse angle is 110°.

Construction Problem.

48.

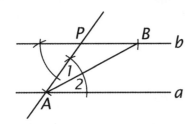

49. $\angle$1 and $\angle$2 seem to be equal.

50. Because $a \parallel b$, $\angle$PBA = $\angle$2 (parallel lines form equal alternate interior angles). Also, PA = PB, and so $\angle$1 = $\angle$PBA (angles opposite equal sides in a triangle are equal). So $\angle$1 = $\angle$2 by substitution.

Set III (page 235)

The context of Kepler's method for determining the shape of Earth's orbit is set forth by David Layzer in *Constructing the Universe* (Scientific American Library, 1984). The parallel lines are in the direction of an arbitrary star. The angle numbered 1 is measured when Mars is "at opposition," that is, when Earth is in alignment between Mars and the sun (M-E-S). The point E' is the position of Earth 1 Martian year later. Layzer explains: "By means of observations of Mars and the sun at intervals separated by one Martian year, Kepler calculated the shape of the earth's orbit and its dimension relative to the reference distance SM."

Plotting successive positions of Earth in this way led to Kepler's realization that the orbits of the planets are elliptical rather than circular and to his three laws of planetary motion to describe this motion.

1. $\angle$4 = $\angle$1 because parallel lines form equal corresponding angles.

2. $\angle$5 = 180° − $\angle$3 because parallel lines form supplementary interior angles on the same side of a transversal.

3. $\angle$6 = $\angle$4 − $\angle$5.

4. $\angle$M is the third angle of $\triangle$MSE'; so $\angle$M = 180° − $\angle$2 − $\angle$6.

5. $\angle$4 = $\angle$1 = 84°.
   $\angle$5 = 180° − $\angle$3 = 180° − 122° = 58°.
   $\angle$6 = $\angle$4 − $\angle$5 = 84° − 58° = 26°.
   $\angle$M = 180° − $\angle$2 − $\angle$6 = 180° − 124° − 26° = 30°.

6.

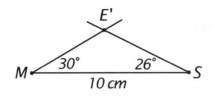

Set I (pages 238–239)

It is strange to think that a steel rod bent into the shape of a triangle would make a good musical instrument and to realize that geometry is useful in describing the sounds produced. They depend in part on where the triangle is struck. According to Thomas Rossing in *The Science of Sound* (Addison-Wesley, 1990), single strokes played on the base create vibrations in the plane of the triangle; tremolos are played between the sides of the vertex angle and create vibrations perpendicular to the plane of the triangle.

The primary appearance of geometry in economics is in graphs. The consumption function example is about the "break even" point. As described in *Economics* by Paul Samuelson and William Nordhaus (McGraw-Hill, 1989), "at any point on the 45° line, consumption exactly equals income . . . . The 45° line tells us immediately, therefore, whether consumption spending is equal to, greater than, or less than the level of income. The point on the consumption schedule where it intersects the 45° line shows us the level of disposable income at which households just break even."

Corollary Proofs.

- 1. The sum of the angles of a triangle is 180°.
- 2. Substitution.
- 3. Substitution.
- 4. Subtraction.
  5. A right angle equals 90°.
  6. Substitution.
  7. Subtraction.
  8. Two angles are complementary if their sum is 90°.
- 9. An equilateral triangle is equiangular.
- 10. An equiangular triangle has three equal angles.
  11. Substitution (and algebra).
  12. Division.
  13. Substitution.

Exterior Angle Theorem.

- 14. Through a point not on a line, there is exactly one line parallel to the line.
  15. Parallel lines form equal alternate interior angles.
- 16. Parallel lines form equal corresponding angles.
  17. Betweenness of Rays Theorem.
  18. Substitution.

Hot Air Balloon Angles.

- 19. $\angle 2 > \angle 1$ and $\angle 2 > \angle A$.
  20. $\angle 2 = \angle 1 + \angle A$.
  21. 28°.

Right and Wrong.

  22. Acute.
  23. The acute angles of a right triangle are complementary; so their sum is a right angle.

Another Kind of Triangle.

- 24. An equilateral triangle.
  25. As a musical instrument.
  26. 60°.

Economics Graph.

- 27. It is a right triangle.
- 28. They are complementary.
  29. 45°.
  30. If two angles of a triangle are equal, the sides opposite them are equal.

Set II (pages 240–241)

The direction of the North Star is treated in exercises 31 through 37 as if it coincides exactly with that of celestial north even though there is a slight discrepancy. Peculiar, a small town near the western border of Missouri, got its name by accident. When the residents applied for a post office, several names that they proposed were rejected because other towns in the state had already taken them. Finally the appointed post master for the town wrote to Washington, D.C., saying, if the names previously submitted would not do, then please assign some "peculiar" name to the town. The Post Office Department did exactly that.

The catch question about the height of the triangle was created by Leo Moser of the University of Alberta.

North Star and Latitude.

31. Parallel lines form equal alternate interior angles.

32. ∠4 and ∠3 are complementary.

33. The acute angles of a right triangle are complementary.

•34. ∠2 and ∠3 are complementary.

35. ∠4 = ∠2.

•36. Complements of the same angle are equal.

37. Substitution. [∠1 = ∠4 (exercise 31) and ∠4 = ∠2 (exercise 35).]

Angle Bisectors.

38.

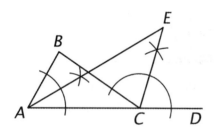

39. (*Student answer.*) (∠B seems to be twice as large as ∠E.)

•40. An exterior angle of a triangle is equal to the sum of the remote interior angles.

•41. Substitution.

42. Multiplication.

43. Substitution.

44. Subtraction.

45. *Proof.*
    (1) In ΔABC and ΔADE, ∠ADE = ∠B. (Given.)
    (2) ∠A = ∠A. (Reflexive.)
    (3) ∠AED = ∠C. (If two angles of one triangle are equal to two angles of another triangle, the third angles are equal.)

46. *Proof.*
    (1) ΔABC is a right triangle with right ∠ABC; ∠ABD and ∠C are complementary. (Given.)
    (2) ∠A and ∠C are complementary. (The acute angles of a right triangle are complementary.)
    (3) ∠ABD = ∠A. (Complements of the same angle are equal.)
    (4) AD = BD. (If two angles of a triangle are equal, the sides opposite them are equal.)
    (5) ΔABD is isosceles. (A triangle is isosceles if it has at least two equal sides.)

Catch Question.

47.

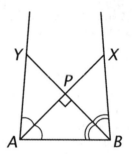

48. Because AX ⊥ BY, ΔABP is a right triangle; so ∠PAB and ∠PBA are complementary. Therefore, ∠PAB + ∠PBA = 90°. Because AX and BY bisect two of the angles, ∠YAP = ∠PAB and ∠XBP = ∠PBA. It therefore follows that ∠YAB + ∠XBA = 180°. If ∠YAB and ∠XBA are supplementary, however, then AY ∥ BX (supplementary interior angles on the same side of a transversal mean that lines are parallel).

The problem is a "catch" question because AY and BX do not intersect to form a triangle.

Set III (page 241)

As this proof without words establishes, the sum of the corner angles of any five-pointed star, no matter how irregular, is 180°. A possible area of exploration would be to draw stars with other numbers of points and try to discover if anything comparable is true of them. (We will return to this problem in our study of circles. Using the fact that an inscribed angle is half the measure of its intercepted arc makes it easy to show for the special case of symmetric stars with an odd number of sides that the sum of the corner angles is 180°.)

1. The sum of the angles that are the corners of a five-pointed star is 180°.

2. A parallel line is drawn through one corner of the star to form angles with one of the sides extended. Applying the exterior angle theorem to two of the large overlapping triangles gives 1 + 3 and 2 + 4 in the smaller triangle on the right. Equal corresponding angles and equal alternate interior angles formed by the parallels reveal that 5 + (2 + 4) + (1 + 3) = 180°.

Chapter 6, Lesson 6

Set I (pages 245–246)

The Rogallo glider is named after Francis Rogallo, a NASA engineer who invented its triangular-winged design in the late 1940s. According to James E. Mrazek in *Hang Gliding and Soaring* (St. Martin's Press, 1976), Rogallos have been tested by the U. S. Army at altitudes of 40 miles and speeds of 2,100 miles per hour and have successfully carried loads as great as three tons! NASA has even considered using radio-controlled Rogallos to bring 50-ton spacecraft boosters back to Earth for reuse.

The Roman alphabet is used almost without exception throughout the world to write the symbols of mathematics, even in such languages as Russian, Hebrew, and Chinese. No matter where you study geometry, the triangles are probably named ABC!

Hang Glider.

•1. ASA.

2. SSS.

3. SAS.

4. AAS.

•5. HL.

Congruent or Not?

•6. Parallel lines form equal corresponding angles.

7. Three.

8. No.

•9. Three.

10. No.

11. Three.

•12. Yes.

13. They are congruent by AAS.

Korean Theorem.

14. That ∠C and ∠C′ are right angles.

15. That ∆ABC and ∆A′B′C′ are congruent.

16. The HL Theorem. If the hypotenuse and a leg of one right triangle are equal to the corresponding parts of another right triangle, the triangles are congruent.

17. Right, hypotenuse, and side.

Triangle Problem.

18.

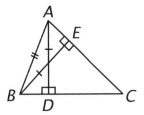

•19. HL.

20. Corresponding parts of congruent triangles are equal.

•21. If two angles of a triangle are equal, the sides opposite them are equal.

22. Isosceles.

Set II (pages 246–248)

Richard G. Pawley explored the question, "Under what conditions can two triangles have five parts of one congruent to five parts of the other, but *not* be congruent?" in an article in *The Mathematics Teacher* (May 1967). He called such triangles "5-con" and proved, among other things, that the sides of such triangles must be in geometric sequence. The angle measures for the two triangles in exercise 52 have been rounded to the nearest degree. Their exact and more precise values are: $\angle A = \text{Arccos } \frac{101}{108} \approx 20.74°$;

$\angle B = \text{Arccos } \frac{-29}{48} \approx 127.17°$; $\angle C = \text{Arccos } \frac{61}{72} \approx$

32.09°. Calculating these values might make an amusing problem for trigonometry students to do.

Mirror Distances.

23.

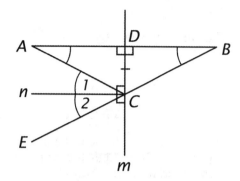

•24. In a plane, two lines perpendicular to a third line are parallel.

25. Parallel lines form equal alternate interior angles.

26. Parallel lines form equal corresponding angles.

•27. Substitution.

28. Reflexive.

•29. AAS.

30. Corresponding parts of congruent triangles are equal.

Angle Bisection.

31.

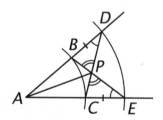

•32. AB = AC, AD = AE, and ∠A = ∠A.

33. SAS.

34. Corresponding parts of congruent triangles are equal.

35. BD = CE (because AD – AB = AE – AC), ∠ADC = ∠AEB, and ∠BPD = ∠CPE.

36. AAS.

37. Corresponding parts of congruent triangles are equal.

38. One set of parts: AD = AE, ∠ADC = ∠AEB, and DP = PE. (Another set of parts: AD = AE, DP = PE, AP = AP.)

39. SAS (or SSS).

•40. Because ∠DAP = ∠EAP (corresponding parts of congruent triangles are equal).

Origami Frog.

41.

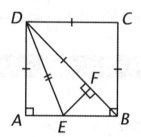

•42. SSS.

43. Corresponding parts of congruent triangles are equal.

•44. 45°.

45. 45°.

46. If two angles of a triangle are equal, the sides opposite them are equal.

47. Right triangles.

48. HL.

49. Corresponding parts of congruent triangles are equal.

•50. Substitution.

More on Equal Parts.

51. Two triangles that have these equal parts must be congruent by SSS.

52.

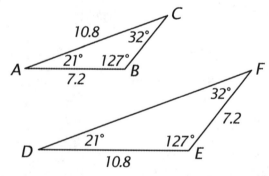

53. Yes. ΔABC and ΔDEF are an example.

54. No. Two triangles that have six pairs of equal parts must be congruent by definition (or SSS).

Set III (page 248)

An entire "Mathematical Games" column by Martin Gardner in *Scientific American* is on the topic of geometrical fallacies. It is included in his *Wheels, Life and Other Mathematical Amusements* (W. H. Freeman and Company, 1983). Gardner reports that Euclid wrote a book on the subject. Called *Pseudaria*, the book is lost and there is no record of its content.

The proof that every triangle is isosceles appeared for the first time in W. W. Rouse Ball's *Mathematical Recreations and Essays*. Lewis Carroll included it in *The Lewis Carroll Picture Book* (1899). As Gardner explains: "The error is one of construction. E is always outside the triangle and at a point such that, when perpendiculars are drawn from E to sides CA and CB, one perpendicular will intersect one side of the triangle but the other will intersect an extension of the other side."

Is Every Triangle Isosceles?

1.

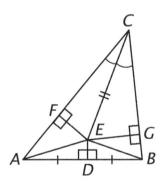

2. AAS.

3. Corresponding parts of congruent triangles are equal.

4. SAS.

5. Corresponding parts of congruent triangles are equal.

6. HL. [EA = EB (exercise 5), EF = EG (exercise 3).]

7. Corresponding parts of congruent triangles are equal.

8. Addition. (FC = GC because ΔCEF ≅ ΔCEG.)

9. Betweenness of Points Theorem.

10. Substitution.

11. The false statements appear at the end. Although it is true for this triangle that A-F-C and so AF + FC = AC, it is not true that B-G-C; so BG + GC ≠ BC. So the last equation, AC = BC, does not follow.

Chapter 6, Review

Set I (pages 250–253)

Exercises 20 through 22 are based on the rather remarkable dissection of an equilateral triangle into five isosceles triangles such that none, one, or two of the triangles are equilateral. Other such dissections are possible.

Flag Symmetries.

•1. Fold the figure along the proposed line of symmetry and see if the two halves coincide.

2. The Cuban flag. The star prevents it from having a horizontal line of symmetry.

3. The Algerian flag has a horizontal line of symmetry. The flag of Barbados has a vertical line of symmetry. The Jamaican flag has two lines of symmetry: one horizontal and the other vertical.

4.

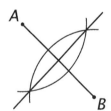

•5. It is the perpendicular bisector of AB.

Lots of Lines.

6. An overhead view of a parking lot.

•7. Equal corresponding angles mean that lines are parallel.

8.

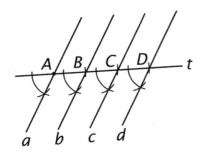

Alphabet.

- •9. To the sum of the remote interior angles.

- 10. Form equal corresponding angles.

- 11. Perpendicular to a third line are parallel.

- •12. They are perpendicular.

- 13. Interior angles mean that lines are parallel.

- •14. Are equal, their sides are perpendicular.

- 15. Are equal.

Ollie's Mistakes.

- •16. In a plane, *two points each* equidistant from the endpoints of a line segment determine the perpendicular bisector of the line segment.

- 17. If two angles *and the side opposite one of them in one triangle* are equal to *the corresponding parts* of another triangle, the triangles are congruent.

- 18. If two *angles* of one triangle are equal to two *angles* of another triangle, the third *angles* are equal.

- 19. *In a plane*, a line perpendicular to one of two parallel lines is also perpendicular to the other.

Triangle Division.

20.

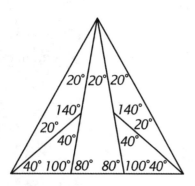

21.

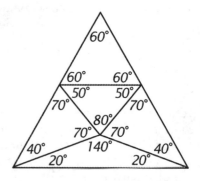

22.

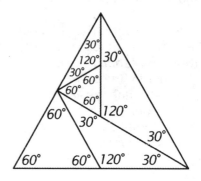

23. They are all isosceles.

24. First figure: no pieces. Second figure: one piece. Third figure: two pieces.

Angle Trisection.

25.

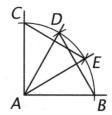

- •26. Equilateral.

- 27. ∠DAB = ∠EAC = 60°. Each angle of an equilateral triangle is 60°.

- 28. ∠EAB = 30°.
  ∠EAB = ∠CAB − ∠DAB = 90° − 60° = 30°.

Surveyor's Triangle.

- 29. Exactly one.

- •30. In a plane, two lines perpendicular to a third line are parallel.

- 31. It would bisect the angle.

32.

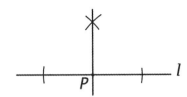

Set II (pages 253–254)

Magnifying Glass.

•33. In a plane, a line perpendicular to one of two parallel lines is also perpendicular to the other.

34. All right angles are equal.

•35. Concurrent.

36. SAS. (BD = DF, ∠BDF = ∠FDP, DP = DP.)

37. Corresponding parts of congruent triangles are equal.

38. They are equal (by addition).

Measuring a Tree.

•39. ∠B = ∠CAD = 45°.

40. BA = AC and AC = CD.

41. If two angles of a triangle are equal, the sides opposite them are equal.

•42. BA = CD by substitution.

Quadrilateral Problem.

43. ∠A and ∠D. (Or ∠B and ∠C.)

44. Parallel lines form supplementary interior angles on the same side of a transversal.

45. No. (This answer may surprise some students. An example of why the answer is "no" is an isosceles trapezoid.)

Isosceles Triangle.

46.

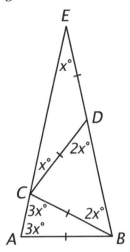

47. ∠ECD = $x°$.

•48. ∠CDB = $2x°$.

49. ∠DBC = $2x°$.

•50. ∠ACB = $3x°$. (∠ACB is an exterior angle of ΔBCE; so ∠ACB = ∠E + ∠EBC = $x° + 2x°$.)

51. ∠A = $3x°$.

52. ∠ABE = $3x°$.

53. ∠E + ∠A + ∠ABE = $x + 3x + 3x = 180$; so $7x = 180$ and $x = \dfrac{180}{7} \approx 25.7$. ∠E $\approx 25.7°$.

Stair Steps.

54.

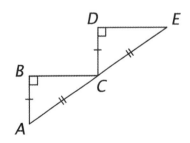

ΔABC ≅ ΔCDE (HL); so ∠BCA = ∠E (corresponding parts of congruent triangles are equal). BC ∥ DE because equal corresponding angles (with a transversal) mean that lines are parallel.

SAT Problem.

55. $q = 40$ because parallel lines form equal alternate interior angles. $p = 140$ because two angles in a linear pair are supplementary. So $p - q = 140 - 40 = 100$.

•1. $\dfrac{3}{6} + \dfrac{4}{6} = \dfrac{7}{6}$.

•2. $\dfrac{5x}{20} - \dfrac{4x}{20} = \dfrac{x}{20}$.

•3. $\dfrac{x+10}{2x}$.

•4. $\dfrac{4x}{x^3} - \dfrac{3}{x^3} = \dfrac{4x-3}{x^3}$.

•5. $\dfrac{24}{36} = \dfrac{2}{3}$.

•6. $\dfrac{x^6}{8}$.

•7. $\dfrac{1}{2} \cdot \dfrac{10}{7} = \dfrac{5}{7}$.

•8. $\dfrac{x+2}{x} \cdot \dfrac{x}{2} = \dfrac{x+2}{2}$.

•9. $\dfrac{12x}{16} = \dfrac{3x}{4}$.

•10. $\dfrac{x+4-x-2}{10} = \dfrac{2}{10} = \dfrac{1}{5}$.

•11. $\dfrac{1}{6}$.

•12. $\dfrac{10}{x} \cdot \dfrac{1}{5} = \dfrac{10}{5x} = \dfrac{2}{x}$.

•13. $\dfrac{4x}{8} + \dfrac{2y}{8} - \dfrac{z}{8} = \dfrac{4x+2y-z}{8}$.

•14. $\dfrac{8\pi r^3}{6} = \dfrac{4\pi r^3}{3}$.

•15. $\dfrac{x}{y} - \dfrac{y}{y} = \dfrac{x-y}{y}$.

•16. $\dfrac{6(x-3)}{3(2x-3)} = \dfrac{2(x-3)}{2x-3}$ or $\dfrac{2x-6}{2x-3}$.

•17. $\dfrac{3x}{(x+1)(x-1)} - \dfrac{2(x-1)}{(x+1)(x-1)} =$

$\dfrac{3x-2x+2}{(x+1)(x-1)} = \dfrac{x+2}{x^2-1}$.

•18. $\dfrac{y}{y(x-y)} + \dfrac{x}{x-y} = \dfrac{1+x}{x-y}$.

•19. $\dfrac{(x+3)(x+5)}{(x+3)(x-2)} \cdot \dfrac{x-2}{x+5} = 1$.

•20. $\dfrac{(x-1)(x^2+x+1)}{x-1} + \dfrac{1}{x-1} = \dfrac{x^3-1+1}{x-1} = \dfrac{x^3}{x-1}$.

Set I (pages 260–261)

The aerial photographer Georg Gerster, the author of *Below from Above* (Abbeville Press, 1986), wrote about his photograph of the freeway overpass covered by carpets:

"Years after having taken this photograph, I am still amazed at this super-scale Paul Klee . . . . In Europe, special mats, a Swiss development, have been adopted. They retain the heat generated as the concrete hardens and make it unnecessary to keep wetting the concrete. The new procedure is more economical, though less colorful . . . ."

Theorem 24.

- •1. Two points determine a line.
- 2. The sum of the angles of a triangle is 180°.
- •3. Addition.
- 4. Betweenness of Rays Theorem.
- 5. Substitution.

Corollary.

- •6. The sum of the angles of a quadrilateral is 360°.
- 7. Division.
- 8. An angle whose measure is 90° is a right angle.
- •9. A quadrilateral each of whose angles is a right angle is a rectangle.
- •10. Each angle of a rectangle is a right angle.
- 11. All right angles are equal.

Carpets.

- 12. Rectangles.
- 13. Right.
- •14. Lines that form right angles are perpendicular.
- 15. In a plane, two lines perpendicular to a third line are parallel.

Rectangles.

- 16. They are equal and they are right angles.
- •17. They seem to be parallel and equal.

18. They seem to be equal (and they seem to bisect each other).

Rhombuses.

- 19. They seem to be equal (and the opposite sides seem to be parallel).
- 20. They seem to be perpendicular (and they seem to bisect each other).

Squares.

- •21. They are equiangular (or each of their angles is a right angle).
- 22. They are equilateral.

Parallelograms.

- 23. They seem to be parallel and equal.
- 24. They seem to be equal.
- 25. They seem to bisect each other.

SAT Problem.

- •26. $2x + 2y = 180$. ($2x + 2y + 180 = 360$.)
- 27. $x + y = 90$.

Set II (pages 262–264)

Martin Gardner dedicated his book of mathematical recreations, *Penrose Tiles to Trapdoor Ciphers* (W. H. Freeman and Company, 1989) to Roger Penrose with these words:

"To Roger Penrose, for his beautiful, surprising discoveries in mathematics, physics, and cosmology; for his deep creative insights into how the universe operates; and for his humility in not supposing that he is exploring only the products of human minds."

Two chapters of Gardner's book are about Penrose tilings, a subject full of rich mathematical ideas. The two tiles in the exercises come from a rhombus with angles of 72° and 108°. John Horton Conway calls them "kites" and "darts." Amazingly, in *all* nonperiodic tilings of these tiles, the ratio of the number of kites to the number of darts is the golden ratio: $\frac{1+\sqrt{5}}{2} = 1.618 \ldots$! There are a number of simple ways to prove that there are infinitely many ways to tile the plane with these two tiles. Even so, any finite region of one tiling appears infinitely many times in every other tiling; so no finite sample of a Penrose tiling can be used to determine which of the infinitely many different tilings it is! Students interested in such mind-boggling ideas will enjoy reading Gardner's material and doing the experiments suggested in it.

The exercises on the angles of polygons are intended to encourage students to realize that memorizing a result such as "each angle of an equiangular n-gon has a measure of $\dfrac{(n-2)180°}{n}$"

is not important. What *is* important is for the student to feel comfortable and confident in being able to reconstruct such a result if it is needed.

Penrose Tiles.

28. The one labeled ABCD.

•29. 144°. (360° – 3 · 72°.)

30.

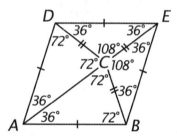

•31. SSS.

32. All are isosceles.

33. They seem to be collinear.

34. They are collinear because ∠ACE = 72° + 108° = 180°.

35. Yes. It appears that it could be folded along the line through A, C, and E so that the two halves would coincide.

36. No. Quadrilateral ABED is equilateral but not equiangular.

Gemstone Pattern.

•37. Triangles, quadrilaterals, and an octagon.

38. Five sides, two diagonals, and three triangles.

39. *Example figure:*

•40. Six sides, three diagonals, and four triangles.

41. *Example figure:*

42. Eight sides, five diagonals, and six triangles.

•43. $n - 3$.

44. $n - 2$.

45. A quadrilateral has $4 - 3 = 1$ diagonal which forms $4 - 2 = 2$ triangles.

•46. 720°. (4 × 180°.)

•47. 120°. ($\dfrac{720°}{6}$.)

48. 1,080°. (6 × 180°.)

49. 135°. ($\dfrac{1,080°}{8}$.)

•50. $(n - 2)180°$.

51. $\dfrac{(n-2)180°}{n}$.

Linkage Problems.

•52. Their sum is 360°.

53. Their sum is 180° if we think of the linkage as a triangle. (It is 360° if we think of the linkage as a quadrilateral, because ∠BCD = 180°.)

•54. They all appear to be smaller.

55. ∠A + ∠B + ∠APB = 180°, and ∠C + ∠D + ∠CPD = 180°; so ∠A + ∠B + ∠APB = ∠C + ∠D + ∠CPD. Because ∠APB = ∠CPD, it follows that ∠A + ∠B = ∠C + ∠D by subtraction.

Quadrilateral Angle Sum.

56. Addition.

57. Substitution.

58. Subtraction.

59. ∠1 and ∠2 are a linear pair; so they are supplementary and their sum is 180°. Likewise for ∠3 and ∠4. (∠1 + ∠2) + (∠3 + ∠4) = S; so 180° + 180° = S, and so S = 360°.

Set III (page 264)

This triangle-to-square puzzle is the puzzle of Henry Dudeney first introduced in the Set III exercise of Chapter 1, Lesson 3. It is perhaps surprising that, by carefully comparing the two figures and knowing that the angles of an equilateral triangle and a square are 60° and 90° respectively, we can find all of the remaining angles from just the one given. (By using the fact that the triangle and square have equal areas and applying some simple trigonometry to the triangle labeled D, we can find the exact values of *all* of the angles in the figure. The 41° angle,

for example, is rounded from Arcsin $\frac{\sqrt[4]{3}}{2}$.)

Triangle into Square.

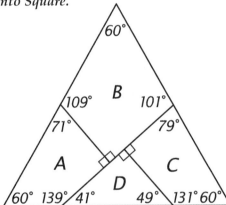

Chapter 7, Lesson 2

Set I (pages 267–268)

1. The first card (the card with the picture of the baseball player).

•2. Point.

•3. See if the figure looks exactly the same when it is turned upside down.

Optical Illusion.

•4. Three.

•5. Yes. The opposite sides of a parallelogram are equal.

6. BC looks longer than CF, but they are actually the same length.

7. The opposite angles of a parallelogram are equal.

•8. An exterior angle of a triangle is greater than either remote interior angle.

9. Substitution.

Theorem 26.

10. The opposite sides of a parallelogram are equal.

•11. The opposite sides of a parallelogram are parallel.

12. Parallel lines form equal alternate interior angles.

13. ASA.

14. Corresponding parts of congruent triangles are equal.

•15. A line segment is bisected if it is divided into two equal segments.

Point Symmetry.

16.

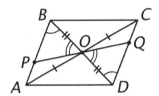

•17. The diagonals of a parallelogram bisect each other.

•18. Two points are symmetric with respect to a point if it is the midpoint of the line segment connecting them.

19. Parallel lines form equal alternate interior angles.

20. Vertical angles are equal.

21. ASA.

22. Corresponding parts of congruent triangles are equal.

23. Two points are symmetric with respect to a point if it is the midpoint of the line segment connecting them.

In encyclopedias and unabridged dictionaries, the entry for "parallelogram" is followed by an entry for "parallelogram law." The *Encyclopedia Britannica* describes it as the "rule for obtaining geometrically the sum of two vectors by constructing a parallelogram. A diagonal will give the requisite vector sum of two adjacent sides. The law is commonly used in physics to determine a resultant force or stress acting on a structure." The excerpt from *Mathematical Principles of Natural Philosophy* is taken from the original Latin edition published in 1687. It immediately follows statements of his three laws of motion.

A specific measure for ∠BAD in the figure for exercises 47 through 50 was given to make the exercises a little easier. Actually, any measure other than 60° or 120° will work. A nice problem for better students would be to rework the exercises letting ∠BAD = x° and to explain what happens when x = 60 or 120.

Parallelogram Rule.

24. Corollary.

25.

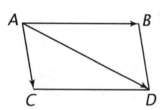

•26. 10. (5 cm represents 50 lb; so 1 cm represents 10 lb.)

27. 25. (2.5 × 10 = 25.)

28. About 6 cm.

29. About 60. (6 × 10 = 60.)

•30. About 24°.

Angle Bisectors.

31.

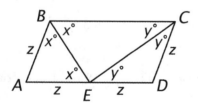

•32. ∠AEB = x°; ∠CED = y°.

33. AB, AE, and ED.

34. Point E is the midpoint of AD.

•35. AD = 2DC.

36. They are supplementary. Parallel lines form supplementary interior angles on the same side of a transversal.

•37. 180.

38. 90.

39. Isosceles.

40. Right. Because x + y = 90, ∠BEC = 90°.

41. BE ⊥ EC.

Two Parallelograms.

42.

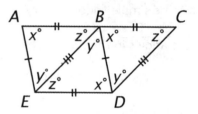

43. Because x, y, and z are the measures of the angles of a triangle, x + y + z = 180. Therefore, ∠ABC is a straight angle; so A, B, and C are collinear.

44. Because ABDE and BCDE are parallelograms, AB ∥ ED and BC ∥ ED. If A, B, and C are not collinear, then through B there are two lines parallel to ED. This contradicts the Parallel Postulate; so A, B, and C must be collinear.

•45. All three (△ABE ≅ △DEB ≅ △BCD).

46. No.

Hidden Triangles.

47.

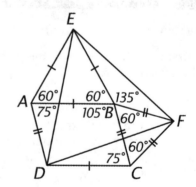

48. △ADE, △CDF, and △EBF.

49. SAS.

50. Three. In addition to ΔABE and ΔBCF, ΔDEF is equilateral because its sides are the corresponding sides of the three congruent triangles.

Set III (page 269)

The four symbols used on modern playing cards were designed in 1392 by the court painter to King Charles VI of France. The four suits represented four classes of French society: the spades, soldiers, the clubs, farmers, the diamonds, artisans, and the hearts, the clergy. According to John Scarne in *Scarne's New Complete Guide to Gambling* (Simon & Schuster, 1974), playing cards first arrived in the New World in 1492 with Columbus and immediately found favor with the Native Americans, who made them of deerskin or sheepskin!

Playing Card Symmetries.

1. So that they look the same whether they are held "right side up" or "upside down."

2. No playing card has line symmetry.

3. The diamonds are more symmetric because the diamond symbol itself has point symmetry, whereas the symbols of the other suits do not. (They do have line symmetry, however.)

4. All of the diamond cards have point symmetry except for the 7. All face cards have point symmetry. In the club, spade, and heart suits, the 2s, 4s, and 10s have point symmetry.

5. The 6s and 8s of clubs, spades, and hearts would have point symmetry if one of the symbols in the middle row were turned upside down. (It is impossible for the odd-numbered cards in these three suits to be redesigned so as to have point symmetry!)

Chapter 7, Lesson 3

Set I (pages 272–273)

The letter *p* is the only letter of the English alphabet that transforms into three other letters by reflections and rotations. Some students may enjoy thinking about other examples of letters that transform into at least one other letter in this way.

Pop-Up Parallelogram.

•1. A quadrilateral is a parallelogram if its opposite sides are equal.

2. The opposite sides of a parallelogram are parallel.

•3. Parallel lines form equal corresponding angles.

•4. In a plane, if a line is perpendicular to one of two parallel lines, it is also perpendicular to the other.

Theorem 28.

5. The sum of the angles of a quadrilateral is 360°.

6. Division.

7. Two angles are supplementary if their sum is 180°.

•8. Supplementary interior angles on the same side of a transversal mean that lines are parallel.

9. A quadrilateral is a parallelogram if its opposite sides are parallel.

Theorem 29.

10. Two points determine a line.

11. Parallel lines form equal alternate interior angles.

12. Reflexive.

•13. SAS.

14. Corresponding parts of congruent triangles are equal.

•15. A quadrilateral is a parallelogram if its opposite sides are equal.

Theorem 30.

16. A line segment that is bisected is divided into two equal parts.

17. Vertical angles are equal.

18. SAS.

19. Corresponding parts of congruent triangles are equal.

•20. Equal alternate interior angles mean that lines are parallel.

•21. A quadrilateral is a parallelogram if two opposite sides are both parallel and equal.

Letter Transformations.

•22. Line symmetry. The figure can be folded along a vertical line so that the two halves coincide.

23. Point symmetry. If the figure is turned upside down, it looks exactly the same.

24. Line symmetry. The figure can be folded along a horizontal line so that the two halves coincide.

Quadrilateral Symmetries.

25. Yes. Point symmetry.

•26. No. Line symmetry (one vertical line).

27. Yes. Point symmetry and line symmetry (two lines, one vertical and one horizontal).

28. Yes. Point symmetry and line symmetry (four lines, one vertical, one horizontal, and two at 45° angles with them).

Economics Graph.

•29. They seem to be parallel.

30. The quadrilateral formed by the four lines is a parallelogram (a quadrilateral is a parallelogram if two opposite sides are equal and parallel); so the lines C and C + I are parallel.

Set II (pages 273–275)

The method for constructing a parallel to a line through a given point not on it is another example of a "rusty compass" construction. Martin Gardner observes that, although the method has long been known, it is still being rediscovered and reported as new.

The grid-bracing illustrations are taken from the section titled "Bracing Structures" in Jay Kappraff's *Connections—The Geometric Bridge between Art and Science* (McGraw-Hill, 1991). Kappraff cites the article "How to Brace a One-Story Building" in *Environment and Planning* (1977) as the original source of this material, an important topic for engineers and architects. Examples of pertinent theorems dealing with this subject in graph theory are: "In any distorted grid all the elements of a row (column) are parallel" and "A bracing of an *n* by *m* grid is a minimum

rigid bracing if and only if the bracing subgraph is a tree."

Parallel Rulers.

31. They are connected so that AB = CD and AC = BD. A quadrilateral is a parallelogram if its opposite sides are equal.

•32. The pairs of opposite angles.

Tent Geometry.

33.

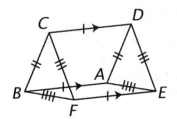

34. Yes. ABCD and CDEF are parallelograms; so AB = CD and CD = EF (the opposite sides of a parallelogram are equal) and AB ∥ CD and CD ∥ EF (the opposite sides of a parallelogram are parallel). It follows that AB = EF (substitution) and AB ∥ EF (two lines parallel to the same line are parallel to each other). Therefore, ABFE is a parallelogram because two opposite sides are parallel and equal.

35. Yes. AD = BC, DE = CF, and AE = BF (the opposite sides of a parallelogram are equal). Therefore, △ADE ≅ △BCF (SSS).

Rope Trick.

36.

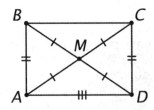

•37. A quadrilateral is a parallelogram if its diagonals bisect each other.

38. The opposite sides of a parallelogram are equal.

•39. SSS.

40. Corresponding parts of congruent triangles are equal.

41. The opposite angles of a parallelogram are equal.

42. Substitution.

•43. A quadrilateral is a rectangle if all of its angles are equal.

Parallel Postulate.

44. Through point P, there is exactly one line parallel to line *l*.

45.

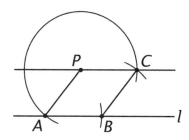

46. PC is parallel to line *l* because ABCP is a parallelogram (a quadrilateral is a parallelogram if its opposite sides are equal).

Flexible Grid.

•47. 12.

48. 4.

49. 12.

50. 4.

51. They remain unchanged because their shapes are determined by their sides (SSS) and their sides do not change.

52. They remain parallel because the opposite sides of the quadrilaterals remain equal; so they are always parallelograms.

•53. They are the sides of the quadrilaterals above and below the braced square.

54. They are the sides of the quadrilaterals to the right of the braced square.

55. Because all of the quadrilaterals remain parallelograms, these sides remain parallel to the horizontal and vertical sides of the braced square.

Set III (page 275)

1. (*The figures below show two possible answers depending on which triangles are constructed inward and outward.*)

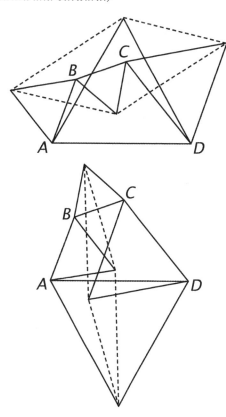

2. The quadrilateral formed seems to be a parallelogram.

Chapter 7, Lesson 4

Set I (pages 277–279)

In his definitive book on the subject, *Dissections—Plane and Fancy* (Cambridge University Press, 1997), Greg N. Frederickson (using radians, 2π radians = 360°) reports:

"We can think of a regular polygon as a union of various rhombuses and isosceles triangles (which are halves of rhombuses). Such a characterization is an *internal structure* of the corresponding polygon. If the regular polygon has an even number of sides, then the internal structure needs only rhombuses .... A regular polygon with $2q$ sides contains q each of $\frac{i\pi}{q}$-rhombuses, for each positive integer i that is less than $\frac{q}{2}$, plus $\frac{q}{2}$ squares if q is even. For example, the dodecagon contains

6 each of $\frac{\pi}{6}$-rhombuses and $\frac{\pi}{3}$-rhombuses, plus 3 squares."

[$\frac{\pi}{6}$ and $\frac{\pi}{3}$ indicate rhombuses having angles of 30° and 60°.] There are several ways in which the rhombuses can be arranged when the dodecagon is divided in this way.

The remark about a square being perfect because all of its angles are "just right" was made by Albert H. Beiler at the beginning of the chapter titled "On The Square" in his book titled *Recreations in the Theory of Numbers—The Queen of Mathematics Entertains* (Dover, 1966).

Regular Dodecagon.

•1. 12.

•2. Three.

3. A square.

4. Three.

•5. 15.

6. Three. (There are six narrow, six medium, and three square.)

Theorem 31.

•7. All of the angles of a rectangle are equal.

8. A quadrilateral is a parallelogram if its opposite angles are equal.

Theorem 32.

9. All of the sides of a rhombus are equal.

•10. A quadrilateral is a parallelogram if its opposite sides are equal.

Theorem 33.

11. All of the angles of a rectangle are equal.

•12. All rectangles are parallelograms.

13. The opposite sides of a parallelogram are equal.

14. Reflexive.

15. SAS.

16. Corresponding parts of congruent triangles are equal.

Theorem 34.

•17. All of the sides of a rhombus are equal.

18. In a plane, two points each equidistant from the endpoints of a line segment determine the perpendicular bisector of the line segment.

Which Parts?

•19. Two consecutive sides.

20. One side.

21. One side and one angle.

22. Two consecutive sides and one angle.

"Just Right."

23. If all of the angles of a triangle were right angles, the sum of their measures would be 270°, rather than 180°.

•24. No. It must be a rectangle.

25. *Shown below are two of many possible examples. (The ambiguity arises because the right angles may be on the inside or the outside of the polygon.)*

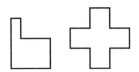

Set II (pages 279–280)

Checking a Wall.

26. No. If the opposite sides of a quadrilateral are equal, it must be a parallelogram but not necessarily a rectangle.

27. No. We know that the diagonals of a rectangle are equal but to conclude that, if the diagonals of a quadrilateral are equal, it is a rectangle would be to make the converse error.

28.

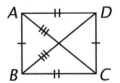

29. SSS.

30. Corresponding parts of congruent triangles are equal.

•31. A quadrilateral is a parallelogram if its opposite sides are equal.

32. The opposite angles of a parallelogram are equal.

33. Substitution.

•34. A quadrilateral all of whose angles are equal is a rectangle.

35. True.

•36. False.

37. True.

38. False.

•39. True.

Square Problem.

40.

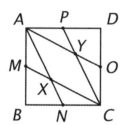

41. Parallelograms.

42. They must be parallelograms because two opposite sides are both parallel and equal. (For example, in AMCO, AM ∥ CO because ABCD is a square and AM = CO because M and O are midpoints of the equal segments AB and CD.)

•43. Its opposite sides are parallel.

44. A rhombus.

45. They seem to be collinear.

46. It appears to have point symmetry and line symmetry (with respect to the lines of the diagonals of the square).

Rhombus Problem.

47.

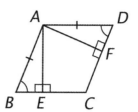

48. Yes. Because ABCD is a rhombus, AB = AD. It is also a parallelogram; so ∠B = ∠D. ΔAEB ≅ ΔAFD (AAS); so AE = AF.

Triangle Problem.

49.

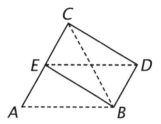

50. Because ABDE is a parallelogram, AB = ED (the opposite sides of a parallelogram are equal). Also, because BDCE is a rectangle, ED = CB (the diagonals of a rectangle are equal). So AB = CB (substitution). Therefore, ΔABC is isosceles. (Additional drawings using the conditions given reveal that it isn't necessarily equilateral.)

Set III (page 280)

Bill Leonard told the story of starting a lesson with a high-school class of remedial mathematics students with the problem of counting the squares in a 4 x 4 grid. He had hoped that the students would think about it for awhile, but it didn't take long for several students to start yelling out "16." Leonard then asked if they had counted squares such as the one shaded in the second figure. The class immediately yelled out in unison, "17!"

Counting Squares.

1. It contains 30 squares in all. There are 16 small squares, 9 2 x 2 squares, 4 3 x 3 squares, and 1 4 x 4 square.

2. It contains 15 squares in all. There are 7 small squares, 6 2 x 2 squares, 1 3 x 3 square, and 1 4 x 4 square.

Set I (pages 282–284)

Hal Morgan in his book titled *Symbols of America* (Viking, 1986), comments on the Chase Manhattan Bank trademark: "After World War II the representational images of the nineteenth century began to give way to abstract designs inspired by the spare, geometric forms of the Bauhaus. Some of these modern designs are quite striking in their clarity and simplicity. . . . The design firm of Chermayeff & Geismar Associates created the Chase Manhattan Bank's distinctive octagon logo in 1960. . . . The abstract, interlocking design seems to have become a sort of model for the designers of other bank and insurance logos." Morgan shows examples of several more-recent logos that are also point-symmetric designs of interlocking solid black congruent polygons.

Quadrilaterals in Perspective.

- •1. A trapezoid.

- •2. That it has exactly one pair of parallel sides.

- 3. Yes. Picnic blankets are usually rectangular. (The perspective alters the blanket's shape.)

- •4. That all of its angles are right angles (or that they are equal).

- •5. Yes. All rectangles are parallelograms.

- 6. Rhombuses and parallelograms.

- 7. Squares, rectangles, rhombuses, and parallelograms.

Geometric Trademark.

- •8. Point.

- 9. It is a square because all of its sides and angles are equal.

- •10. They seem to be congruent.

- •11. Trapezoids.

- 12. CD and JE.

- •13. No. The legs of a trapezoid must be equal for it to be isosceles.

- 14. No. One is a right angle and the other is acute.

- 15. An octagon.

Theorem 35.

- •16. The bases of a trapezoid are parallel.

- •17. Through a point not on a line, there is exactly one line parallel to the line.

- 18. A quadrilateral is a parallelogram if its opposite sides are parallel.

- •19. The opposite sides of a parallelogram are equal.

- 20. The legs of an isosceles trapezoid are equal.

- 21. Substitution.

- •22. If two sides of a triangle are equal, the angles opposite them are equal.

- 23. Parallel lines form equal corresponding angles.

- 24. Substitution.

- •25. Parallel lines form supplementary interior angles with a transversal.

- 26. The sum of two supplementary angles is 180°.

- 27. Substitution.

- 28. Subtraction.

Theorem 36.

- 29. The legs of an isosceles trapezoid are equal.

- •30. The base angles of an isosceles trapezoid are equal.

- 31. Reflexive.

- 32. SAS.

- 33. Corresponding parts of congruent triangles are equal.

Set II (pages 284–285)

Stepladder.

34.

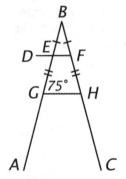

35. An isosceles trapezoid.

•36. 75°.

37. The base angles of an isosceles trapezoid are equal.

38. Isosceles.

39. 75°.

•40. Parallel lines form equal corresponding angles.

•41. 105°.

42. 30°.

Regular Pentagon.

43. Examples are ABCD, ABCE, ACDE, and ABDE. Their bases are parallel (they form supplementary interior angles on the same side of a transversal) and their legs are equal (given).

44. Examples are ABCH, ABGE, and AFDE. Their opposite sides are parallel (established in exercise 43).

45. Examples are ABCH, ABGE, and AFDE. They are rhombuses because all of their sides are equal. In regard to ABCH, for example, AB = BC (given), AB = HC, and AH = BC (the opposite sides of a parallelogram are equal); so HC = AH (substitution).

Trapezoid Diagonals.

•46. AC and DB bisect each other.

•47. A quadrilateral is a parallelogram if its diagonals bisect each other.

48. The opposite sides of a parallelogram are parallel.

•49. A trapezoid has only one pair of opposite sides parallel.

50. AC and DB bisect each other.

51. AC and DB do not bisect each other.

52. Yes. One example of such a trapezoid is shown below.

Set III (page 285)

The Moscow Puzzles is the English language edition of Boris Kordemsky's *Mathematical Know-how,* which, according to Martin Gardner, is the best and most popular puzzle book ever published in the Soviet Union. Kordemsky taught mathematics at a high school in Moscow for many years.

Gardner reported Andrew Miller's discovery of a second solution to Kordemsky's trapezoid dissection problem in the February 1979 issue of *Scientific American* and posed it as a new puzzle for the readers of that magazine.

Congruence Puzzle.

Kordemsky's solution is shown below.

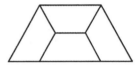

The solution thought of by Andrew Miller is shown below. The three cuts intersect the upper base at points $\frac{1}{8}$, $\frac{3}{4}$, and $\frac{7}{8}$ of the way from its left endpoint.

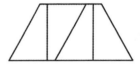

Chapter 7, Lesson 6

Set I (pages 288–289)

The "state capital" quadrilateral was chosen partly for fun but primarily to show that the original quadrilateral can have any shape. It doesn't even have to be convex. Not only can the quadrilateral be concave, it can even be "self-intersecting" or skew!

Portuguese Theorem.

1. Midpoints.

•2. A midsegment.

3. Other side.

4. A midsegment of a triangle is parallel to the third side and half as long.

•5. DE ∥ AC and DE = $\frac{1}{2}$AC.

State Capitals Problem.

6.

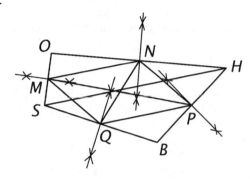

7. MNPQ looks like a parallelogram.

•8. A midsegment of a triangle is parallel to the third side. (MN and QP are midsegments of ΔSOH and ΔSBH.)

•9. In a plane, two lines parallel to a third line are parallel to each other.

•10. A midsegment of a triangle is half as long as the third side.

11. Substitution.

12. That it is a parallelogram.

•13. MNPQ is a parallelogram because two opposite sides are both parallel and equal.

14. MP and NQ seem to bisect each other.

15. The diagonals of a parallelogram bisect each other.

Midpoint Quadrilateral.

16. A midsegment of a triangle is half as long as the third side.

•17. Addition.

•18. Substitution.

19. Its perimeter.

•20. The sum of its diagonals.

21. The perimeter of the midpoint quadrilateral is equal to the sum of the diagonals of the original quadrilateral.

Two Midsegments.

22.

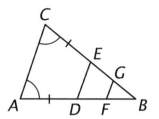

23. ADEC is an isosceles trapezoid because DE ∥ AC (a midsegment of a triangle is parallel to the third side) and AD = CE.

•24. ∠A = ∠C because the base angles of an isosceles trapezoid are equal.

25. ΔABC is isosceles because, if two angles of a triangle are equal, the sides opposite them are equal.
(Alternatively, AB = 2AD = 2CE = CB .)

26. FG = $\frac{1}{4}$AC because FG = $\frac{1}{2}$DE and

DE = $\frac{1}{2}$AC [so FG = $\frac{1}{2}(\frac{1}{2}$AC) = $\frac{1}{4}$AC].

Also, FG ∥ AC because FG ∥ DE and DE ∥ AC (in a plane, two lines parallel to a third line are parallel to each other).

Set II (pages 289–291)

Exercises 27 through 36 are intended to review some basics of coordinate geometry, including the distance formula, and to get students to thinking about how midpoints and the directions of lines might be treated.

The Midsegment Theorem is, in a way, a "special case" of the fact that the line segment connecting the midpoints of the legs of a trapezoid is parallel to the bases and half their sum.

The edition of Euclid's *Elements* with the fold-up figures was its first translation into English, by Henry Billingsley. After a career in business, Billingsley became sheriff of London and was elected its mayor in 1596. An excerpt from *The Elements of Geometrie*, including an actual fold-up model, is reproduced in Edward R. Tufte's wonderful book *Envisioning Information* (Graphics Press, 1990).

In *Mathematical Gems II* (Mathematical Association of America, 1976), Ross Honsberger wrote: "Solid geometry pays as much attention to the tetrahedron as plane geometry does to the triangle. Yet many elementary properties of the tetrahedron are not very well known." Exercises 37 through 41 establish that any triangle can be folded along its midsegments to form a tetrahedron. Unlike tetrahedra in general, however, a tetrahedron formed in this way is always isosceles; that is, each pair of opposite edges are equal. All the faces of an isosceles tetrahedron are congruent triangles and consequently have the same area; furthermore, the sum of the face angles at each vertex is 180°. More information on this geometric solid can be found in chapter 9 of Honsberger's book.

Midpoint Coordinates.

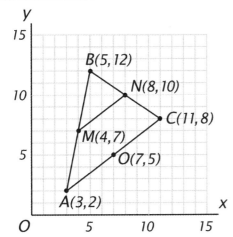

27.

•28. $(4, 7)$.

•29. $(8, 10)$.

30. $(7, 5)$.

31. Yes. The coordinates of the midpoint are "midway between" the respective coordinates of the endpoints. (They are their averages.)

•32. $MN = \sqrt{(8-4)^2 + (10-7)^2} = \sqrt{4^2 + 3^2} = \sqrt{16+9} = \sqrt{25} = 5$.

33. $AC = \sqrt{(11-3)^2 + (8-2)^2} = \sqrt{8^2 + 6^2} = \sqrt{64+36} = \sqrt{100} = 10$.

34. Yes. $MN = \frac{1}{2}AC$ as the Midsegment Theorem states.

•35. They should be parallel.

36. Yes. They seem to be parallel because they have the same direction; each line goes up 3 units as it goes 4 units to the right.

Tetrahedron.

37.

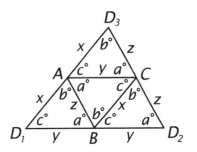

•38. They are congruent by SSS.

39.

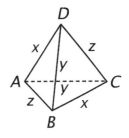

40. They are equal.

41. 180°. (The measures of the three angles that meet at each vertex are a, b, and c.)

More Midpoint Quadrilaterals.

42. *Example figure:*

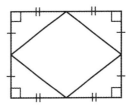

•43. A rhombus.

44. The line segments divide the rectangle into four right triangles and the quadrilateral. The triangles are congruent by SAS; so their hypotenuses are equal. Because the hypotenuses are the sides of the quadrilateral, it is equilateral and must be a rhombus.

45. *Example figure:*

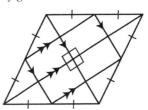

46. A rectangle.

47. The sides of the new quadrilateral are parallel to the diagonals of the rhombus because of the Midsegment Theorem. Consequently, they form four small parallelograms. Each parallelogram has a right angle because the diagonals of a rhombus are perpendicular. Because the opposite angles of a parallelogram are equal, it follows that the four angles of the new quadrilateral are right angles; so it is a rectangle.

48. Another square.

Base Average.

•49. MN seems to be parallel to AB and DC, the bases of the trapezoid.

50.

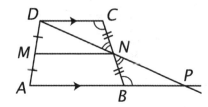

51. DC ∥ AP; so ∠C = ∠NBP (parallel lines form equal alternate interior angles).
CN = NB because N is the midpoint of CB.
∠CND = ∠BNP (vertical angles are equal).
So △DCN ≅ △PBN (ASA).
   [An alternative proof could use ∠CDN = ∠BPN (equal alternate interior angles) in place of either pair of angles above, in which case the triangles would be congruent by AAS.]

52. DN = NP (corresponding parts of congruent triangles are equal); so N is the midpoint of DP. Therefore, MN is a midsegment of △DAP and so MN ∥ AP.

53. MN ∥ AP and DC ∥ AP; so MN ∥ DC. (In a plane, two lines parallel to a third line are parallel to each other.)

54. $MN = \frac{1}{2}AP$ by the Midsegment Theorem.
Because AP = AB + BP and DC = BP (corresponding parts of congruent triangles are equal), $MN = \frac{1}{2}(AB + BP) = \frac{1}{2}(AB + DC)$ (substitution).

Set III (page 291)

The problem of the nested triangles is a preview of the idea of a limit, a topic considered later in the treatment of measuring the circle. The sum of the lengths of all of the sides of the triangles is the sum of the geometric series:

$$12 + \frac{1}{2}12 + (\frac{1}{2})^2 12 + (\frac{1}{2})^3 12 + \cdots = 2 \times 12$$

Students who tackle this problem will be interested to know that they will explore it further in second-year algebra.

Infinite Series.

The perimeter of the largest triangle is 12 in. From the Midsegment Theorem, we know that the sides of the next triangle are half as long; so its perimeter is 6 in. Reasoning in the same way, we find that the perimeters in inches of successive triangles are 3, 1.5, 0.75, 0.375, 0.1875, and so on.

Starting with the first two triangles, we find that the successive sums of the sides are 18, 21, 22.5, 23.25, 23.625, and so on. The sums seem to be getting closer and closer to 24 inches and (as students who study infinite series will learn to prove) the sum of *all* of the sides of the triangles in the figure is 2 feet.

Chapter 7, Review

Set I (pages 292–294)

A question such as that in exercise 11 is intended to arouse the student's curiosity. According to Jearl Walker, who posed it in his book titled *The Flying Circus of Physics* (Wiley, 1977), "If the ratio of the bar's density to the fluid's density is close to 1 or 0, the bar floats in stable equilibrium as in the first figure. If the ratio is some intermediate value, then the bar floats in stable equilibrium with its sides at 45° to the fluid's surface. In each case, the orientation of stable equilibrium is determined by the position in which the potential energy of the system is least." This floating-bar problem is a good illustration of how scientists arrive at conclusions by using inductive reasoning (performing experiments with floating bars to see what happens) and how they then "explain" the results by using deductive reasoning (calculating the potential energy and its minimum value by using mathematics, including *geometry*).

The frequent appearance in this chapter of theorems whose converses also are theorems can easily mislead students into forgetting that a statement and its converse are not logically equivalent. The intent of exercises 17 through 30 is to remind students of this fact and to encourage them to draw figures to test unfamiliar conclusions (or even to remind themselves of facts that they have proved). Knowledge of the properties of the various quadrilaterals should not be as much a matter of memorization as of looking at pictures and applying simple common sense.

Palm Strand Rhombus.

1.

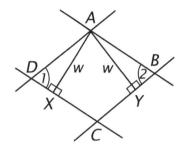

•2. Both pairs of its opposite sides are parallel (because the strips have parallel sides).

3. The opposite angles of a parallelogram are equal.

•4. AAS.

5. Corresponding parts of congruent triangles are equal.

6. The opposite sides of a parallelogram are equal.

•7. Substitution.

8. All of its sides are equal.

Floating Bar.

•9. Four.

10. Yes.

11. *(Student answer.)* (The bar can float either way depending on it and the liquid.)

12.

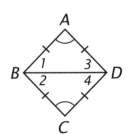

•13. All of the sides and angles of a square are equal.

14. SAS (or SSS).

15. Corresponding parts of congruent triangles are equal.

•16. If a line divides an angle into two equal parts, it bisects the angle.

Related Statements.

17. It is the converse of the first statement.

18. No. The first statement is true; the first figure shows that the triangles are congruent by SSS. The second figure shows that the second statement is false.

19. True.

20. True.

•21. True.

22. False.
Example of a quadrilateral that has two pairs of equal angles but is not an isosceles trapezoid:

23. False.
Example of a parallelogram that is not a rhombus:

24. True.

25. True.

26. True.

•27. True.

28. False.
Example of a quadrilateral with perpendicular diagonals that is not a rhombus:

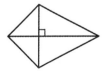

29. False.

Example of a parallelogram whose diagonals are not equal:

30. False.

Example of a quadrilateral whose diagonals are equal that is not a parallelogram:

Jumping Frog.

•31. The sum of the angles of a quadrilateral is 360°.

32. They are rhombuses and all rhombuses are parallelograms. (Or, both pairs of their opposite sides are equal.)

•33. The opposite angles of a parallelogram are equal.

34. They are parallelograms because their diagonals bisect each other.

35. The opposite sides of a parallelogram are parallel and, in a plane, two lines parallel to a third line are parallel to each other.

Set II (pages 294–295)

Only the right angles in the pieces of the puzzle of exercises 36 through 50 actually have measures that are integers. From $\triangle ABG$ it is evident that

$\tan A = \dfrac{2c}{c} = 2$; so $\angle A = \text{Arctan } 2 \approx 63.43°$.

Although part of basic vocabulary in books of the past, words such as "trapezium" and "rhomboid" are now close to being obsolete. Strangely, in British usage, the word "trapezium" is used to mean "trapezoid."

Even though the quadrilaterals in exercises 54 through 57 are drawn to scale, it is almost impossible to recognize them merely by their appearance. It is tempting to conclude that, because quadrilateral ABCD is a trapezoid whose base angles are equal, it is isosceles, which, of course, is the converse error. To prove that it is isosceles would require adding an extra line to the figure

to form a parallelogram and a triangle; the proof is comparable to that of Theorem 35, outlined in exercises 16 through 28 of Lesson 5. The additional fact that the base opposite the larger base angles of an isosceles trapezoid is longer than the other base is the basis for the magic trick of exercise 58. Like the question about the floating bar in Set I, the question about the magic trick is included for fun and not meant to be taken too seriously.

Dissection Puzzle.

•36. D, E, and F.

37. D.

38. D and F.

39. C and D.

40. A, B, D, E, and F.

41.

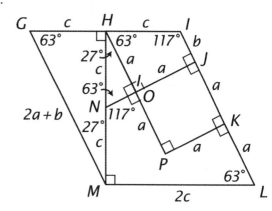

•42. GI ⊥ HM and ML ⊥ HM. In a plane, two lines perpendicular to a third line are parallel.

43. A parallelogram.

44. A quadrilateral is a parallelogram if its opposite sides are parallel.

•45. 2c.

46. $2a + b$. [Some students may recognize that GM can also be expressed in terms of c by using the Pythagorean Theorem. $GM^2 = c^2 + (2c)^2 = 5c^2$; so $GM = c\sqrt{5}$.]

47. A square.

48. All of its sides and angles are equal.

•49. A trapezoid.

50. It has exactly one pair of parallel sides. (HO ∥ IJ because, in a plane, two lines perpendicular to a third line are parallel.)

Words from the Past.

•51. No, because a trapezoid has two parallel sides.

52. Yes, because, other than squares, rhombuses are parallelograms that have no right angles.

53. No, because a rectangle has four right angles.

Rectangles Not.

54. ABCD is a trapezoid because BC ∥ AD (they form supplementary interior angles on the same side of a transversal). (It is also possible to show that ABCD is isosceles.)

55. EFGH is a parallelogram because both pairs of opposite angles are equal.

•56. IJKL is a trapezoid because IJ ∥ LK.

57. From its angles we can conclude that MNOP is neither a parallelogram nor a trapezoid; so it is not any of the special types of quadrilaterals that we have studied.

58. While you look at the card, the magician rotates the pack 180° so that the bottom edges of the cards are at the top. When your chosen card is put back, it is easy to slide it up because it is now wider at the top than any of the others.

Another Midpoint Quadrilateral.

59.

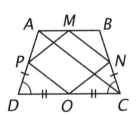

60. MN ∥ AC and PO ∥ AC (a midsegment of a triangle is parallel to the third side); so MN ∥ PO. MN = $\frac{1}{2}$AC and PO = $\frac{1}{2}$AC (a midsegment of a triangle is half as long as the third side); so MN = PO. MNOP is a parallelogram because two opposite sides are both parallel and equal.

61. It appears to be a rhombus. ΔPDO ≅ ΔNCO (SAS); so PO = NO. Because MN = PO and MP = NO, MN = MP; so all of the sides of MNOP are equal.

•1. 2.236.

•2. 22.36.

•3. 223.6.

•4. 7.07.

•5. 70.7.

•6. $\sqrt{500} = \sqrt{100 \cdot 5} = \sqrt{100}\sqrt{5} = 10\sqrt{5}$.

•7. $\sqrt{50{,}000} = \sqrt{100 \cdot 500} = \sqrt{100}\sqrt{500} = 10(10\sqrt{5}) = 100\sqrt{5}$.

•8. $\sqrt{50} = \sqrt{25 \cdot 2} = \sqrt{25}\sqrt{2} = 5\sqrt{2}$.

•9. $\sqrt{5{,}000} = \sqrt{100 \cdot 50} = \sqrt{100}\sqrt{50} = 10(5\sqrt{2}) = 50\sqrt{2}$.

•10. $\sqrt{25x} = \sqrt{25}\sqrt{x} = 5\sqrt{x}$.

•11. $\sqrt{x^5} = \sqrt{x^4 \cdot x} = \sqrt{x^4}\sqrt{x} = x^2\sqrt{x}$.

•12. $\sqrt{\pi x^2} = \sqrt{\pi}\sqrt{x^2} = x\sqrt{\pi}$.

•13. $\sqrt{12x^{12}} = \sqrt{4 \cdot 3x^{12}} = \sqrt{4}\sqrt{3}\sqrt{x^{12}} = 2x^6\sqrt{3}$.

•14. $\sqrt{147} + \sqrt{3} = \sqrt{49 \cdot 3} + \sqrt{3} = \sqrt{49}\sqrt{3} + \sqrt{3} = 7\sqrt{3} + \sqrt{3} = 8\sqrt{3}$.

•15. $\sqrt{147 + 3} = \sqrt{150} = \sqrt{25 \cdot 6} = \sqrt{25}\sqrt{6} = 5\sqrt{6}$.

•16. $\sqrt{700} - \sqrt{175} = \sqrt{100 \cdot 7} - \sqrt{25 \cdot 7} = \sqrt{100}\sqrt{7} - \sqrt{25}\sqrt{7} = 10\sqrt{7} - 5\sqrt{7} = 5\sqrt{7}$.

•17. $\sqrt{700 - 175} = \sqrt{525} = \sqrt{25 \cdot 21} = \sqrt{25}\sqrt{21} = 5\sqrt{21}$.

•18. $\sqrt{20 + 20 + 20} = \sqrt{60} = \sqrt{4 \cdot 15} = \sqrt{4}\sqrt{15} = 2\sqrt{15}$.

•19. $\sqrt{20} + \sqrt{20} + \sqrt{20} = 3\sqrt{20} = 3\sqrt{4 \cdot 5} = 3\sqrt{4}\sqrt{5} = 6\sqrt{5}$.

•20. $\sqrt{6^2} + \sqrt{8^2} = 6 + 8 = 14$.

•21. $\sqrt{6^2 + 8^2} = \sqrt{36 + 64} = \sqrt{100} = 10$.

•22. $\sqrt{14}\sqrt{21} = \sqrt{14 \cdot 21} = \sqrt{2 \cdot 7 \cdot 3 \cdot 7} = 7\sqrt{6}$.

•23. $\dfrac{\sqrt{242}}{\sqrt{2}} = \sqrt{\dfrac{242}{2}} = \sqrt{121} = 11$.

•24. $(4\sqrt{5})(6\sqrt{2}) = 24\sqrt{10}$.

•25. $(3\sqrt{7})^2 = 9 \cdot 7 = 63$.

•26. $\dfrac{12\sqrt{15}}{4\sqrt{3}} = 3\sqrt{5}$.

•27. $(5 - \sqrt{2}) + (5 + \sqrt{2}) = 10$.

•28. $(5 - \sqrt{2})(5 + \sqrt{2}) = 25 - 2 = 23$.

•29. $\sqrt{3}(\sqrt{27} + 1) = \sqrt{81} + \sqrt{3} = 9 + \sqrt{3}$.

•30. $\sqrt{3} + (\sqrt{27} + 1) = \sqrt{3} + \sqrt{9 \cdot 3} + 1 = \sqrt{3} + 3\sqrt{3} + 1 = 4\sqrt{3} + 1$.

•31. $(\sqrt{x} + \sqrt{y}) + (\sqrt{x} + \sqrt{y}) = 2\sqrt{x} + 2\sqrt{y}$.

•32. $(\sqrt{x} + \sqrt{y})^2 = (\sqrt{x})^2 + 2(\sqrt{x}\sqrt{y}) + (\sqrt{y})^2 = x + 2\sqrt{xy} + y$.

Set I (pages 300–302)

In their book titled *Symmetry—A Unifying Concept* (Shelter Publications, 1994), István and Magdolna Hargittai point out that it was Louis Pasteur who first discovered that otherwise identical crystals can be mirror images of one another. Except for glycine, all amino acids can exist in opposite-handed forms, but only the left-handed version occurs naturally. The Hargittais write: "Many biologically important chemical compounds exist in left-handed and right-handed forms, and the biological activity of the two forms may be very different. . . . Humans metabolize only right-handed glucose. Left-handed glucose, although still sweet, passes through the system untouched." Martin Gardner's book titled *The New Ambidextrous Universe* (W. H. Freeman and Company, 1990) is a good source of further information on the subject.

At first glance, it might appear that Roger Shepard's figure (exercises 18 through 21) also illustrates rotations because the arrows point in opposite directions. The arrows pointing to the left, however, are not congruent to the arrows pointing to the right. The figure is featured on the cover of Al Seckel's excellent book titled *The Art of Optical Illusions* (Carlton Books, 2000).

Transformations in Art.

• 1. A translation.

2. A rotation.

• 3. A rotation.

4. No.

5. No.

Mirror Molecules.

• 6. A reflection.

7. "Left-handed" and "right-handed."

Reflections.

8.

9.

A | A

10.

11.

12.

E | Ǝ

13.

Z | Ƨ

14.

15.

• 16. Exercises 9, 11, and 14.

• 17. Each has a vertical line of symmetry.

Down the Stairs.

18. A translation.

19. A one-to-one correspondence between two sets of points.

• 20. Yes.

21. A transformation that preserves distance and angle measure.

Peter Jones.

22. A rotation.

23. The image is produced from the original figure by rotating the figure 90° clockwise.

24. S.

Set II (pages 302–304)

As the Adobe Illustrator User Guide explains, the program "defines objects mathematically as vector graphics." Exactly how vector graphics combines geometry with linear algebra and matrix theory is explained in detail in *The Geometry Toolbox for Graphics and Modeling* by Gerald E. Farin and Dianne Hansford (A. K. Peters, 1998).

Escalator Transformations.

•25. A translation.

26. A rotation.

27.

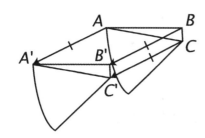

•28. They are parallelograms because they have two sides that are both parallel and equal.

29. The opposite sides of a parallelogram are parallel.

30. It is a parallelogram.

•31. The opposite sides of a parallelogram are equal.

32. SSS.

33. Corresponding parts of congruent triangles are equal.

34. A transformation that preserves distance and angle measure.

35.

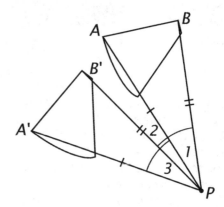

•36. Betweenness of Rays Theorem.

•37. Substitution.

38. Subtraction.

39. SAS.

40. Corresponding parts of congruent triangles are equal.

41. Distances.

Triangle Construction.

42.

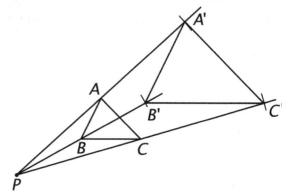

•43. A dilation.

•44. They seem to be twice as long.

45. They seem to be equal.

46. No. It is not an isometry, because it doesn't preserve distance.

Computer Geometry.

47. A translation.

48. A rotation.

49. A dilation.

50.

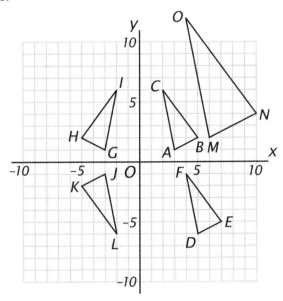

51. A(3, 1) → D(3 + 2, 1 – 7), or D(5, –6).
    B(5, 2) → E(5 + 2, 2 – 7), or E(7, –5).
    C(2, 6) → F(2 + 2, 6 – 7), or F(4, –1).

•52. A translation.

53. A(3, 1) → G(–3, 1).
    B(5, 2) → H(–5, 2).
    C(2, 6) → I(–2, 6).

54. A reflection.

55. A(3, 1) → J(–3, –1).
    B(5, 2) → K(–5, –2).
    C(2, 6) → L(–2, –6).

•56. A rotation.

57. A(3, 1) → M(6, 2).
    B(5, 2) → N(10, 4).
    C(2, 6) → O(4, 12).

58. A dilation.

Set III (page 304)

Toothpick Puzzle.

1. (There are two possible ways to solve the puzzle, one of which is a reflection of the other. The three toothpicks moved are shown as dotted lines in the figures below.)

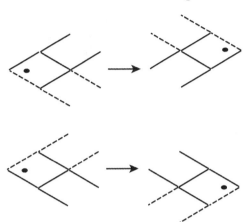

2. Yes. The reversed fish is a rotation image of the original fish. The center of rotation can be the midpoint of either toothpick forming the fish's back.

Chapter 8, Lesson 2

Set I (pages 308–309)

Capital letters in many typefaces do not have the simple symmetries suggested by exercises 10 through 13. For example, the horizontal bar of an H doesn't always connect the midpoints of the side bars. The upper part of an S is frequently smaller than the lower part.

David Moser, the creator of the "China" transformation, has designed some other amazing figures of this sort. Two are included in the book of visual illusions titled *Can You Believe Your Eyes* by J. R. Block and Harold E. Yuker (Brunner/Mazel, 1992). One changes "England" from Chinese into English and the other does the same thing with "Tokyo"! Both word transformations are accomplished, like the example in the text, by a 90° rotation.

A Suspicious Cow.

1. The water from which the cow is drinking.

2. It is upside down. The cow and barn that we see "above" the water are actually reflections.

3.

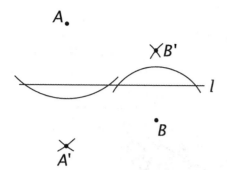

•4. 2*x* units.

•5. It is the segment's perpendicular bisector.

Double Reflections.

6. *b*.

•7. A rotation.

8. 90°.

•9. 180°.

10.

a

W | W

———————— b

W

11.

a

K | K

———————— b

K

12.

a

H | H

———————— b

H

13.

a

S | S

———————— b

S

•14. Its image looks the same.

15. Its image looks the same.

16. Its image looks the same.

17. In exercise 12. The figure has both a vertical and a horizontal line of symmetry.

SAT Problem.

•18. B.

19. A reflection.

20. Measure the distances from B and P to the dotted line and see if they are the same.

Can You Read Chinese?

21. A reflection.

•22. A reflection.

23. A rotation.

•24. Yes, because it is the composite of two reflections in intersecting lines.

25. It is a translation of the word "China" from Chinese into English.

Set II (pages 309–311)

Ethologist Niko Tinbergen carried out a number of interesting laboratory studies concerning animal vision, including one with the "goose-hawk" and an experiment with circular disks interpreted by newly hatched blackbirds as their "mother." These examples and others are included by cognitive scientist Donald D. Hoffman in *Visual Intelligence—How We Create What We See* (Norton, 1998).

Kaleidoscope Patterns.

26. B, D, and F.

•27. C and E.

•28. 120°.

29. Three. The three mirror lines. (The monkey faces are almost mirror symmetric; so the lines that bisect the angles formed by the mirrors almost look like lines of symmetry.)

•30. No. The figure does not look exactly the same upside down. (The monkey's left nostril is closer to its left eye than its right nostril is to its right eye; so it is possible to tell the difference.)

Scaring Chickens.

31. B.

•32. E.

33. D.

34. A.

•35. E.

36. A.

•37. A translation.

38. A translation is the composite of two reflections in parallel lines.

39. Seeing the bird at C flying to A.

Boomerang.

40.

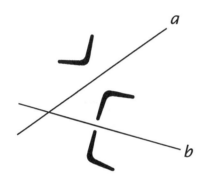

•41. A rotation.

42. A rotation is the composite of two reflections in intersecting lines.

43. At the point in which the two lines intersect.

Triangle Reflections.

•44. If a point is reflected through a line, the line is the perpendicular bisector of the segment connecting the point and its image.

45. A translation.

•46. Its magnitude.

47. It is twice as long; AA" = 2XY.

•48. SAS.

49. Corresponding parts of congruent triangles are equal.

50. A rotation.

51. Its magnitude.

52. It is twice as large; ∠AOA" = 2∠XOY.

Set III (page 311)

A. E. W. Mason, the author of *The House of the Arrow* (1924), is best known for his novels *The Four Feathers* and *Fire Over England*. *The House of the Arrow*, which featured his detective Inspector Hanaud investigating the murder of a French widow, was made into a movie three times. Following the excerpt quoted in the Set III exercise, the story continues:

"It was exactly half-past one; the long minute hand pointing to six, the shorter hour hand on the right-hand side of the figure twelve, half-way between the one and the two. With a simultaneous movement they all turned again to the mirror; and the mystery was explained. The shorter hour-hand seen in the mirror was on the left-hand side of the figure twelve, and just where it would have been if the hour had been half-past ten and the clock actually where its reflection was. The figures on the dial were reversed and difficult at a first glance to read."

What Time Was It?

1.

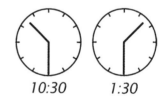

10:30   *1:30*

2. Examples suggest that the reflection of a clock face in a vertical mirror always looks like an actual time. This is not true for the reflection of a clock face in a horizontal mirror. For example, a horizontal reflection of a clock face reading 10:30 would not look like an actual time because the hour hand would be in the wrong position when the minute hand is pointing to the top of the clock.

10:30

???

Set I (pages 314–315)

The number "eight" seems to have a special connection to sports in which synchronization is important. One of the definitions given in *The American Heritage Dictionary of the English Language* for the word "eight" is "an eight-oared racing shell." Team routines in synchronized swimming consist of eight swimmers.

Prevaricator.

1. He has "liar" written all over his face!

2. A translation.

•3. A reflection.

4. A rotation.

•5. A glide reflection.

6. Yes.

7. If two figures are congruent, there is an isometry such that one figure is the image of the other.

Synchronized Oars.

•8. ∠PAB and ∠PCD.

9. If the oars are assumed to be identical, AB = CD. A quadrilateral is a parallelogram if two opposite sides are both parallel and equal.

•10. The opposite sides of a parallelogram are equal.

11. A translation.

12. A glide reflection.

13. Synchronized swimming.

Swing Isometries.

•14. A reflection.

15. A rotation.

16. Its center.

•17. The magnitude of the rotation.

•18. The lines bisect these angles.

19. They are the perpendicular bisectors of these line segments.

20. It is twice as large.

Quadrilateral Reflections.

21. Two points determine a line.

•22. They are the perpendicular bisectors of these line segments.

23. They appear to be parallel.

24. A translation.

•25. Two figures are congruent if there is an isometry such that one figure is the image of the other.

Set II (pages 315–316)

The symmetry of the illustration for the piano-moving problem suggests that the distances from A and J to the corner of the room are equal. If this is the case, then it is easy to prove that ΔDGP in the figure below is an isosceles right triangle. It follows that ∠CDG = ∠DGH = 135°. The piano, then, has been rotated 270° in all. If the symmetry doesn't exist, ΔDGP will still be a right triangle with complementary acute angles; so, even though the measures of ∠CDG and ∠DGH will change, their sum will remain the same.

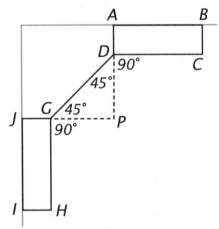

Moving a Piano.

26. A rotation.

•27. D.

28. E.

•29. G.

30. D.

31. Another rotation about point G.

32. 270°.

Bulldogs.

33. Two figures are congruent if there is an isometry such that one figure is the image of the other.

34. A translation.

•35. No, because the two dogs are not mirror images of each other.

36. Yes. A translation is the composite of two reflections through parallel lines.

37. No, because, if there were three reflections, one dog would be a mirror image of the other.

38. A glide reflection.

•39. A translation and a reflection.

40. Three reflections.

Grid Problem.

41.

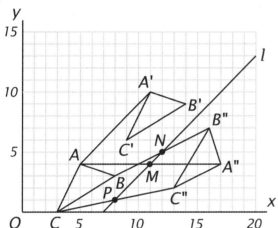

•42. B'(14, 9), C'(9, 6).

•43. $6\sqrt{2}$. (AA' $= \sqrt{(11-5)^2 + (10-4)^2} = \sqrt{36+36} = \sqrt{36 \cdot 2} = 6\sqrt{2}$.)

44. AA' appears to be parallel to line *l*.

45. A glide reflection, because it is the composite of a translation and a reflection in a line parallel to the direction of the translation.

•46. M(11, 4), N(12, 5), P(8, 1).

47. The *y*-coordinate is 7 less than the *x*-coordinate.

48. They lie on line *l*.

Set III (pages 317–318)

The irregular rubber stamp figure was chosen to try to impress upon the student's mind that the number of reflections needed to show that two figures are congruent has nothing to do with their complexity. David Henderson suggests a nice experiment in his book titled *Experiencing Geometry—In Euclidean, Spherical and Hyperbolic Spaces* (Prentice Hall, 2001). He says: "Cut a triangle out of an index card and use it to draw two congruent triangles in different orientations on a sheet of paper. . . . Now, can you move one triangle to the other by three (or fewer) reflections? You can use your cutout triangle for the intermediate steps." This leads to proving that "on the plane, spheres, or hyperbolic planes, every isometry is the composition of one, two, or three reflections." Chapter 11 of Henderson's book is a good resource for students (and teachers) wanting to learn more about isometries.

Stamp Tricks.

1. Ollie stamped one of the images on the back of the tracing paper.

2. One figure is now the mirror image of the other and two reflections (of an asymmetric figure) cannot produce a mirror image.

3. Yes. The figure below shows one of the many ways in which it can be done.

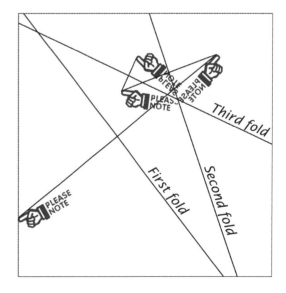

Set I (pages 321–322)

Two wonderful sources of ambigrams are Scott Kim's *Inversions* (Byte Books, 1981) and John Langdon's *Wordplay* (Harcourt Brace Jovanovich, 1992). In the introduction to his book, Langdon wrote:

"Ambigrams come in a number of forms, limited only by the ambigrammist's imagination, and usually involve some kind of symmetry. This book is made up of three types:
1. words with rotational symmetry . . .
2. words that have bilateral, or mirror-image symmetry . . .
3. chains. These are ambigrams that cannot stand alone as single words, but depend on being linked to the preceding and ensuing words."

In other words, ambigrams are based on the three basic types of symmetry in the plane: rotation, reflection, and translation.

The fact that games normally have two opponents or two opposing teams requires that almost all playing fields and courts have two lines of symmetry. Baseball and its related versions can use a field with just one line of symmetry because the teams regularly interchange their positions in the game.

It is interesting that, although Washington had the extra window painted solely to "complete the symmetry," the two windows on either side of the front door, with the windows above them, are not quite in the right places.

Ambigrams.

• 1. Rotation (or point) symmetry.

2. See if it coincides with its rotation image. (Or, for point symmetry, see if it looks the same upside down.)

3. Reflection (line) symmetry.

4. See if it coincides with its reflection image or fold it to see if the two halves coincide.

Sport Symmetry.

5. Baseball.

• 6. It has reflection (line) symmetry with respect to a line through home plate and second base.

7. Basketball.

8. It has reflection (line) and rotation (point) symmetry. It has two lines of symmetry and 2-fold rotation symmetry.

9. The same type as the basketball court.

10. *(Student answer.)* (It is so that each team has the same view of the other side.)

Mount Vernon.

11. To make his house look more symmetric.

Symmetries of Basic Figures.

• 12. The point itself. (Not its "center," because a point does not have a center!)

13. Yes. A point is symmetric with respect to every line that contains it (because a point on a reflection line is its own image.)

• 14. A line looks the same if it is rotated 180°.

15. Any point on the line can be chosen as its center of symmetry.

• 16. A line has reflection symmetry because it can be reflected (folded) onto itself.

17. Infinitely many. The line itself and every line that is perpendicular to it.

18. Yes. A line can be translated any distance along itself and still look the same.

• 19. That rays OA and OD are opposite rays and that rays OB and OC are opposite rays.

20. If ∠AOB is rotated 180° about point O, it coincides with ∠COD.

21. Vertical angles are equal.

• 22. In a plane, two points each equidistant from the endpoints of a line segment determine the perpendicular bisector of the line segment.

23. A and C.

24. If two sides of a triangle are equal, the angles opposite them are equal.

• 25. They bisect each other.

26. CD.

27. The opposite sides of a parallelogram are equal.

28. No. BD cannot be a rotation image of AC because they do not have the same length.

Set II (pages 322–324)

For the piano keyboard, it is interesting to note that all of the translation images of a given key have the same letter name. The key that is the first translation image to the right of a given key is one octave higher and has a frequency twice as great. Without the translation pattern of the black keys, pianists would have trouble keeping their place!

In his book titled *Reality's Mirror* (Wiley, 1989), Bryan Bunch explains the connection of odd and even wave functions to such topics as cold fusion and the Pauli Exclusion Principle. Bunch writes: "It would be only a slight exaggeration to say that symmetry accounts for all the observable behavior of the material world."

Piano Keyboard.

29. *Example answer:*

•30. Translation.

31. No.

Water Wheel.

32. Rotation and point symmetry.

•33. 22.5°. ($\frac{360°}{16}$.)

•34. No. ($\frac{125°}{22.5°}$ is not an integer.)

35. Yes. ($\frac{225°}{22.5°} = 10$.)

•36. 16.

37. No.

Wave Functions.

•38. The y-axis.

39. Reflection (line).

40. The origin.

41. Rotation (or point).

•42. Even.

43. Neither.

44. Odd.

Cherry Orchard.

•45. Because it can be translated (in various directions) and look exactly the same.

46. The distance between any pair of neighboring trees.

47.

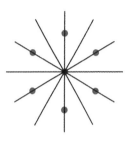

48. Because it can be reflected (in various lines) and look exactly the same.

49. Because it can be rotated (about various points) and look exactly the same.

50. Any tree. Also, any point centered between three neighboring trees. (Also, the midpoints of the segments between neighboring trees are centers of 2-fold rotation symmetry.)

51. 60°.

Set III (page 324)

Short Story.

1.

in the summer, *sis* goes to the beach and *suns* herself, adds up *sums* of numbers, does *stunts* and collects *tin* cans. The ride home is *hilly*, and that makes her head feel *fizzy*.

2. The figure can be read as either Scott's first name or his last name.

Set I (pages 325–327)

In *Visual Intelligence—How We Create What We See* (Norton, 1998), Donald Hoffman reports that perhaps the earliest version of the "vase-faces" illusion appeared in a picture puzzle in 1795. Perceptual psychologists first began to study the illusion in 1915. Of the many variations that have appeared since, surely the most clever is its use in the vase pictured in the text.

In 1941–42, Escher filled a large notebook with notes and drawings that he titled *Regular Division of the Plane with Asymmetrical Congruent Polygons*. This notebook is reproduced in its entirety in *Visions of Symmetry—Notebooks, Periodic Drawings, and Related Work of M. C. Escher*, by Doris Schattschneider (W. H. Freeman and Company, 1990). She remarks about this work: "It is impossible to look at this notebook and not conclude that in this work, Escher was a research mathematician."

Remarkably, the first book to be published on Escher's work was written by Caroline H. Macgillavry, a professor of chemical crystallography. In *Symmetry Aspects of M. C. Escher's Periodic Drawings* (The International Union of Crystallography, 1965), Macgillavry wrote that Escher's notebook had been a "revelation" to her, commenting: "It is no wonder that x-ray crystallographers, confronted with the ways in which nature solves the same problem of packing identical objects in periodic patterns, are interested in Escher's work."

In lecturing on his work, Escher explicitly discussed his use of the four isometries of the plane in creating his mosaics. For more on this, see the section titled "The geometric rules" (p. 31ff) in *Visions of Symmetry*.

Amazing Vase.

1. No. The left and right sides of the upper part of the vase do not quite look like mirror images of each other.

2. The profiles of Prince Philip and Queen Elizabeth can be seen in silhouette.

Double Meanings.

3. A one-to-one correspondence between two sets of points.

4. A transformation that preserves distance and angle measure.

5. A transformation in which the image is an enlargement or reduction of the original.

•6. No.

Monkey Rug.

7. A translation.

•8. A glide reflection.

•9. A rotation.

10. A reflection.

Clover Leaves.

•11. That it can be rotated so that it looks the same in three positions.

12. 120°.

13. No. It doesn't look the same upside down.

14.

15.

16. 90°.

•17. Yes.

18. Good.

Musical Transformations.

19. A translation.

•20. A reflection.

21. A translation.

22. A rotation.

Fish Design.

23. No. There seem to be two shapes of fish in the mosaic. The differences in their noses and tails are the most obvious.

•24. Two figures are congruent if there is an isometry such that one figure is the image of the other.

25. A glide reflection.

26. A rotation.

27. Yes. Translations.
   Example answer:

Set II (pages 327–329)

Keep Your Eye on the Ball—The Science and Folklore of Baseball, by Robert G. Watts and A. Terry Bahill (W. H. Freeman and Company, 1990) is a great source of examples for use in illustrating the application of algebra, geometry, and physics to the analysis of baseball. The authors describe the figure used for exercises 28 through 32: "The swing of a baseball bat exhibits two types of motion: translational and rotational. . . . To move the bat from position A to position C, one can first *rotate* the bat about the center of mass and then *translate* the center of mass."

John Langdon wrote that his "past" and "future" ambigrams "are intended to represent the interminable conveyor belt of time, and the individual letters and words are the experiences, the 'life bites' of our existence."

Batter's Swing.

28. A rotation.

•29. A translation.

30. For a rotation, the two reflection lines intersect. For a translation, they are parallel.

•31. About 47°.

32. About 24 in. (In the figure, the length of the bat is 42 mm and the magnitude of the translation is about 24 mm.)

33. *Example figure:*

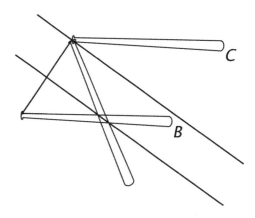

Construction Problem.

34.

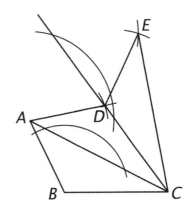

35. The image of AB is AD and the image of BC is DC.

•36. Because △ADC is the reflection image of △ABC.

•37. It has reflection (line) symmetry with respect to line AC.

38. A rotation.

From N to Z.

39.

40. A rotation.

•41. 45°.

42. 90°.

43. The image of H looks like I. The image of W looks somewhat like E. (And, the image of Z looks like N.)

Past and Future.

44. *Example answers:*

45. Translation.

46. So that the E's look "correct" in both directions.

•47. No.

48. The letters A and T in PAST.
The letters T, U, and E in FUTURE.

49. Yes. The letter E.

Dice Symmetries.

50. *Group 1:* 1, 4, and 5.
These faces have 4 lines of symmetry and 4-fold rotation (as well as point) symmetry.
Group 2: 2, 3, and 6.
These faces have 2 lines of symmetry and 2-fold rotation (as well as point) symmetry. In regard to 2 and 3, the symmetry lines contain the diagonals of the face; in regard to 6, they are parallel to the edges and midway between them.

Set I (pages 330–333)

Chill Factor.

1. A conditional statement.

2. If you don't put your coat on.

Losers, Sleepers.

3. An Euler diagram.

4. If you snooze, you lose.

5. You lose.

Marine Logic.

6. A syllogism.

7. If one of its premises were false.

Finding Truth.

8. Theorems.

9. Postulates.

Only in Geometry.

10. Points that lie on the same line.

11. The side opposite the right angle in a right triangle.

12. A triangle or trapezoid that has two equal legs.

13. A quadrilateral all of whose sides are equal.

Why Three?

14. Three noncollinear points determine a plane.

15. Point, plane.

What Follows?

16. A line.

17. 180°.

18. Bisects it.

19. Are equal.

20. Equiangular.

21. Either remote interior angle.

22. The perpendicular bisector of the line segment.

23. Are parallel.

24. Are parallel.

25. Also perpendicular to the other.

26. The remote interior angles.

27. Equal.

28. Are equal.

Formulas.

29. The area of a circle is π times the square of its radius.

30. The area of a rectangle is the product of its length and width.

31. This is the Distance Formula. The distance between two points is the square root of the sum of the squares of the differences of their x-coordinates and their y-coordinates.

32. The perimeter of a triangle is the sum of the lengths of its sides.

33. The perimeter of a rectangle is the sum of twice its length and twice its width.

Protractor Problems.

34. 84°. (123 – 39.)

35. 42°. ($\frac{1}{2}$84°.)

36. 81. (39 + 42 or 123 – 42.)

37. 48°. (90° – 42°.)

38. 171. (123 + 48 = 171.)

39. Yes. OD-OC-OB because 171 > 81 > 39.

Metric Angles.

40. Two angles are supplementary iff their sum is 200 grades.

41. An angle is obtuse if it is greater than 100 grades but less than 200 grades.

42. The sum of the angles of a quadrilateral is 400 grades.

43. Each angle of an equilateral triangle is $66\frac{2}{3}$ grades.

Linear Pair.

44. That the figure contains two opposite rays.

45. No. If the angles are equal, they would be right angles.

46. They are supplementary.

47. Each angle is a right angle.

Polygons.

48. $\sqrt{1125}$ or $15\sqrt{5}$.

   ($c^2 = 15^2 + 30^2 = 225 + 900 = 1125$,
   $c = \sqrt{1125} = 15\sqrt{5}$.)

49. 30. (It can't be 15, because 15 + 15 = 30.)

50. 30°, 150°, and 150°. (A rhombus is a parallelogram. The opposite angles of a parallelogram are equal and the consecutive angles are supplementary.)

51. 30°, 150°, and 150°. (The base angles of an isosceles trapezoid are equal. The parallel bases form supplementary angles with the legs.)

Bent Pyramid.

52. 126°. (180° – 54°.)

53. 94°. [180° – 2(43°).]

54. 54°. (∠E = ∠A.)

55. 169°. (126° + 43°.)

Six Triangles.

56. ΔAFP and ΔDEP.

57. By SAS. (ED = AP because AB = ED and AB = AP.)

58. By HL. (AF = PB because AF = CD and CD = PD.)

Italian Theorem.

59. ∠B > ∠A.

60. If two sides of a triangle are unequal, the angle opposite the greater side is greater than that opposite the smaller side. (If two sides of a triangle are unequal, the angles opposite them are unequal in the same order.)

61. Major and minor.

Impossibly Obtuse.

62. ∠A + ∠B + ∠C > 270°.

63. The fact that the sum of the angles of a triangle is 180°.

64. It shows that what we supposed is false.

65. Indirect.

Converses.

66. If two angles are complementary, then they are the acute angles of a right triangle. False.

67. If the diagonals of a quadrilateral bisect each other, then the quadrilateral is a parallelogram. True.

Construction Exercises.

68.

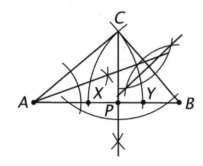

69. No.

70. Point P.

71. CA, CP, and CB.

Set II (pages 333–336)

Grid Exercise.

72.

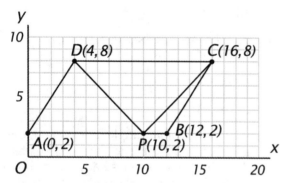

73. A parallelogram.

74. AB = DC = 12; AD = BC = $\sqrt{52}$ or $2\sqrt{13}$.
(AD = $\sqrt{(4-0)^2+(8-2)^2}$ = $\sqrt{16+36}$ = $\sqrt{52}$ = $2\sqrt{13}$.
BC = $\sqrt{(16-12)^2+(8-2)^2}$ = $\sqrt{16+36}$ = $\sqrt{52}$ = $2\sqrt{13}$.

75. *(Student answer.)* (PD looks longer to many people.)

76. PD = $\sqrt{(4-10)^2+(8-2)^2}$ = $\sqrt{36+36}$ = $\sqrt{72}$ = $6\sqrt{2}$.
PC = $\sqrt{(16-10)^2+(8-2)^2}$ = $\sqrt{36+36}$ = $\sqrt{72}$ = $6\sqrt{2}$.

Roof Truss.

77. ∠A = ∠L. ∠A and ∠L are corresponding parts of ΔAGF and ΔLGF, which are congruent by SSS.

78. BC ∥ FG. BC ∥ DE and DE ∥ FG because equal alternate interior angles mean that lines are parallel. BC ∥ FG because, in a plane, two lines parallel to a third line are parallel to each other.

79. Yes. CD ∥ EF because they form equal alternate interior angles with DE.

80. No. If ∠AGF and ∠GFA were supplementary, GA would be parallel to FA (supplementary angles on the same side of a transversal mean that lines are parallel). GA and FA intersect at A.

Angles Problem 1.

81.

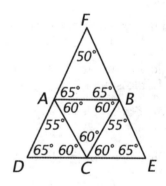

82. ΔACD ≅ ΔBCE by ASA. AC = BC because ΔABC is equilateral, ∠DAC = 55° = ∠EBC, and ∠DCA = 60° = ∠ECB.

Irregular Star.

83.

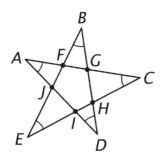

84. True. ΔEFC is isosceles because EF = FC because ∠C = ∠E.

85. True. FGHIJ is equiangular because each of its angles is an angle of one of the five overlapping isosceles triangles, whose base angles are the equal angles at the corners of the star. If two angles of one triangle are equal to two angles of another triangle, the third angles are equal.

86. False. (This is evident from the figure.)

87. True. There are 10 isosceles triangles in all. The base angles of the triangles that are acute (ΔBFG is an example) are supplementary to the equal angles of pentagon FGHIJ; so they are equal.

Angle Trisector.

88. These triangles are congruent by SSS.

89. ∠1 = ∠2 and ∠2 = ∠3 because they are corresponding parts of the congruent triangles; so ∠1 = ∠3.

Quadrilateral Problem.

90. $x + 2x + 3x + 4x = 360$, $10x = 360$, $x = 36$.

91. 144°. [4(36°).]

92. ∠B = 72°, ∠C = 108°, AB ∥ DC, ABCD is a trapezoid.

Angles Problem 2.

93.

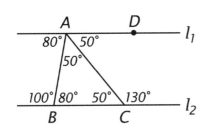

94. ΔABC is isosceles. ∠BAC = 50° = ∠ACB; so AB = BC.

A, B, C.

95. $a^2 + b^2 = c^2$.

96. $a + b > c$, $a + c > b$, $b + c > a$.

97. No. The sum of two of its sides would equal the third side, which would contradict the Triangle Inequality Theorem.

Midsegments.

98.

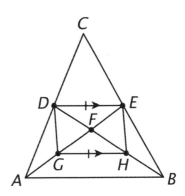

99. Applying the Midsegment Theorem to ΔABC gives DE = $\frac{1}{2}$AB and applying it to ΔABF gives GH = $\frac{1}{2}$AB; so DE = GH by substitution.

100. Again, by the Midsegment Theorem, DE ∥ AB and GH ∥ AB; so DE ∥ GH. In a plane, two lines parallel to a third line are parallel to each other.

101. DEHG is a parallelogram because two opposite sides, DE and GH, are both parallel and equal.

102. AG = GF and BH = HF because G and H are the midpoints of AF and BF. GF = FE and HF = FD because the diagonals of a parallelogram bisect each other. So AG = GF = FE and BH = HF = FD.

On the Level.

103. The design of the swing is based on a parallelogram. The supports of the plank are equal, and the part of the plank between the supports is equal to the distance between the supports at the top. The top remains level with the ground. The plank, the opposite side of the parallelogram, is always parallel to the top; so it also stays level with the ground.

Folding Experiment.

104. *Example figure (the student's figure will depend on the point and corner chosen):*

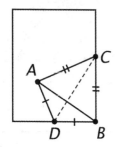

105. It appears to be the perpendicular bisector of AB.

106. When B falls on A, CA = CB and DA = DB. In a plane, two points each equidistant from the endpoints of a line segment determine the perpendicular bisector of the line segment.

Earth Measurement.

107. 20,520 mi. (360 × 57.)

108. About 3,300 mi. ($c = 2\pi r$; so

$$r = \frac{c}{2\pi} = \frac{20{,}520}{2\pi} \approx 3{,}300.)$$

Not a Square.

109. Yes. ABCD could be a rhombus if all of its sides were equal.

110. Yes. If ABCD is a rhombus, its diagonals will be perpendicular.

111. Yes. ABCD is a parallelogram and the diagonals of a parallelogram bisect each other.

112. No. If ABCD were a rectangle, ∠DAB would be equal to ∠ABC.

113. No. If AC = DB, then ΔDAB ≅ ΔCBA by SSS (DA = CB because ABCD is a parallelogram.) If ΔDAB ≅ ΔCBA, then ∠DAB = ∠ABC but ∠DAB ≠ ∠ABC.

Construction Exercise.

114. *Example figure:*

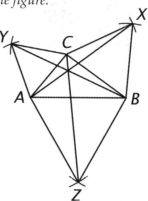

115. They seem to be concurrent.

SAT Problem.

116. A rotation.

117. A translation.

118. A reflection.

119. A reflection.

120. Figure E.

Quilt Patterns.

121. Rotation (point) symmetry (4-fold).

122. Rotation (point) symmetry and reflection (line) symmetry (4 lines).

123. Reflection (line) symmetry (1 line).

Dividing a Lot.

124.

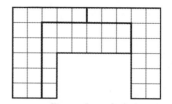

125.

Chapter 9, Lesson 1

Set I (pages 340–341)

The Amtrak trademark first appeared in 1971 with the beginning of the government corporation in charge of the nation's passenger trains. According to Hal Morgan in *Symbols of America* (Viking, 1986), "the arrow design was chosen to convey the image of speed and purpose of direction."

The geometry of drums is not well known and rather surprising. In 1911, Hermann Weyl proved that the overtones of a drum determine its area. In 1936, another mathematician proved that a drum's overtones also determine its perimeter! In 1966, Mark Kac published a paper titled "Can One Hear the Shape of a Drum?" In 1991, Carolyn Gordon gave a lecture at Duke University about drum geometry. She and her husband David Webb have subsequently discovered a number of pairs of sound-alike drums, including the pair pictured in the exercises. The drums in each pair are polygonal in shape, have at least eight sides, and the same perimeter and area. All of this information is from the chapter titled "Different Drums" in Ivars Peterson's *The Jungles of Randomness* (Wiley, 1998). The book also includes a page of color photographs showing the standing waves of the first few normal modes of these drums.

Train Logo.

 1.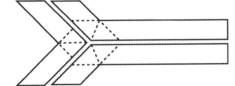

(The dotted lines in this figure illustrate the possible choices of ways to divide the figure into six quadrilaterals.)

2. Trapezoids and parallelograms.

•3. Its area is equal to the sum of the areas of its nonoverlapping parts.

Star in Square.

4. The star because of the Triangle Inequality Theorem.

5. The square because of the Area Postulate. (The area of the square is equal to the area of the "star" plus the positive areas of the four white triangular regions.)

6. No.

Area Relations.

•7. True, because $\triangle ADC \cong \triangle ABC$.

8. True, because $\triangle 1 \cong \triangle 4$ and $\triangle 2 \cong \triangle 3$.

•9. True, because $\alpha EPHD = \alpha \triangle ADC - \alpha \triangle 1 - \alpha \triangle 2 = \alpha \triangle ABC - \alpha \triangle 4 - \alpha \triangle 3 = \alpha GBFP$.

10. True, because $\alpha AGHD = \alpha EPHD + \alpha AGPE = \alpha GBFP + \alpha AGPE = \alpha ABFE$.

11. True, because $\alpha EFCD = \alpha EPHD + \alpha PFCH = \alpha GBFP + \alpha PFCH = \alpha GBCH$.

12. False. $\alpha AGPE > \alpha PFCH$.

Flag Geometry.

13. x square units.

•14. $2x$ square units.

15. $6x$ square units.

•16. $(x + y)$ square units.

17. $(2x - 2y)$ square units.

18. $(x + y)$ square units.

19. Yes. $(x + y) + (2x - 2y) + (x + y) = 4x$.

Drum Polygons.

•20. They are both concave octagons.

21. They have equal areas. ($3.5b^2$.)

22. They have equal perimeters.

•23. $3a + 6b$.

24. No.

Set II (pages 341–342)

Exercises 35 through 39 are a nice example of how some areas can be easily compared. The areas of the regular hexagons circumscribed about a circle and inscribed in it are related in a surprisingly simple way: their ratio is 4 to 3.

Plato's method for doubling the square is based on the same simple idea of counting triangles. It nicely circumvents the fact that, if the side of the original square is 1 unit, the side of the doubled square is an irrational number.

According to Sir Thomas Heath in his commentary on Euclid's *Elements*, Proclus stated that Euclid coined the word "parallelogram." Euclid first mentions parallelograms in Proposition 34 of Book I. It is the next theorem, Proposition 35, that is demonstrated in exercises 46 through 55. (The figure is the one that Euclid used; it illustrates just one of three possible cases relating the upper sides of the parallelograms.) Euclid does not actually mention area, stating the theorem in the form "Parallelograms which are on the same base and in the same parallels are equal to one other." Heath remarks: "No *definition* of equality is anywhere given by Euclid; we are left to infer its meaning from the few *axioms* about 'equal things.'" Previously to the theorem, Euclid had used equality to mean congruence exclusively.

Midsegment Triangle.

25.

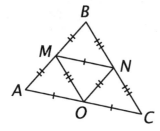

•26. They are congruent (SSS) and therefore equal in area.

27. $\alpha\triangle MNO = \frac{1}{4}\alpha\triangle ABC$.

28. They are parallelograms.

29. Their opposite sides are parallel (or, their opposite sides are equal). Both follow from the Midsegment Theorem.

•30. No.

•31. Yes. Each one contains two of the four triangles, all of whose areas are equal.

32. Trapezoids.

33. No.

34. Yes. Each one contains three of the four triangles, all of whose areas are equal.

Circle Area.

•35. $\frac{1}{6}$ square unit. ($\frac{3}{18}$.)

36. $\frac{1}{2}$ square unit. [$3(\frac{1}{6})$.]

•37. 2 square units. [$12(\frac{1}{6})$.]

38. 4 square units. [$24(\frac{1}{6})$.]

39. Roughly 3.5 square units. (The area of the circle appears to be about halfway between the areas of the two hexagons. Using methods that the students won't know until later, we can show that the area of the circle is $\frac{2\pi}{\sqrt{3}} \approx 3.63\ldots$)

Doubling a Square.

40. SAS.

41. Congruent triangles have equal areas.

•42. It is equilateral and equiangular. (Both follow from the fact that corresponding parts of congruent triangles are equal.)

•43. BFED contains four of the congruent triangles, and ABCD contains two of them.

SAT Problem.

•44. 4 square units. ($\frac{100-84}{4} = 4$.)

45. 116 square units. [$100 + 4(4) = 116$.]

Comparing Parallelograms.

46. The opposite sides of a parallelogram are equal.

47. The opposite sides of a parallelogram are parallel.

48. Parallel lines form equal corresponding angles.

•49. They are both equal to AB.

•50. Because DF = DC + CF = CF + FE = CE.

51. SAS.

52. Congruent triangles have equal areas.

53. Substitution. ($\alpha\triangle ADF = \alpha 1 + \alpha 3$ and $\alpha\triangle BCE = \alpha 3 + \alpha 4$.)

54. Subtraction.

55. $\alpha 1 + \alpha 2 = \alpha 2 + \alpha 4$ (addition); so $\alpha ABCD = \alpha ABEF$ (substitution).

According to Jerry Slocum and Jack Botermans (*New Book of Puzzles*, W. H. Freeman and Company, 1992), the magic playing card puzzle first appeared in a version known as "The Geometric Money" in *Rational Recreations*, by William Hooper, published in 1794.

Unfortunately, it isn't possible to make this puzzle out of an ordinary playing card, because the back of the card, before it is cut up, would have to be redesigned for the purpose of the trick. The figure below shows what it would look like for the card shown in the text.

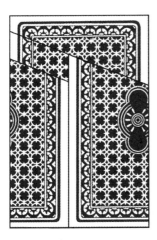

Magic Playing Card.

1. No. If they were, the two would have equal areas. Including the "hole," the second rectangle has a greater area.

2. The widths are obviously the same; so the lengths must be different.

3. The second figure is slightly longer than the first. The extra area is equal to the area of the "hole."

Chapter 9, Lesson 2

Set I (pages 345–347)

The "Abstract Painting, 1960–61" pictured in the text is minimal artist Ad Reinhardt's most famous work. It is actually composed of nine congruent squares, each painted in a slightly different shade of black. At the time of this writing, there is a site on the Internet where another of his pictures, "Black Painting, 1960–66" can be "viewed". The site, which appropriately has the word "piffle" in its name, is . . . black.

In Book II of the *Elements*, Euclid used geometric methods to prove a series of algebraic identities. Proposition 1 deals with the distributive property. Proposition 4, which concerns the square of a binomial, is stated as: "If a straight line be cut at random, the square on the whole is equal to the squares on the segments and twice the rectangle contained by the segments." All this to say: $(a + b)^2 = a^2 + b^2 + 2ab$. In his *History of Mathematics* (Allyn & Bacon, 1985), David Burton wrote: "By Euclid's time, Greek geometric algebra had reached a stage of development where it could be used to solve simple equations." The equations, both linear and quadratic, were solved by making geometric constructions.

Abstract Art.

•1. 20 ft.

2. 240 in.

3. 25 ft^2.

•4. 3,600 in^2.

Tile Pattern.

5. (*Student answer.*)

6. Blue region: 36^2 units; green region: 28^2 units; yellow region: 36^2 units. ($6^2 = 36$; $8^2 - 6^2 = 28$; $10^2 - 8^2 = 36$.)

Fishing Nets.

7. The sides of the small squares are $\frac{1}{2}$ inch long.

•8. Four.

9. 16.

10. To allow the smaller, younger fish to escape.

Distributive Property.

•11. $a(b + c)$.

•12. ab and ac.

13. $a(b + c) = ab + ac$.

Binomial Square.

•14. $(a + b)^2$.

15. a^2, ab, ab, and b^2.

16. $(a + b)^2 = a^2 + 2ab + b^2$.

Difference of Two Squares.

17. $a^2 - b^2 = (a + b)(a - b)$.

Area Connection.

18. $A = ab$.

•19. Two congruent triangles.

20. The Area Postulate: Congruent triangles have equal areas.

21. $A = \frac{1}{2}ab$.

Two Squares.

•22. $\sqrt{26}$.

23. No. ($\sqrt{26} = 5.0990195\ldots$)

24. 25.999801.

•25. No. (5.0990195^2 is a long decimal ending in 5, not the integer 26.)

26. No.

27. Irrational. (As students may recall from algebra, an irrational number is a number that cannot be written as the quotient of two integers. If an integer is not the square of an integer, its square roots are irrational.)

Set II (pages 347–349)

In his book titled *The Cosmological Milkshake* (Rutgers University Press, 1994), physics professor Robert Ehrlich remarks that "lying on a bed of nails need be no more painful than lying on your own bed, provided that the nail spacing is small enough, so that the fraction of your weight supported by any one nail is not too large." He also provides a simple explanation of why a spacing of one nail per square inch is practical.

Lumber measurements are interesting in that the three dimensions of a board are not all given in the same units: a 10-foot "2-by-4", for example, has one dimension given in feet and the other two in inches (a "2-by-4" was originally just that but is now actually 1.5-by-3.5). Another example of combining different units is the "board foot." A board foot is a unit of cubic measure that is the volume of a piece of wood measuring 1 *foot* by 1 *foot* by 1 *inch*. (Because the grading of a piece of lumber is based on cutting it into pieces of the same thickness as the lumber, the calculations

can be done in terms of area rather than volume.)

Bed of Nails.

•28. 2,592 nails. ($6 \times 12 \times 3 \times 12$.)

29. (*Student answer.*) (Standing on it would hurt the most.)

Surveyor's Chain.

•30. 66 ft. ($\frac{100 \times 7.92}{12}$.)

31. One chain by ten chains.

32. Ten square chains.

33. 5,280 ft. (80×66.)

•34. 6,400 square chains. (80^2.)

35. 640. ($\frac{6,400}{10}$.)

Map Reading.

36. 480 cm^2.

•37. 84 cm^2.

•38. 17.5%. ($\frac{84}{480} = 0.175$.)

39. $33\frac{1}{3}$%. (The area of the awkward region is now 160 cm^2; $\frac{160}{480} = \frac{1}{3}$.)

Wallpaper Geometry.

•40. 21.6 ft long. ($36 \text{ ft}^2 = 5,184 \text{ in}^2$; $\frac{5,184 \text{ in}^2}{20 \text{ in}} = 259.2 \text{ in} = 21.6 \text{ ft}$.)

41. 16 ft long. ($\frac{5,184 \text{ in}^2}{27 \text{ in}} = 192 \text{ in} = 16 \text{ ft}$.)

42. The wallpaper strips have to be placed so that the pattern matches, which means that some paper will be wasted.

•43. The total surface area of the walls.

44. In rounding to the nearest number, if we round down, we may not have enough paper. (For example, if the result is 21.4, 21 rolls may not be enough.)

45. It is about taking into account the area of windows and doors.

46. It would reduce it. (The basic rule is to *subtract 1 roll for every 50 square feet of door or window opening.*)

Cutting a Board.

•47. 1,728 in². (12 in × 144 in.)

•48. 189 in². (3.5 in × 54 in.)

49. 243 in². (4.5 in × 54 in.)

50. 459 in². (8.5 in × 54 in.)

51. 408 in². (6 in × 68 in.)

52. Yes. The total area of the four cuttings is

1,299 in² and $\dfrac{1,299}{1,729} \approx 75\%$.

The Number 17.

53.

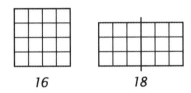

16   18

54. The number of linear units in the perimeter of each figure is equal to the number of square units in its area.

55. [This is a challenging problem. Letting x and y be the dimensions of such a rectangle, $2x + 2y = xy$. Finding solutions to this equation is easiest if we first solve for one variable in terms of the other. $2x = xy - 2y$; so $y(x - 2) = 2x$, and so $y = \dfrac{2x}{x-2}$. Letting $x = 5$ gives $y = \dfrac{10}{3}$; so another rectangle is a 5 by $3\dfrac{1}{3}$ rectangle. Of course, x does not have to be an integer. Any value of $x > 2$ will do.]

Total Living Area.

56. 1,290 ft².
Example figure (answers will vary):

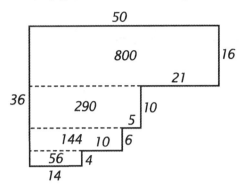

57. *Example figure (answers will vary):*

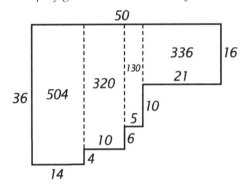

Set III (page 350)

According to David Wells (*The Penguin Dictionary of Curious and Interesting Geometry*, Penguin Books, 1991), a Russian mathematician in the early part of the twentieth century claimed that the "squaring the square" problem couldn't be done. In 1939, Roland Sprague discovered a solution using 55 squares and, in 1978, Dutch mathematician A. J. W. Duijvestijn found the simplest solution, consisting of 21 squares. Martin Gardner's "Mathematical Games" column in the November 1958 issue of *Scientific American* contained William T. Tutte's account of how he and his fellow Cambridge students solved the problem of squaring the square, which was the reason for the topic being featured on the cover of the magazine. The figure on the cover was discovered by Z. Morón in 1925, and it was proved in 1940 that it is the simplest possible way in which to square a rectangle. Tutte's account is charming and very instructive. It shows what fun and progress can be made if you are persistent, systematic, and lucky. Mrs. Brooks, the mother of R. L. Brooks, one of the four Cambridge students who did this work, put together a jigsaw puzzle that her son had made of a squared rectangle but, to everybody's astonishment, she got a rectangle of different dimensions. Such a thing had never been seen before. It was a key observation and unexpected demonstration of the fact that making a real model and playing with it can lead to real progress. More recent discoveries are reported by Gardner in chapter 11 of his book titled *Fractal Music, Hypercards and More* (W. H. Freeman and Company, 1992).

Dividing into Squares.

1.

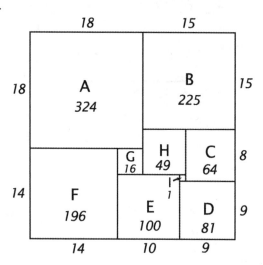

2. A, 324; B, 225; C, 64; D, 81; E, 100; F, 196; G, 16; H, 49; I, 1.

3. No. Its base is 33 and its altitude is 32.

Chapter 9, Lesson 3

Set I (pages 353–355)

Matchstick puzzles such as the one in exercises 1 through 3 first became popular in the nineteenth century, when matches were universally used to light lamps and stoves. As a followup to these exercises, it might be amusing to place 12 matches in the 3-4-5 right triangle arrangement on an overhead projector, observe that they enclose an area of 6 square units, and consider how they might be rearranged to get areas of 5 and 4 square

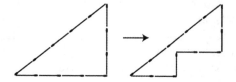

units (the obvious solutions are "corollaries" to the solution for getting an area of 3 square units). Even better would be to pose the puzzle of what areas can be enclosed by 12 matches as a problem for interested students to explore. In one of his early "Mathematical Games" columns for *Scientific American*, Martin Gardner posed the "triangle to area of 4" version of the problem. Several readers of the magazine pointed out that, if 12 matches are arranged to form a six-pointed star, the widths

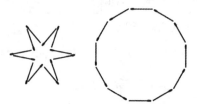

of the star's points can be adjusted to produce any area between 0 and 3 cot 15° (11.196 . . .), the area of a regular dodecagon (*The Scientific American Book of Mathematical Puzzles and Diversions*, Martin Gardner, Simon & Schuster, 1959).

In the days before students had calculators, the problem of finding the area of the four-sided field required a lot more work. The value of knowing how to factor is evident if the problem of calculating

$$\frac{1}{2}(14.36)(8.17) + \frac{1}{2}(14.36)(5.74)$$

is compared with the problem of calculating

$$\frac{1}{2}(14.36)(8.17 + 5.74),$$

as it would have been done more than a century ago.

Match Puzzle.

• 1. 6 square units.

2. Yes. 3 square units have been removed; so 3 square units are left.

3. They are the same: 12 units.

Isosceles Right Triangle.

4. $\frac{1}{2}a^2$.

• 5. $\frac{1}{4}c^2$.

Theorem 39.

- •6. The area of a right triangle is half the product of its legs.

- •7. The area of a polygonal region is equal to the sum of the areas of its nonoverlapping parts (the Area Postulate).

- 8. Substitution.

- •9. Substitution.

- 10. The area of a right triangle is half the product of its legs.

- 11. The area of a polygonal region is equal to the sum of the areas of its nonoverlapping parts (the Area Postulate).

- 12. Subtraction.

- 13. Substitution.

- 14. Substitution.

Enlargement.

- •15. They seem to be equal.

- 16. They are twice as long.

- •17. ΔABC, 40 units; ΔDEF, 80 units.

- 18. It is twice as large.

- 19. ΔABC, 60 square units; ΔDEF, 240 square units.

- 20. It is four times as large.

- •21. They seem to stay the same.

- 22. It is doubled.

- 23. It is multiplied by four.

Chinese Parallelogram.

24.

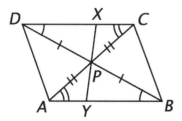

- 25. Yes. The triangles bounding them are congruent. (The diagonals of a parallelogram bisect each other, parallel lines form equal alternate interior angles, and the pairs of vertical angles at P are equal.)

- 26. They have equal areas. (They are also congruent.)

Four-Sided Field.

- •27. αABCD ≈ 99.87 square chains. [αΔABD + αΔCBD =

$$\frac{1}{2}BD \cdot AE + \frac{1}{2}BD \cdot CF = \frac{1}{2}BD(AE + CF) =$$

7.18(8.17 + 5.74) ≈ 99.87.]

- 28. About 9.987 acres.

Set II (pages 355–357)

The key to the puzzle of the surfer is Viviani's Theorem, named after an obscure Italian mathematician (1622–1703): In an equilateral triangle, the sum of the perpendiculars from any point P to the sides is equal to the altitude of the triangle. The proof outlined in exercises 29 through 34 works for the sum of the perpendiculars from any point P in the interior of a convex equilateral *n*-gon to its *n* sides. The pentagon case appeared as the "Star Geometry" problem in Chapter 6, Lesson 1 (page 216, exercises 30 through 32).

A practical application of Viviani's Theorem is in "trilinear charts." A trilinear chart is useful in presenting data belonging to three sets of numbers, each set of which sums to 100%. A. S. Levens describes them in his book titled *Graphics* (Wiley, 1962): "These charts are in the shape of an equilateral triangle. It can be shown that the sum of the perpendiculars, drawn from a point within the triangle, to the sides of the triangle is equal to the altitude of the triangle. Now, when we graduate each side of the triangle in equal divisions ranging from 0% to 100% and regard each side as representing a variable, we have a means for determining the percentages of each of the three variables that will make their sum equal 100%."

The kite problem lends itself to further exploration. For what other quadrilaterals would taking half the product of the lengths of the diagonals give the area? Do the diagonals have to be perpendicular? Does one diagonal have to bisect the other? Does the diagonal formula work for squares? Rhombuses? Rectangles? Are there any trapezoids for which it works? And so on.

In his book titled *Mathematical Encounters of the Second Kind* (Birkhäuser, 1997), Philip J. Davis relates an amusing story about a phone call from a man who identified himself as a roofer and who had a roof whose area he wanted to estimate. As Davis tells it: "He recalled that there were mathematical ways of doing this and had known them in high school, but he had forgotten them and would I help him. It seemed to me that any roofer worthy of his trade could at a single glance estimate the area of a roof to within two packages of shingles, whereas any professional mathematician would surely make an awful hash of it." After finally getting out of the man the fact that the roof consisted of four triangles, Davis asked: "'These four triangles, what shape are they? Are they equilateral triangles like one sees in some high pitched garage roofs or are they isosceles triangles? I've got to know.' No reply from my caller. 'Well, let's look at it another way: You've got four triangles. Find the area of each triangle and add them up.' 'How do you do that?' 'The area of a triangle is one half the base times its altitude.' 'My triangles don't have an altitude. . . .'"

Surfer Puzzle.

•29. The area of a triangle is half the product of any base and corresponding altitude.

30. The area of a polygonal region is equal to the sum of the areas of its nonoverlapping parts (the Area Postulate).

•31. Substitution.

32. Substitution.

•33. Division.

34. It proves that the sum of the lengths of the three paths is equal to the altitude of the triangle.

Three Equal Triangles.

•35. 84 square units.

(α right $\Delta DEF = \frac{1}{2}7 \cdot 24 = 84$.)

•36. 11.2 units. ($\frac{1}{2}15h = 84$.)

37. 6.72 units. ($\frac{1}{2}25h = 84$.)

38. 24 units. (EF = 24.)

39. 16.8 units. ($\frac{1}{2}10h = 84$.)

Kite Geometry.

40.

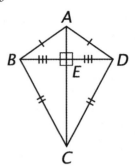

•41. In a plane, two points each equidistant from the endpoints of a line segment determine the perpendicular bisector of the line segment.

•42. Betweenness of Points Theorem.

43. The area of a triangle is half the product of any base and corresponding altitude.

44. Area Postulate.

45. Substitution.

Dividing a Cake.

46. Because it has less icing on its sides.

47.

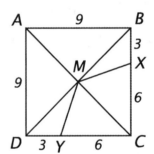

•48. $\alpha\Delta AMB = \alpha\Delta AMD = 20.25$ in²; $\alpha\Delta BMX = \alpha\Delta DMY = 6.75$ in²; $\alpha\Delta CMX = \alpha\Delta CMY = 13.5$ in².

49. Yes. The area of the top of each piece is 27 in² and each piece has the same amount of icing.

50.

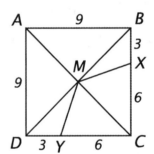

Divide the perimeter of the cake, 36, by 5 to get 7.2. Starting at A, mark off points around the square at intervals of 7.2 in. Cut from the center of the cake through these points.

Shingle Roof.

51. On the scale drawing, the lengths of the base and corresponding altitude are 2.5 in and 1.5 in. Because 1 in represents 8 ft, these lengths on the roof are 20 ft and 12 ft. Therefore, the area of the section is 120 ft².

•52. About 1,200 shingles.

53. 4.8, or about 5 bundles. (Because 4 bundles contain 1,000 shingles, 1 bundle contains 250 shingles. $\frac{1,200}{250} = 4.8$.)

54. 3.6 pounds. ($\frac{120}{100} \times 3 = 3.6$.)

Grid Exercise.

55.

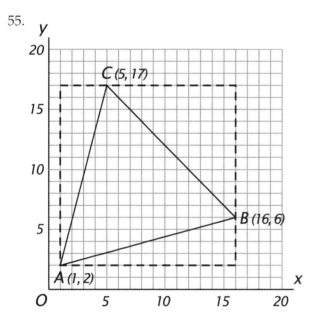

56. It is isosceles. By the Distance Formula,
$$AB = \sqrt{15^2 + 4^2} = \sqrt{241} \approx 15.52;$$
$$BC = \sqrt{11^2 + 11^2} = \sqrt{242} \approx 15.56;$$
$$AC = \sqrt{4^2 + 15^2} = \sqrt{241} \approx 15.52.$$

57. The exact area is 104.5 square units.
($15^2 - 2 \cdot \frac{1}{2} \cdot 4 \cdot 15 - \frac{1}{2} 11^2$.)

58. Irrational.

59. Rational.

Set III (page 357)

Although Heron's name is attached to the method for finding the area of a triangle from the lengths of its sides, Archimedes is thought to have actually discovered it. Heron's proof of the method is included in his book titled *Metrica*, which was lost until a manuscript of it was discovered in Constantinople in 1896. J. L. Heilbron remarks: "A good example of the difference in approach between the geometer and the algebraist may be drawn from proofs of the famous formula of Heron for the area of a triangle." Heilbron includes Heron's (or Archimedes's) geometric proof based on similar triangles, as well as an elegant trigonometric proof, in the "Tough Knots" chapter of his *Geometry Civilized* (Clarendon Press, 1998). An algebraic proof is included in the preceding editions of my *Geometry* (1974 and 1987). The algebra can be checked by symbol manipulation programs but is otherwise more challenging than instructive.

Heron's Theorem.

1. *(Student answer.)* ("Triangle 3" with sides 5, 5, and 10 seems "the most obvious.")

2. Triangle 1: perimeter 16; so $s = 8$.
   Area = $\sqrt{8(8-5)(8-5)(8-6)} = \sqrt{8 \cdot 3 \cdot 3 \cdot 2} = \sqrt{144} = 12$.
   Triangle 2: perimeter = 18; so $s = 9$.
   Area = $\sqrt{9(9-5)(9-5)(9-8)} = \sqrt{9 \cdot 4 \cdot 4 \cdot 1} = \sqrt{144} = 12$.
   Triangle 3: perimeter = 20; so $s = 10$.
   Area = $\sqrt{10(10-5)(10-5)(10-10)} = \sqrt{10 \cdot 5 \cdot 5 \cdot 0} = \sqrt{0} = 0$.

3. "Triangle 3" can't be a triangle, because, if it were, the lengths of its sides would contradict the Triangle Inequality Theorem.

4. Triangle 4 because "triangle 5" can't be a triangle.

5. Perimeter = 18; so $s = 9$.
   Area = $\sqrt{9(9-4)(9-6)(9-8)} = \sqrt{9 \cdot 5 \cdot 3 \cdot 1} = \sqrt{135}$, or $3\sqrt{15}$, or approximately 11.6.

Set I (pages 360–362)

The two arrangements of the automobiles in exercises 1 through 5 have something in common with D'Arcy Thompson's fish transformation. Both are examples of a "shear" transformation, a transformation not usually discussed in geometry but one that is a basic tool of computer graphics. For example, some computer programs use shear transformations to generate slanted italic fonts from standard ones. Although these transformations are not isometries, they do preserve area.

An interesting story of the construction and problems of the Hancock Tower, headquarters of the John Hancock Mutual Life Insurance Company in Boston, is told by Mattys Levi and Mario Salvadori in their book titled *Why Buildings Fall Down* (Norton, 1992). Although now altered to be more structurally sound, the tower at the beginning had problems with the wind because of its unusual nonrectangular cross section. Among other things, all of the more than 10,000 panels of reflective glass covering its walls had to be replaced with stronger panels!

From Above.

•1. That they are equal.

•2. 2,800 ft². [$x = (5)(14) = 70$, $y = (5)(8) = 40$; $xy = 2,800$.]

3. 220 ft. [$2(x + y) = 2(70 + 40) = 220$.]

4. The perimeter of the second figure is greater.

5. No.

Tax Assessor Formula.

•6. Area $= \frac{1}{4}(s + s)(s + s) = \frac{1}{4}(2s)(2s) = s^2$.

7. Yes. Area $= \frac{1}{4}(b + b)(h + h) = \frac{1}{4}(2b)(2h) = bh$.

•8. $\frac{1}{4}(20 + 20)(13 + 13) = \frac{1}{4}(40)(26) =$

260 square units.

9. No. The area of a parallelogram is the product of any base and corresponding altitude; so the area = (20)(12) = 240 square units.

10. $\frac{1}{4}(3 + 24)(10 + 17) = \frac{1}{4}(27)(27) =$

182.25 square units.

11. No. The area of a trapezoid is half the product of its altitude and the sum of its bases; so the area $= \frac{1}{2}(8)(3 + 24) = 4(27) =$ 108 square units.

12. $\frac{1}{4}(20 + 24)(7 + 15) = \frac{1}{4}(44)(22) =$

242 square units.

13. No. The area of each right triangle is half the product of its legs; so the area $=$ $\frac{1}{2}(7)(24) + \frac{1}{2}(20)(15) = 84 + 150 =$ 234 square units.

•14. Area $= \frac{1}{4}(37 + 37)(37 + 37) = \frac{1}{4}(74)(74) =$

1,369 square units.

15. No. The diagonals of a rhombus are perpendicular and bisect each other; so the area of this rhombus is $4(\frac{1}{2})(12)(35) = 4(210) =$ 840 square units.

16. Because the formula seems to always give an area either equal to or larger than the correct answer.

Pegboard Quadrilaterals.

•17. A parallelogram. (Two of its opposite sides are both parallel and equal.)

•18. 1 unit. $A = bh = (1)(1) = 1$.

19. A parallelogram.

20. 1 unit.

21. A parallelogram.

22. 1 unit. It can be divided by a horizontal diagonal into two triangles, each with a base and altitude of 1; $2(\frac{1}{2})(1)(1) = 1$.

Skyscraper Design.

•23. 30,794 ft². [$(300)(104) - 2(\frac{1}{2})(14)(29) =$

31,200 − 406 = 30,794.]

24. 237,000 ft². [(300)(790) = 237,000.]

Shuffleboard Court.

25. 3 ft². (Triangle: $\frac{1}{2}(2)(3) = 3$.)

•26. 4.5 ft². (Trapezoid: $\frac{1}{2}(3)(1 + 2) = 4.5$.)

27. 7.5 ft². (Trapezoid: $\frac{1}{2}(3)(2 + 3) = 7.5$.)

•28. 8.25 ft². (Trapezoid: $\frac{1}{2}(1.5)(5 + 6) = 8.25$.)

29. 35.25 ft².

Set II (pages 362–364)

Students who do exercises 47 through 49 in which the trapezoidal rule is used to find the approximate area of the region between a parabola and the *x*-axis may be intrigued to know that the actual area is represented by the expression

$$\int_0^3 \frac{x^2}{2}\,dx$$

and that those who take calculus will learn how to find its exact value, 4.5.

The SAT problem is a good example of how important it is to "look before you leap." The method of subtracting the shaded area from the rectangle to get the unshaded region is tempting but, even if the areas of all of the regions could be found, would be more time consuming than would simply adding the areas of the two unshaded regions. Time taken to consider alternative approaches to solving a problem is often time well spent.

Inaccessible Field.

30. The entire figure (ECBG) is a trapezoid. Subtract the areas of trapezoid ECDF, ΔFDA, and ΔABG from its area.

31.

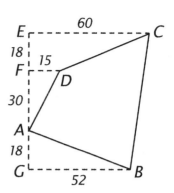

•32. 675 m². [$\frac{1}{2}(18)(60 + 15) = 675$.]

33. 225 m². [$\frac{1}{2}(15)(30) = 225$.]

34. 468 m². [$\frac{1}{2}(18)(52) = 468$.]

•35. 3,696 m². [$\frac{1}{2}(60 + 52)(18 + 30 + 18) = 3,696$.]

36. 2,328 m². (3,696 − 675 − 225 − 468 = 2,328.)

A Fold-and-Cut Experiment.

37.

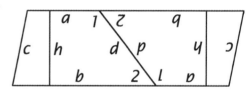

38. The bases line up because ∠1 and ∠2 are supplementary (parallel lines form supplementary interior angles on the same side of a transversal); together ∠1 and ∠2 form a straight angle.

39. A parallelogram.

40. A quadrilateral is a parallelogram if its opposite sides are equal. (Alternatively, a quadrilateral is a parallelogram if two opposite sides are both parallel and equal.)

•41. Its area is $(a + b)h$.

42. The parallelogram consists of two congruent trapezoids; so the area of each is $\frac{1}{2}(a + b)h$, or $\frac{1}{2}h(a + b)$.

Trapezoidal Rule.

•43. In a plane, two lines perpendicular to a third line are parallel.

44. $\alpha ABGH = \frac{1}{2}x(y_1 + y_2)$. $\alpha BCFG = \frac{1}{2}x(y_2 + y_3)$.

$\alpha CDEF = \frac{1}{2}x(y_3 + y_4)$.

45. $\alpha ABGH + \alpha BCFG + \alpha CDEF =$

$\frac{1}{2}x(y_1 + y_2) + \frac{1}{2}x(y_2 + y_3) + \frac{1}{2}x(y_3 + y_4) =$

$\frac{1}{2}x(y_1 + y_2 + y_2 + y_3 + y_3 + y_4) =$

$\frac{1}{2}x(y_1 + 2y_2 + 2y_3 + y_4)$.

•46. 165 square units.

$[\frac{1}{2}(5)(8 + 2(15) + 2(11) + 6) = \frac{1}{2}(5)(66) = 165.]$

47. B(2, 2), C(3, 4.5).

48. 4.75. $[\frac{1}{2}(1)(0 + 2(0.5) + 2(2) + 4.5).]$

49. It is larger.

SAT Problem.

•50. ΔEBH, FHCG (and FHCD).

51. We don't know exactly where point H is; so we don't know the lengths of BH and HC.

52. 9 square units.

53. The unshaded region consists of two triangles, ΔEFD and ΔEFH, with a common base of 2. The corresponding altitudes are

5 and 4; so $\alpha EDFH = \frac{1}{2}(2)(5) + \frac{1}{2}(2)(4) =$

$5 + 4 = 9$.

Set III (page 364)

The chessboard puzzle was discovered by a German mathematician named Schlömilch and published in 1868 in an article titled "A Geometric Paradox." Good references on it and similar puzzles can be found in the second chapter on "Geometrical Vanishes" in Martin Gardner's *Mathematics, Magic and Mystery* (Dover, 1956) and the chapter titled "Cheated, Bamboozled, and Hornswoggled" in *Dissections—Plane and Fancy,*

by Greg N. Frederickson (Cambridge University Press, 1997). A generalization of the chessboard paradox was made by Lewis Carroll; an algebraic analysis of it can be found in J. L. Heilbron's *Geometry Civilized* (Clarendon Press, 1998).

Chessboard Mystery.

1. The area of each right triangle is

$\frac{1}{2}(3)(8) = 12$ units. The area of each

trapezoid is $\frac{1}{2}(5)(3 + 5) = 20$ units.

The four areas add up to 64 units, as we would expect for a square board consisting of 64 unit squares.

2.

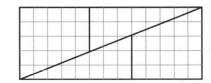

3. 65 units.

4. According to the Area Postulate, the area of a polygonal region is equal to the sum of the areas of its nonoverlapping parts. But 65 ≠ 64.

(The explanation is that the pieces don't quite fill the rectangle. The exaggerated drawing below shows where the extra unit comes from.)

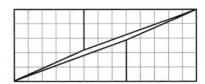

Chapter 9, Lesson 5

Set I (pages 367–368)

J. L. Heilbron in *Geometry Civilized* (Clarendon Press, 1998), wrote: "The Pythagorean Theorem and its converse bring the first book of the *Elements* to a close. They therefore mark a climax, the high point toward which all the apparatus of definitions, postulates, and propositions was aimed. The diagram of the windmill [the figure used by Euclid to prove this theorem] may be taken as a symbol of geometry, and hence of Greek thought, which generations . . . have regarded as the bedrock of their culture."

Note that in exercises 1 through 3, dealing with the special case of an isosceles right triangle, the areas of the squares on its sides are expressed in terms of the area of the triangle rather than the length of one of its sides.

The proof of the converse of the Pythagorean Theorem is, in essence, that of Euclid. The only difference is that Euclid chose to construct the second triangle so that it shared one of its shorter sides with the original triangle.

In thinking about exercise 32, it is helpful to consider the squares representing a^2 and b^2 as being hinged together at one of their vertices. If the angle of the hinge is adjusted to 90°, we have the Pythagorean Theorem with $c^2 = a^2 + b^2$. As the angle is made smaller, the distance c shrinks and c^2 gets smaller; so, for acute triangles, $c^2 < a^2 + b^2$. If instead, the angle is made larger, the distance c increases and c^2 grows; so, for obtuse triangles, $c^2 > a^2 + b^2$.

Batik Design.

1. *Example figure:*

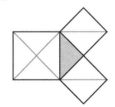

•2. 2, 2, and 4 units.

3. 2 + 2 = 4. (The square on the hypotenuse of a right triangle is equal to the sum of the squares on its legs.)

Theorem 43.

•4. The Ruler Postulate.

•5. The Protractor Postulate.

6. The Ruler Postulate.

7. Two points determine a line.

•8. The square of the hypotenuse of a right triangle is equal to the sum of the squares of its legs.

•9. Substitution.

10. SSS.

11. Corresponding parts of congruent triangles are equal.

•12. Substitution.

13. A 90° angle is a right angle.

14. A triangle with a right angle is a right triangle.

Squares on the Sides.

•15. 12.

•16. 1,225.

17. 1,369.

•18. 37.

19. 10.

20. 26.

21. 576.

22. 24.

23. 484.

24. 16.

25. 15.

•26. No. 256 + 225 = 481 ≠ 484.

27. 2,304.

28. 3,025.

29. 73.

30. Yes. 2,304 + 3,025 = 5,329. (If the square of one side of a triangle is equal to the sum of the squares of the other two sides, the triangle is a right triangle.)

Ollie's Triangles.

31. *Example figure:*

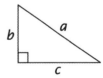

32. *Example figure:*

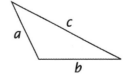

Frank J. Swetz and T. I. Kao, the authors of *Was Pythagoras Chinese? An Examination of Right Triangle Theory in Ancient China* (Pennsylvania State University Press, 1977), report that, although estimates of the date of origin of the *Chou Pei Suan Ching* date as far back as 1100 B.C., much of the material in it seems to have been written at the time of Confucius in the sixth century B.C. It is interesting to observe that, although there is no proof of the Pythagorean Theorem given in the *Chou Pei*, the figure, based on the 3-4-5 right triangle, may have been the inspiration for the figure on which the proof in the text of Lesson 5 is based.

In *The Mathematical Universe* (Wiley, 1994), William Dunham tells the story of James A. Garfield and his proof of the Pythagorean Theorem. After completing his college education, Garfield began teaching mathematics at Hiram College in Ohio. Within a few years, he was elected to the Ohio Senate and then joined the Union Army; he served for 17 years in the House of Representatives before being elected president in 1880. At the time of its publication in *The New England Journal of Education* in 1876, it was reported that Garfield had discovered his proof of the Pythagorean Theorem during "some mathematical amusements and discussions with other congressmen."

The figure illustrating exercises 50 and 51 is another nice example of a "proof without words." The fact that the bisector of the right angle of a right triangle bisects the square on the hypotenuse is clearly connected to the second figure's rotation symmetry.

Was Pythagoras Chinese?

•33. 6 units. $[\frac{1}{2}(3)(4).]$

34. 24 units. (4×6.)

•35. 25 units. ($24 + 1$.)

36. $\frac{1}{2}ab$.

•37. $2ab$. $[4(\frac{1}{2}ab).]$

•38. $(b - a)^2$.

39. $\alpha ABCD = 2ab + (b - a)^2$ and $\alpha ABCD = c^2$.

40. $2ab + (b - a)^2 = c^2$; so $2ab + b^2 - 2ab + a^2 = c^2$, and so $a^2 + b^2 = c^2$.

Garfield's Proof.

41. SAS.

42. Corresponding parts of congruent triangles are equal.

•43. The acute angles of a right triangle are complementary (so the sum of these angles is 90°).

44. A right angle.

•45. A trapezoid.

46. $\alpha ABDE = \frac{1}{2}(a + b)(a + b)$ and

$\alpha ABDE = \frac{1}{2}ab + \frac{1}{2}ab + \frac{1}{2}c^2$.

47. $\frac{1}{2}(a + b)^2 = \frac{1}{2}ab + \frac{1}{2}ab + \frac{1}{2}c^2$; so $(a + b)^2 = 2ab + c^2$; so $a^2 + 2ab + b^2 = 2ab + c^2$, and so $a^2 + b^2 = c^2$.

Angle-Bisector Surprise.

48.

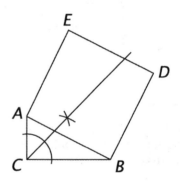

•49. It appears to bisect ABDE (that is, divide it into two congruent parts having equal areas).

50. It bisects $\angle ACB$. $\triangle FCG \cong \triangle HCG$; so $\angle FCG = \angle HCG$.

51. It bisects ABDE.

Set III (page 370)

Twenty Triangles.

1. *Example figure:*

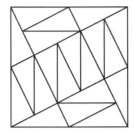

[The areas of the three squares are 4, 16, and 20 square inches; so the sides of the large square are $\sqrt{20} \approx 4.5''$ (or exactly, $\sqrt{20} = 2\sqrt{5}''$). We can't get this length by using the legs of the triangles; so we have to use their hypotenuses: $1^2 + 2^2 = c^2$, and so $c = \sqrt{5} \approx 2.2$. Two hypotenuses are needed to form each side of the large square.]

2. It is tempting to try to use the right angles of the triangles to form the right angles of the third square. This won't work, however, because the legs of the triangles have to be *inside* the square.

Chapter 9, Review

Set I (pages 371–373)

In the section on Babylonian mathematics in *The History of Mathematics* (Allyn & Bacon, 1985), David Burton wrote: "In the ancient world, the error was widespread that the area of a plane figure depended entirely on its perimeter; people believed that the same perimeter always confined the same area. Army commanders estimated the number of enemy soldiers according to the perimeter of their camp, and sailors the size of an island according to the time for its circumnavigation. The Greek historian Polybius tells us that in his time unscrupulous members of communal societies cheated their fellow members by giving them land of greater perimeter (but less area) than what they chose for themselves; in this way they earned reputations for unselfishness and generosity, while they really made excessive profits."

An excellent source of information on tangrams are the two chapters on them by Martin Gardner in *Time Travel and Other Mathematical Bewilderments* (W. H. Freeman and Company, 1988). Although many references report that tangrams date back about 4,000 years, historians now believe that they actually originated in China in about 1800. An ornate set of carved ivory tangrams once owned by Edgar Allan Poe is now in the New York Public Library.

Olympic Pools.

- •1. 125 m². (50 × 2.5.)

2. No. There is room left on the sides. The area of the pool is 1,050 m² and the total area of the 8 lanes is 1,000 m² (the width of the pool is 21 m and the total width of the 8 lanes is 20 m.)

- •3. 525. ($\frac{1,050}{2}$.)

Enemy Camps.

- •4. Camp A, 68 paces; camp B, 64 paces; camp C, 60 paces.

5. Camp A, 208 square paces; camp B, 220 square paces; camp C, 216 square paces.

6. Camp A because it has the greatest perimeter.

7. No. Camp B has the greatest area.

Tangrams.

- •8. 16 units.

9. 4 units.

10. The other triangle, 8 units; the square, 8 units; the parallelogram, 8 units.

11. 64 units.

- •12. The area of a polygonal region is equal to the sum of the areas of its nonoverlapping parts.

Altitudes and Triangles.

13.

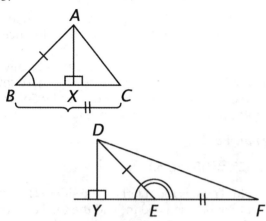

•14. ∠DEY.

15. △ABX ≅ △DEY.

•16. AAS.

17. Congruent triangles have equal areas.

•18. Corresponding parts of congruent triangles are equal.

19. △ABC and △DEF.

•20. Triangles with equal bases and equal altitudes have equal areas.

Suriname Stamp.

21. The square of the hypotenuse of a right triangle is equal to the sum of the squares of its legs.

22. If the square on one side of a triangle is equal to the sum of the squares of the other two sides, the triangle is a right triangle.

•23. 11.

24. 49.

25. 169.

26. No. $121 + 49 = 170 \neq 169$.

Moroccan Mosaic.

•27. $4a$.

28. a^2.

•29. $4a + 4b$.

•30. $a(a + 2b)$ or $a^2 + 2ab$.

31. $4a + 2b$.

32. $b(a + b)$ or $ab + b^2$. [$\frac{1}{2}b(a + a + 2b)$.]

Set II (pages 373–375)

The Daedalus project was named after the character in Greek mythology who constructed wings from feathers and wax to escape from King Minos. The planes built in the project were designed and built at M.I.T. Daedalus 88, the plane shown in the photograph, was flown 74 miles from Crete to the island of Santorini in the Mediterranean in 3 hours, 54 minutes, all by human power alone! According to an article in the October 1988 issue of *Technology Review*, the plane weighed 70 pounds when empty. Its wing span and area are worked out in the exercises. The plane had a flight speed ranging from 14 to 17 miles per hour and its human "engine" power was about a quarter of a horsepower!

The problems about a crack forming in a stretched bar are based on information in *The New Science of Strong Materials* (Princeton University Press, 1976), by J. E. Gordon, who writes: "When a crack appears in a strained material it will open up a little so that the two faces of the crack are separated. This implies that the material immediately behind the crack is relaxed and the strain energy in that part of the material is released. If we now think about a crack proceeding inwards from the surface of a stressed material we should expect the area of material in which the strain is relaxed to correspond roughly to the two shaded triangles. Now the area of these triangles is roughly l^2, where l is the length of the crack. The relief of strain energy would thus be expected to be proportional to the square of the crack length, or rather depth, and in fact this rough guess is confirmed by calculation. Thus a crack two microns deep releases four times as much strain energy as one one micron deep and so on."

The exercises on the area of an irregular tract reveal that the trapezoidal rule is used in surveying as well as in calculus.

George Biddle Airy was the astronomer royal at the Greenwich Observatory from 1836 to 1881. A nice treatment of dissection proofs of the Pythagorean Theorem and their connection to superposing tessellations is included by Greg Frederickson in chapter 4 of *Dissections—Plane and Fancy* (Cambridge University Press, 1997).

Courtyard Design.

33.

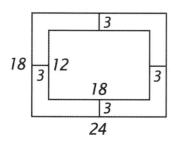

•34. 216 ft². (12 × 18.)

35. 216 ft². (18 × 24 − 216.)

36. Yes.

Daedalus Wing.

37. 112 ft.

•38. 332.5 ft². $[2 \cdot \frac{1}{2}(14)(1.25 + 2.5) +$

$2 \cdot \frac{1}{2}(28)(2.5 + 3.75) + 28(3.75) =$

$52.5 + 175 + 105 = 332.5.]$

Crack Formation.

39. $a + b = 2l$. The colored area is $\frac{1}{2}al + \frac{1}{2}bl = l^2$;

so $\frac{1}{2}a + \frac{1}{2}b = l$, and so $a + b = 2l$.

•40. It would be four times as great [because $(2l)^2 = 4l^2$].

Triangle Comparisons.

41.

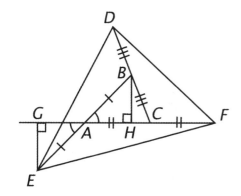

42. AAS.

43. Corresponding parts of congruent triangles are equal.

•44. α△AEF = 2x (because AF = 2AC and BH = EG).

45. α△DEF = 7α△ABC. (By the same reasoning, α△BDE = 2x and α△CFD = 2x. Because α△DEF = α△ABC + α△AEF + α△BDE + α△CFD, α△DEF = x + 2x + 2x + 2x = 7x = 7α△ABC.)

SAT Problem.

•46. $4x^2 - xy$.

•47. $4x - y$.

48. $8x$. $[4(2x).]$

49. $10x - 2y$. $\{2[x + (4x - y)] = 2(5x - y).\}$

Surveying Rule.

50. $\frac{1}{2}x(a + b) + \frac{1}{2}x(b + c) + \frac{1}{2}x(c + d) +$

$\frac{1}{2}x(d + e)$ or $\frac{1}{2}x(a + 2b + 2c + 2d + e)$.

•51. It would be rectangular.

52. $\frac{1}{2}x(a + 2a + 2a + 2a + a) = \frac{1}{2}x(8a) = 4xa$.

Pythagorean Proof.

53. In the first arrangement, the three pieces form the squares on the legs of the right triangle and their total area is $a^2 + b^2$. In the second arrangement, the three pieces form the square on the hypotenuse of the right triangle and their total area is c^2; so $c^2 = a^2 + b^2$.

54. It refers to the area of the red piece. [In the first figure, the area of the red piece is clearly the sum of the areas of the two smaller squares minus the areas of the two triangles: $a^2 + b^2 - 2(\frac{1}{2}ab)$. When the two triangles "stand on" the red piece, the square on the hypotenuse is formed and, when the red piece "stands on" the two triangles, the squares on the legs are formed.]

•1. 500. $[x = 10(50) = 500.]$

•2. $\dfrac{1}{5}$ or 0.2. $(3 = 15x, x = \dfrac{3}{15} = \dfrac{1}{5} = 0.2.)$

•3. 2.5. $[6 = 4(x - 1), 6 = 4x - 4, 10 = 4x,$
$x = \dfrac{10}{4} = \dfrac{5}{2} = 2.5.]$

•4. $\pm 10.$ $(x^2 = 100, x = \pm 10.)$

•5. $-3.$ $(11 + 2x = 5, 2x = -6, x = -3.)$

•6. 42. $(2x = 3x - 42, x = 42.)$

•7. 15. $(x - 12 = 3, x = 15.)$

•8. 20. $(5x = 4x + 20, x = 20.)$

•9. 41. $[7(x + 5) = 2(4x - 3), 7x + 35 = 8x - 6,$
$41 = x.]$

•10. 3.5. $[\dfrac{4}{x - 3} = 8, 4 = 8(x - 3), 4 = 8x - 24,$
$28 = 8x, x = \dfrac{28}{8} = \dfrac{7}{2} = 3.5.]$

•11. 0. $[15 - 2(x + 3) = 3(x + 3),$
$15 - 2x - 6 = 3x + 9, 9 - 2x = 3x + 9, 5x = 0,$
$x = 0.]$

•12. 8. $(\dfrac{8x}{x} = x, x = 8.)$

•13. 5. $[3(15) + 1(x + 4) = 2(3)(x + 4),$
$45 + x + 4 = 6x + 24, 49 + x = 6x + 24, 25 = 5x,$
$x = 5.]$

•14. 5. $[5(3x + 1) - 4x = 20(x - 2),$
$15x + 5 - 4x = 20x - 40, 11x + 5 = 20x - 40,$
$45 = 9x, x = 5.]$

•15. $-36.$ $[4(x + 8) - (x - 4) = 2x,$
$4x + 32 - x + 4 = 2x, 3x + 36 = 2x, x = -36.]$

Set I (pages 380–382)

According to Benno Artmann in *Euclid—The Creation of Mathematics* (Springer, 1999), the architects of the Parthenon, Ictinos and Callikrates, based its dimensions on the square numbers 4 and 9. Starting with a height of 4^2, they chose $4 \cdot 9$ for the width and 9^2 for the length, so that the width was the geometric mean between the height and the length.

Euclid developed the theory of proportion with a series of 25 theorems in Book V of the *Elements*. His definition of *ratio* as "a sort of relation in respect of size between two magnitudes of the same kind" is, as one commentator remarked, unusual for Euclid because it is so vague as to be of no practical use.

The numbers connected with the *Turtles Forever* picture illustrate the fact that any three consecutive terms of a geometric sequence are proportional (as are *any* three evenly spaced terms). The term in the middle is consequently the geometric mean between the other two terms.

Parthenon Architecture.

•1. 2.25.

2. 2.25.

3. $81 \times 16 = 1{,}296$ and $36 \times 36 = 1{,}296$.

•4. The geometric mean.

•5. $\dfrac{9}{6} = \dfrac{6}{4}$.

6. $1.5 = 1.5$.

7. $101\dfrac{1}{3}$ ft.

•8. $h \approx 45$ ft. (*Example proportion:* $\dfrac{228}{h} = \dfrac{81}{16}$.)

•9. $\dfrac{2}{7}$ (or $0.285714\ldots$). ($\dfrac{2d}{7d}$.)

10. $\dfrac{4}{5}$ (or 0.8). ($\dfrac{8d}{10d}$.)

11. $\dfrac{10}{1}$ (or 10). ($\dfrac{10d}{d}$.)

Turtles Forever.

12. $\dfrac{16}{32} = \dfrac{32}{64}$.

13. $\dfrac{2}{16} = \dfrac{16}{128}$.

14. 4 and 16, 2 and 32, 1 and 64.

Ratio.

•15. Division.

16. The second.

17. Euclid referred to "magnitudes of the same kind," which suggests that numbers of the same kind of units are being compared.

Four Rectangles.

18. Rectangle 1, ab; rectangle 2, ad; rectangle 3, cd; rectangle 4, bc.

•19. Rectangles 2 and 4. ($ad = bc$ because $\dfrac{b}{a} = \dfrac{d}{c}$.)

20. Rectangles 1 and 3. ($\dfrac{a}{c} = \dfrac{b}{d}$ because $ad = bc$.)

21. Rectangles 1 and 3.

Enlargement.

•22. 30.

23. 36.

24. Correct. (Both ratios are equal to $\dfrac{13}{14}$.)

25. Correct. ($\dfrac{126}{84} = \dfrac{42}{28} = \dfrac{3}{2} = 1.5$.)

26. Correct. [$\dfrac{\frac{1}{2} \cdot 42 \cdot 36}{\frac{1}{2} \cdot 28 \cdot 24} = 2.25$ and $(\dfrac{39}{26})^2 = 2.25$.]

Set II (pages 382–384)

The official design of the current United States flag was signed into law by President Eisenhower in 1959, soon after Alaska became the 50th state. The first chapter of *Slicing Pizzas, Racing Turtles, and Further Adventures in Applied Mathematics* (Princeton University Press, 1999), by Robert Banks, deals with the geometry of the flag.

Concerning the ratio $\frac{19}{10}$ Banks writes: "Why the relative length of the flag is precisely 1.9—or indeed why the relative length of the union is exactly 0.76 (could it be 1776?)—is not known. It just is."

As Thomas Rossing remarks in *The Science of Sound* (Addison Wesley, 1990), "the word *scale* is derived from a Latin word (*scala*) meaning *ladder* or *staircase*. A musical scale is a succession of notes arranged in ascending or descending order." In chapter 9 of his book, Rossing includes complete discussions of the scales based on just intonation and equal temperament as well as the Pythagorean scale on which exercises 41 through 47 are based. Another good reference on the structure of musical scales is *Connections—The Geometric Bridge between Art and Science*, by Jay Kappraff (McGraw-Hill, 1991). It is interesting to see from the relative lengths of the strings producing the notes of the Pythagorean scale that they are all derived from the primes 2 and 3:

$$1 \quad \frac{2^3}{3^2} \quad \frac{2^6}{3^4} \quad \frac{3}{2^2} \quad \frac{2}{3} \quad \frac{2^4}{3^3} \quad \frac{2^7}{3^5} \quad \frac{1}{2}$$

In the currently used even-tempered scale, the ratios $\frac{C}{D}$, $\frac{D}{E}$, $\frac{F}{G}$, $\frac{G}{A}$, and $\frac{A}{B}$ are $\left(\sqrt[12]{2}\right)^2 \approx 1.122$, in contrast with the Pythagorean ratio of $\frac{9}{8} = 1.125$.

The "even tempered" ratios $\frac{E}{F}$ and $\frac{B}{C'}$ are $\sqrt[12]{2} \approx 1.059$ in contrast with the Pythagorean ratio, $\frac{256}{243} \approx 1.053$.

The fact that the means of a proportion can be interchanged is proved in Book V of the *Elements* as Proposition 16: "If four magnitudes be proportional, they will also be proportional alternately." The addition property proved in exercises 51 and 52 appears in the *Elements* as Proposition 18 of the same book.

Arm Spans.

27. No. The units of length being compared are not the same; so the result is meaningless.

•28. 36 in; $\frac{36}{15} = 2.4$.

29. 1.25 ft; $\frac{3}{1.25} = 2.4$.

30. No.

United States Flag.

•31. $\frac{19}{10}$.

32. $\frac{7}{13}$.

33. $\frac{7}{6}$.

34. The length of the flag.

35. 304 feet.

•36. The width of one stripe.

•37. Approximately 12.3 ft. ($\frac{160}{13} \approx 12.3$.)

38. Approximately 86 ft. ($7 \times 12.3 \approx 86$.)

•39. Approximately 122 ft. ($\frac{b}{160} = 0.76$, $b = 121.6$.)

40. No. $\frac{a}{b} \approx \frac{86}{121.6} \approx 0.7$ and $\frac{w}{l} = \frac{10}{19} \approx 0.53$.

Pythagorean Tuning.

•41. $\dfrac{D}{E} = \dfrac{\frac{8}{9}}{\frac{64}{81}} = \dfrac{8}{9} \cdot \dfrac{81}{64} = \dfrac{9}{8}$.

•42. $\dfrac{E}{F} = \dfrac{\frac{64}{81}}{\frac{3}{4}} = \dfrac{64}{81} \cdot \dfrac{4}{3} = \dfrac{256}{243}$.

43. $\dfrac{F}{G} = \dfrac{\frac{3}{4}}{\frac{2}{3}} = \dfrac{3}{4} \cdot \dfrac{3}{2} = \dfrac{9}{8}$.

44. $\dfrac{G}{A} = \dfrac{\frac{2}{3}}{\frac{16}{27}} = \dfrac{2}{3} \cdot \dfrac{27}{16} = \dfrac{9}{8}$.

45. $\dfrac{A}{B} = \dfrac{\frac{16}{27}}{\frac{128}{243}} = \dfrac{16}{27} \cdot \dfrac{243}{128} = \dfrac{9}{8}$.

46. $\dfrac{\text{B}}{\text{C}'} = \dfrac{\frac{128}{243}}{\frac{1}{2}} = \dfrac{128}{243} \cdot \dfrac{2}{1} = \dfrac{256}{243}.$

47. G is the geometric mean between F and A;
    A is the geometric mean between G and B.
    $\left(\dfrac{\text{F}}{\text{G}} = \dfrac{\text{G}}{\text{A}} = \dfrac{9}{8} \text{ and } \dfrac{\text{G}}{\text{A}} = \dfrac{\text{A}}{\text{B}} = \dfrac{9}{8}.\right)$

Geometric Proportions.

•48. The means are interchanged.

49. Multiplication (or, in a proportion, the product of the means is equal to the product of the extremes).

•50. Division.

51. Addition.

52. Substitution (and algebra).

Sinker or Floater.

53. Alice is a floater because her density is
    $\dfrac{120 \text{ pounds}}{1.95 \text{ cubic feet}} \approx 61.5$ pounds per cubic foot.

54. Let x = Ollie's weight in pounds. His
    density is $\dfrac{x \text{ pounds}}{2.3 \text{ cubic feet}}$. Because he is a
    sinker, $\dfrac{x}{2.3} > 62.4$ and $x > 143.52$. Ollie
    weighs more than 143 pounds.

Set III (page 384)

The White Horse of Uffington was created by removing sod from the chalky rock lying underneath. A book written as far back as the fourteenth century ranked it second only to Stonehenge among the tourist attractions of Britain. The horse is in excellent condition owing to its being maintained by the National Trust.

The White Horse.

1. (*Student answer.*) (The length of the horse in the photograph is about 95 mm and its
   height is about 24 mm: $\dfrac{l}{h} = \dfrac{x \text{ m}}{28 \text{ m}} = \dfrac{95 \text{ mm}}{24 \text{ mm}}$,
   $x \approx 110.8$. The horse is about 110 meters long.)

2. (*Student answer.*) (The scale is about $\dfrac{1}{1,200}$.
   $\dfrac{24 \text{ mm}}{28,000 \text{ mm}} \approx \dfrac{1}{1,167}$.)

Chapter 10, Lesson 2

Set I (pages 387–389)

According to Robert Bauval and Adrian Gilbert (*The Orion Mystery*, Crown, 1994), the apparent relative positions of the three pyramids at Giza not only match those of the three stars in Orion's belt but their orientation with respect to the Nile matches Orion's apparent orientation with respect to the Milky Way! The arrangement of the pyramids and their relative sizes (seemingly based on the apparent sizes of the stars) seem to indicate that the Egyptians were trying to build a replica of heaven on Earth.

A "plane table" is a drawing board and ruler mounted on a tripod, used by surveyors in the field to sight and map data. More details on how it works can be found in J. L. Heilbron's *Geometry Civilized* (Clarendon Press, 1998).

According to Robert C. Yates in *The Trisection Problem* (Franklin Press, 1942; N.C.T.M., 1971), Alfred Kempe was "one of the cleverest amateur mathematicians of the [eighteenth] century." Kempe gave a historic lecture on linkages to a group of science teachers in London in the summer of 1876. This lecture, titled "How to Draw a Straight Line," and published in 1877, was reprinted by N.C.T.M. in 1977. The linkage invented by Kempe for trisecting angles is also discussed by Yates in the chapter on mechanical trisectors in *The Trisection Problem* and by Martin Gardner in the chapter titled "How to Trisect an Angle" in *Mathematical Carnival* (Knopf, 1975). As Kempe himself observed, his idea can be extended to produce a linkage that can divide an angle into any number of equal parts.

The Pyramids and Orion.

•1. They are proportional.

•2. They are equal.

Plane Table.

•3. $\angle$DAC = $\angle$FBC.

4. $\angle$AEC = $\angle$BGC.

5. $\angle$GCB = $\angle$ECA.

•6. $\dfrac{AD}{BF} = \dfrac{DC}{FC}$.

7. $\dfrac{GC}{EC} = \dfrac{BC}{AC}$.

8. $\dfrac{BC}{AC} = \dfrac{BG}{AE}$.

Kempe's Linkage.

•9. OE = 4 and OG = 8. ($\dfrac{1}{2} = \dfrac{2}{OE} = \dfrac{OE}{OG}$.)

10. They are always equal. Corresponding angles of similar triangles are equal.

11. It trisects ∠AOG.

Billboards.

12. Their corresponding angles are equal because billboards are rectangular and all right angles are equal. Their corresponding sides, however, are not proportional, because $\dfrac{5}{11} \neq \dfrac{11}{23}$. As a calculator shows, $\dfrac{5}{11} \approx 0.45$, whereas $\dfrac{11}{23} = 0.47826\ldots$.

•13. 24.2. ($\dfrac{5}{11} = \dfrac{11}{x}$, $5x = 121$, $x = 24.2$.)

14. $\dfrac{121}{23}$ (or $5\dfrac{6}{23}$ or approximately 5.26).

($\dfrac{x}{11} = \dfrac{11}{23}$, $23x = 121$, $x = \dfrac{121}{23} = 5\dfrac{6}{23}$.)

Dilation Problem.

•15. The center of the dilation.

16. AB = 5, BC = 4, AC = 6, A′B′ = 2.5, B′C′ = 2, and A′C′ = 3.

17. They are half as long.

•18. $\dfrac{1}{2}$.

19. They are their midpoints.

Quadrilateral Conclusions.

•20. Corresponding sides of similar polygons are proportional.

21. Multiplication. (In a proportion, the product of the means is equal to the product of the extremes.)

22. They have equal areas.

23. They also have equal areas.
αADGH = αAEFH + αEDGF and
αEDCI = αEDGF + αFGCI.
Because αADGH = αEDCI (exercise 22),
αAEFH + αEDGF = αEDGF + αFGCI
(substitution); so αAEFH = αFGCI
(subtraction).

Set II (pages 389–391)

A good article on the instruments of the violin family is "The Physics of Violins" by Carleen Maley Hutchins, published in the November 1962 issue of *Scientific American* and reprinted in the *Scientific American Resource Library* (W. H. Freeman and Company, 1978).

The International Standard of paper sizes is now in common use in Europe. As the exercises reveal, they are based on a sheet with an area of 1 square meter that can be cut in half to produce a second sheet similar in shape. Further sizes are produced in the same way. As the exercises also reveal, this pattern requires that all sizes have the ratio $\dfrac{\sqrt{2}}{1}$. The sequence beginning with size A0 continues to size A10 (a tiny 26 mm by 37 mm sheet). Computer programs usually offer sizes A4 and A5 in their menus of choices for printing formats.

Similar-Triangles Proof.

•24. A midsegment of a triangle is parallel to the third side.

25. Parallel lines form equal corresponding angles.

26. Reflexive.

27. A midsegment of a triangle is half as long as the third side.

•28. Division.

29. The midpoint of a line segment divides it into two equal segments.

•30. Betweenness of Points Theorem.

31. Substitution.

32. Division.

33. Substitution (they are all equal to $\frac{1}{2}$).

•34. Two triangles are similar if their corresponding angles are equal and their corresponding sides are proportional.

Violin Family.

•35. OC = 56 mm. $56 = r_2\,28$; so $r_2 = \dfrac{56}{28} = 2.0$.

36. OD = 87 mm. $87 = r_3\,28$; so $r_3 = \dfrac{87}{28} \approx 3.1$.

•37. BF = 24 mm. $\dfrac{BF}{AE} = \dfrac{24}{21} \approx 1.1$.

38. CG = 42 mm. $\dfrac{CG}{AE} = \dfrac{42}{21} = 2.0$.

39. DH = 65 mm. $\dfrac{DH}{AE} = \dfrac{65}{21} \approx 3.1$.

40. Viola: $1.1 \times 24 \approx 26$ inches.
    Cello: $2.0 \times 24 = 48$ inches.
    Bass: $3.1 \times 24 \approx 74$ inches.

Paper Sizes.

41. $\dfrac{1188}{x} = \dfrac{x}{594}$.

•42. Approximately 840 mm. ($x^2 = 1188 \times 594$, $x \approx 840$.)

43. The geometric mean.

•44. Approximately 1 m². (1188 mm × 840 mm = 997,920 mm² ≈ 1 m².)

45. Approximately 420 mm. ($\dfrac{x}{2} \approx \dfrac{840}{2} = 420$).

46. $\dfrac{r}{1} = \dfrac{1}{\frac{r}{2}}$.

•47. $\dfrac{r^2}{2} = 1$, $r^2 = 2$, $r = \sqrt{2}$; so the exact ratio of the length to width is $\sqrt{2}$.

Picture Frames.

48. $\dfrac{28 + 2w}{28} = \dfrac{20 + 2w}{20}$.

•49. $\dfrac{28 + 2w}{28} = \dfrac{20 + 2w}{20}$, $\dfrac{14 + w}{14} = \dfrac{10 + w}{10}$, $140 + 10w = 140 + 14w$, $0 = 4w$, $w = 0$.

50. $\dfrac{a + 2w}{a} = \dfrac{b + 2w}{b}$, $ab + 2bw = ab + 2aw$, $2bw = 2aw$, $bw = aw$. If $w \neq 0$, then $a = b$.

51. They mean that it is only possible for a rectangular picture to be surrounded by a frame of constant width whose outer edge is similar to it if the picture is square.

Set III (page 391)

The Villa Foscari, also called "La Malcontenta," is located on the Brenta River near Venice. It was commissioned by the brothers Nicolo and Alvise Foscari and is still in the possession of the Foscari family.

Similar Rectangles.

1. *Ratio 1:2.*

   Dimensions 1 × 2:
   HION, JKQP, NOUT, PQWV.

   Dimensions 1.5 × 3:
   ACIG, BDJH, CEKI, DFLJ.

   Dimensions 2 × 4:
   GHTS, IJVU, KLXW, HKQN, NQWT.

   Dimensions 4 × 8:
   GLXS.

2. *Ratio 1:4.*

   Dimensions 1 × 4:
   HIUT, JKWV.

   Dimensions 1.5 × 6:
   AEKG, BFLH.

   Dimensions 2 × 8:
   GLRM, MRXS.

3. *Ratio 2:3.*

   Dimensions 1 × 1.5:
   BCIH, DEKJ.

   Dimensions 2 × 3:
   GIOM, HJPN, IKQO, JLRP, MOUS, NPVT, OQWU, PRXV.

   Dimensions 4 × 6:
   GKWS, HLXT.

4. *Ratio 3:4.*

   Dimensions 1.5 × 2:
   ABHG, CDJI, EFLK.

   Dimensions 3 × 4:
   GIUS, HJVT, IKWU, JLXV.

Chapter 10, Lesson 3

Set I (pages 394–396)

Mathematics is truly a universal language. Although the words in the Turkish version of the Side-Splitter Theorem are incomprehensible to someone who does not know Turkish, the figure and symbolic statements about it are recognizable everywhere.

Descartes's method for multiplying two numbers geometrically appeared in Book I, "Problems the Construction of Which Requires Only Straight Lines and Circles," of his *Géométrie*, published as Appendix I to his *Discours de la méthode* (1637). As Descartes described the construction: "Taking one line which I shall call the unit in order to relate it as closely as possible to numbers, and which can in general be chosen arbitrarily, and having given two other lines, to find a fourth line which shall be to one of the given lines as the other is to the unit (which is the same as multiplication)." More on this subject can be found in *A Source Book in Mathematics, 1200–1800*, edited by D. J. Struik (Harvard University Press, 1969).

In his book titled *Perspective in Perspective* (Routledge & Kegan Paul, 1983), Lawrence Wright describes the various ways in which artists have drawn a cube, ranging from a mere square to "spherical perspective." Although the "two-point perspective" version used for exercises 22 through 25 was the basis for perspective drawings in the Renaissance, Wright observes: "None of the three faces shown is square, the overall shape is not even partly square, none of the twelve angles seen is a right angle, and of the three sets of parallels only the verticals are shown as such."

Turkish Theorem.

1. "Paralel" and "hipotez" obviously mean "parallel" and "hypothesis." "Distan" seems to mean "distance." ("Hüküm" evidently means "conclusion," but it is unrecognizable as such.)

•2. If a line parallel to one side of a triangle intersects the other two sides in different points.

3. It divides the sides in the same ratio.

Errors of Omission.

4. *Example answer:*

5. *Example answer:*

6. *Example answer:*

Picturing Products.

•7. The area of the rectangle.

•8. As lengths: AC = 3, AD = 2, and AE = 6.

9. $\dfrac{1}{a} = \dfrac{b}{c}$.

10. $c = ab$.

11.

3^2

12.

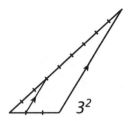

3^2

Supply and Demand.

•13. $\dfrac{AC}{CO} = \dfrac{AP}{PB}$.

14. $\dfrac{CO}{AO} = \dfrac{PB}{AB}$.

15. $\dfrac{OD}{DB} = \dfrac{AP}{PB}$.

16. $\dfrac{OD}{OB} = \dfrac{AP}{AB}$.

•17. Substitution. (Both are equal to $\dfrac{AP}{PB}$.)

Side-Splitter Practice.

•18. $x = 2.4$. $(\dfrac{x}{4} = \dfrac{3}{5}, 5x = 12, x = 2.4.)$

19. $x = 8$. $(\dfrac{x}{4} = \dfrac{16}{x}, x^2 = 64, x = 8.)$

20. $x = 7.5$. $(\dfrac{3}{x} = \dfrac{4}{10}, 4x = 30, x = 7.5.)$

21. $x = 10$. $(\dfrac{x}{24} = \dfrac{15}{36}, 36x = 360, x = 10.)$

Two-Point Perspective.

•22. Reasonable because BE ∥ CF in ΔOCF (the Side-Splitter Theorem).

23. Not reasonable, because lines AD and BC intersect in O.

24. Reasonable because of division (BC = CD and EF = FG).

25. Reasonable because DG ∥ CF in ΔPCF (the corollary to the Side-Splitter Theorem).

Set II (pages 396–398)

The Dames Point Bridge is the longest cable-stayed bridge in the United States and the only one to feature the "harp," or parallel-stay, arrangement on two vertical planes. Completed in 1989, it is 2 miles long.

Parallelogram Exercise.

•26. It is a parallelogram because DF ∥ EB and DF = EB.

27. XE ∥ YB in ΔAYB and FY ∥ DX in ΔCDX. If a line parallel to one side of a triangle intersects the other two sides in different points, it divides the sides in the same ratio.

28. AX = XY = YC. Because AE = EB, $\dfrac{AE}{EB} = 1$; so $\dfrac{AX}{XY} = 1$ and AX = XY. Because DF = FC, $\dfrac{DF}{FC} = 1$; so $\dfrac{XY}{YC} = 1$ and XY = YC.

Corollary to Theorem 44.

•29. If a line parallel to one side of a triangle intersects the other two sides in different points, it divides the sides in the same ratio.

30. Multiplication. (In a proportion, the product of the means is equal to the product of the extremes.)

31. Addition.

32. Substitution (factoring).

33. Division (and substitution).

Parallels Path.

34. If a line parallel to one side of a triangle intersects the other two sides in different points, it cuts off segments proportional to the sides.

35. $\dfrac{AM}{MN} = \dfrac{BM}{MO} = \dfrac{CN}{NO} = \dfrac{DN}{MN} = \dfrac{EO}{MO} = \dfrac{FO}{NO} = \dfrac{GM}{MN}$.

•36. EO. (Because $\dfrac{BM}{MO} = \dfrac{EO}{MO}$.)

37. FO. (Because $\dfrac{CN}{NO} = \dfrac{FO}{NO}$.)

38. Because $\dfrac{AM}{MN} = \dfrac{GM}{MN}$, AM = GM by multiplication.

•39. Point G is the same point as point A.

40. It would retrace itself.

Bridge Cables.

•41. EF = FG; so $\dfrac{EF}{FG} = 1$ by division.

42. $\dfrac{EH}{HI} = 1$. In ΔIEG, HF ∥ IG; so, by the Side-Splitter Theorem, $\dfrac{EH}{HI} = \dfrac{EF}{FG}$.

•43. In a plane, two lines perpendicular to a third line are parallel.

44. ABHI and BCEH are parallelograms (both pairs of their opposite sides are parallel).

•45. The opposite sides of a parallelogram are equal.

46. $\dfrac{BC}{AB} = 1$. Because $\dfrac{EH}{HI} = 1$, BC = EH and AB = HI, $\dfrac{BC}{AB} = 1$ by substitution.

47. Because $\dfrac{BC}{AB} = 1$, AB = BC by multiplication.

Original Proofs.

48. *Proof. (One possibility.)*
    (1) In △ABC, AD bisects ∠BAC. (Given.)
    (2) ∠BAD = ∠DAC. (If an angle is bisected, it is divided into two equal angles.)
    (3) AE = ED. (Given.)
    (4) ∠EDA = ∠DAC. (If two sides of a triangle are equal, the angles opposite them are equal.)
    (5) ∠BAD = ∠EDA. (Substitution.)
    (6) ED ∥ AB. (Equal alternate interior angles mean that lines are parallel.)
    (7) $\dfrac{AE}{EC} = \dfrac{BD}{DC}$. (If a line parallel to one side of a triangle intersects the other two sides in different points, it divides them in the same ratio.)

49. *Proof.*
    (1) In △ABC, AB = AC and DE ∥ BC. (Given.)
    (2) $\dfrac{AD}{AB} = \dfrac{AE}{AC}$. (If a line parallel to one side of a triangle intersects the other two sides in different points, it cuts off segments proportional to the sides.)
    (3) $\dfrac{AD}{AB} = \dfrac{AE}{AB}$. (Substitution.)
    (4) AD = AE. (Multiplication.)

Set III (page 398)

According to an article on the reproducing pantograph by Jack W. Jacobsen, it "first saw use as an early copying machine, making exact duplicates of written documents. Artists soon adopted its

use to duplicate drawings. It is known that da Vinci used one to make duplicates of his drawings and possibly to duplicate those drawings onto canvas.
  . . . It was not long before sculptors and carvers adapted the pantograph's use for tracing drawings onto blocks of marble or wood. They would then use the reproduced lines as guidelines for carving. True advancement in the pantograph design came about late in the 18th century with the advent of typeset printing. A pantograph was used to cut out the typeset letters. . . ."

Pantograph.

1.

2. ABCD is always a parallelogram because both pairs of its opposite sides are equal.

3. ∠2, ∠4, and ∠6.

4. ∠1 = ∠3 = ∠5 = ∠7 = $\left(\dfrac{180 - x}{2}\right)°$ or $\left(90 - \dfrac{x}{2}\right)°$.

5. 180°.

6. 180°. (∠4 = ∠2 and ∠5 = ∠1; so ∠3 + ∠4 + ∠5 = ∠3 + ∠2 + ∠1 by substitution).

7. It is always a straight angle.

8. They are always collinear.

9. $\dfrac{PD}{DE}$ always stays the same because $\dfrac{PD}{DE} = \dfrac{PA}{AB} = \dfrac{BC}{CE}$ and PA, AB, BC, and CE are fixed lengths.

10. $\dfrac{PE}{PD} = \dfrac{PB}{PA} = \dfrac{35\ cm}{20\ cm} = 1.75$.

Chapter 10, Lesson 4

Set I (pages 401–403)

It is a remarkable thing that, if two similar triangles are in *any* position but have the same orientation, the "average" triangle formed by taking the midpoints of the line segments connecting their corresponding vertices is similar to them. David Wells points out in *The Penguin Dictionary of Curious and Interesting Geometry* (Penguin Books, 1991) that "the same is true of polygons in general. It is also true if, instead of taking the midpoints of the lines, they are just divided in the same ratio."

The apple on the stamp honoring Sir Isaac Newton symbolizes the story that an apple falling from a tree led Newton to the idea of universal gravitation by getting him to thinking that, if the earth pulls on an apple, it also pulls on more distant objects such as the moon. (The moon doesn't fall as does the apple, because it falls *around* the earth. Its sideways motion makes it miss.) The geometric figure on the stamp is from Newton's *Principia Mathematica*. As William Dunham remarks in *The Mathematical Universe* (Wiley, 1994), "This work presented Newtonian mechanics in a precise, careful, and mathematical fashion. In it he introduced the laws of motion and the principle of universal gravitation and deduced, mathematically, everything from tidal flows to planetary orbits. *Principia Mathematica* is regarded by many as the greatest scientific book ever written."

The figure for exercises 9 through 11 is more remarkable than it might seem. As recently as 1972, C. Stanley Ogilvy observed in *Tomorrow's Math* (Oxford University Press) that, although a way to dissect a right triangle into five triangles similar to itself was known, a way to do this for nonright triangles was not. The division of the 30°-30°-120° triangle on which exercises 9 through 11 are based was discovered soon afterward and has been proved to be the only such solution. Zalman Usiskin and S. G. Wayment reported this result in an article published in *Mathematics Magazine* titled "Partitioning a Triangle into Five Triangles Similar to It." All of this is reported by Martin Gardner in *Wheels, Life and Other Mathematical Amusements* (W. H. Freeman and Company, 1983).

The chessboard puzzle was first introduced in the Set III exercises of Chapter 9, Lesson 4.

Two entire chapters of J. V. Field's *The Invention of Infinity* (Oxford University Press, 1997) deal with Piero della Francesca's mathematics and his treatise on perspective, a work strongly influenced by Euclid.

Thales' Method.

•1. Parallel lines form equal corresponding angles.

2. All right angles are equal.

•3. AA.

4. Corresponding sides of similar triangles are proportional.

Triangle Average.

5. They appear to be their midpoints.

•6. Yes. Two triangles similar to a third triangle are similar to each other.

Newton's Figure.

7. ΔPRI ~ ΔPDF and ΔPQI ~ ΔPES. (Also, ΔPDF ~ ΔSEF, from which it follows that ΔPRI ~ ΔSEF.)

8. AA. (All these triangles are right triangles. The triangles in the first two pairs share acute angles at P. The triangles in the third pair have equal vertical angles at F. The fourth pair follows from the fact that two triangles similar to a third triangle are similar to each other.)

Nine Triangles.

9. ΔABC, ΔADC, ΔBCE, ΔCDF, ΔCEF, ΔDEF.

•10. 30°, 30°, and 120°. (It is easiest to see this by starting with the three equal angles at F.)

11. ΔCDE is equiangular (and equilateral). ΔACE and ΔBCD are right triangles with acute angles of 30° and 60°.

Chessboard Puzzle.

12. The square has an area of 64 units and the other arrangement seems to have an area of 65 units.

13. AA. [∠CAE = ∠DAF (reflexive), and ∠AEC = ∠AFD (all right angles are equal.]

•14. Corresponding sides of similar triangles are proportional.

15. EC = 3 and FD = 5.

16. No. $\frac{8}{13} \neq \frac{3}{5}$.

•17. They are not collinear.

18. Indirect.

Piero's Theorem.

•19. That they are parallel.

20. That lines contain the same point.

•21. $\triangle AHK \sim \triangle ABD$ and $\triangle AKL \sim \triangle ADE$.

22. $\dfrac{HK}{BD} = \dfrac{AK}{AD}$ and $\dfrac{AK}{AD} = \dfrac{KL}{DE}$.

•23. $\dfrac{HK}{BD} = \dfrac{KL}{DE}$.

24. $\dfrac{LM}{EF} = \dfrac{MN}{FG} = \dfrac{NI}{GC}$.

Set II (pages 403–405)

Students who have solved "work problems" in algebra may be interested in knowing that the figure illustrating two electrical resistances in parallel also illustrates problems with two workers working simultaneously. The equation $R = \dfrac{R_1 R_2}{R_1 + R_2}$ is equivalent to the equation $\dfrac{1}{R_1} + \dfrac{1}{R_2} = \dfrac{1}{R}$, a form more familiar in the work-problem context: R_1 and R_2 become the times needed for the two workers to complete the job individually and R the time for them to complete it working together.

Exercises 31 through 36 are based on an interesting fact that is not well known. Although the angles of a 3-4-5 triangle are not related in any interesting way, the largest angle of a 4-5-6 triangle is exactly twice as large as its smallest angle.

Mascheroni's work *Geometria del Compasso*, in which he proved that every straightedge and compass construction can be done with an adjustable compass alone, was published in 1797. For more on Mascheroni constructions, see chapter 17 of Martin Gardner's *Mathematical Circus* (Knopf, 1979).

Electrician's Formula.

25. AA.

26. Corresponding sides of similar triangles are proportional.

•27. Addition.

•28. Multiplication.

29. Substitution (factoring).

30. Division.

Sides and Angles.

•31. $\triangle ABD \sim \triangle CBA$.

32. $\dfrac{AB}{CB} = \dfrac{BD}{BA} = \dfrac{AD}{CA}$.

33. $\dfrac{12}{18} = \dfrac{x}{12} = \dfrac{y}{15}$.

•34. $x = 8$, $y = 10$, and $z = 10$. ($z = 18 - x$.)

35. $\angle 2 = \angle C$. ($y = 10 = z$.)

36. $\angle BAC = 2\angle C$. ($\angle BAC = \angle 1 + \angle 2$, $\angle 1 = \angle C$, $\angle 2 = \angle C$; so $\angle BAC = \angle C + \angle C$.)

Mascheroni Construction.

37.

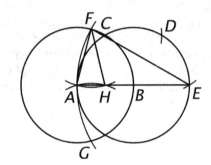

38. Isosceles.

•39. AA ($\angle A = \angle EFA$ and $\angle FHA = \angle A$).

40. Corresponding sides of similar triangles are proportional.

41. It is equal to $\dfrac{1}{2}$. $FH = FA = AB = \dfrac{1}{2}AE$; so $\dfrac{FH}{AE} = \dfrac{1}{2}$.

42. $\dfrac{AH}{FA} = \dfrac{FH}{EA} = \dfrac{1}{2}$ (exercises 40 and 41); so $\dfrac{AH}{FA} = \dfrac{1}{2}$ and $AH = \dfrac{1}{2}FA$. Because $FA = AB$, $AH = \dfrac{1}{2}AB$ by substitution.

Deja Vu.

43. AA ($\angle A = \angle A$ and $\angle C = \angle ADE$).

44. $\dfrac{BC}{ED} = \dfrac{AC}{AD}$.

•45. $\dfrac{a}{s} = \dfrac{b}{b-s}$.

46. $ab - as = bs$; so $ab = as + bs$; so $s(a + b) = ab$;

so $s = \dfrac{ab}{a+b}$.

47. It represents a single resistance to which resistances a and b are equivalent.

Original Proofs.

48. *Proof.*
    (1) $\triangle ACD$ with BE $\parallel$ CD. (Given.)
    (2) $\angle ABE = \angle C$ and $\angle AEB = \angle D$. (Parallel lines form equal corresponding angles.)
    (3) $\triangle ABE \sim \triangle ACD$. (AA.)

49. *Proof.*
    (1) Trapezoid ABCD with bases AB and DC and diagonals AC and BD meeting at E. (Given.)
    (2) AB $\parallel$ DC. (The bases of a trapezoid are parallel.)
    (3) $\angle BAC = \angle ACD$ and $\angle ABD = \angle BDC$. (Parallel lines form equal alternate interior angles.)
    (4) $\triangle AEB \sim \triangle CED$. (AA.)
    (5) $\dfrac{AE}{CE} = \dfrac{EB}{ED}$. (Corresponding sides of similar triangles are proportional.)
    (6) $AE \times ED = BE \times EC$. (Multiplication.)

50. *Proof.*
    (1) $\triangle ABC$ with midsegments MN, MO, and NO. (Given.)
    (2) MN $\parallel$ AB, NO $\parallel$ BC, and MO $\parallel$ AC. (A midsegment of a triangle is parallel to the third side.)
    (3) ANMO and BMNO are parallelograms. (A quadrilateral is a parallelogram if both pairs of opposite sides are parallel.)
    (4) $\angle OMN = \angle A$ and $\angle ONM = \angle B$. (The opposite angles of a parallelogram are equal.)
    (5) $\triangle MNO \sim \triangle ABC$. (AA.)

Set III (page 406)

Dividing Line.

1. 4.5 square units each.

2. The area of the purple region is

$\dfrac{1}{2}(4 \times 2) + 1 = 5$ square units; so the area of

the yellow region is $9 - 5 = 4$ square units.

3. AA ($\angle DBQ = \angle CBA$ and $\angle QDB = \angle C$).

4. Corresponding sides of similar triangles are proportional.

5. $QD = \dfrac{1}{4}$. Because $\dfrac{QD}{AC} = \dfrac{DB}{CB}$, $\dfrac{QD}{1} = \dfrac{1}{4}$.

6. The area of the purple region is

$\dfrac{1}{2}(4 \times 1.75) + 1 = 4.5$ square units.

7. Yes.

Chapter 10, Lesson 5

Set I (pages 409–410)

Exercises 25 through 30 are a good basis for further exploration. After confirming the answers to exercises 29 and 30, it might be fun to experiment with other linear transformations of the coordinates; that is, $(a, b) \to (ca + e, db + f)$, to see what choices of c, d, e, and f produce dilations and what sorts of transformations are produced when $c \neq d$.

Fish Story.

•1. CG and CF.

2. In a plane, two lines perpendicular to a third line are parallel.

3. $\triangle CDE \sim \triangle CAB$ by AA ($\angle CDE = \angle A$ and $\angle CED = \angle B$ because parallel lines form equal corresponding angles.)

•4. Corresponding altitudes.

5. $\dfrac{CG}{CF} = \dfrac{DE}{AB}$.

6. Corresponding altitudes of similar triangles have the same ratio as that of the corresponding sides.

•7. CG, the distance of the fish from the camera.

8. The fish is 1 foot long. ($\dfrac{2}{12} = \dfrac{DE}{6}$, DE = 1.)

9. The center.

•10. 7.5 cm.

11. 3 cm.

•12. 2.5. ($r_1 = \dfrac{BE}{DE} = \dfrac{7.5}{3}$.)

13. It indicates that the larger fish is 2.5 times as long as the smaller fish.

14. AB = 5 cm. CD = 2 cm. $\dfrac{AB}{CD} = \dfrac{5}{2} = 2.5$.

•15. 0.4. $\left(r_2 = \dfrac{DE}{BE} = \dfrac{3}{7.5}.\right)$

16. It indicates that the smaller fish is 0.4 times as long as the larger fish.

17. $\dfrac{CD}{AB} = \dfrac{2}{5} = 0.4$.

•18. r_2 is the reciprocal of r_1.

Drawing Conclusions.

19. No. $\triangle AGB$ appears to be larger than $\triangle DGE$. Also, we do not know that any sides are equal.

20. Yes. The triangles are similar by AA. (The right angles at G and the alternate interior angles formed by AB and ED.)

•21. Yes. Corresponding sides of similar triangles are proportional.

22. Yes. Because $\dfrac{AG}{DG} = \dfrac{GB}{GE}$, AG × GE = GB × DG.

•23. No. These areas cannot be equal if $\triangle AGB$ is larger.

24. Yes. Adding $\alpha\triangle DGE$ to each side of $\alpha AGEF = \alpha BGDC$ gives $\alpha ADEF = \alpha BEDC$.

Grid Exercise.

25.

•26. B(5, 3) → B′(2 · 5 − 3, 2 · 3 − 4), or B′(7, 2).
C(10, 6) → C′(2 · 10 − 3, 2 · 6 − 4), or C′(17, 8).

•27. $\triangle A'B'C'$ appears to be an enlargement of $\triangle ABC$. (Also, the corresponding sides of the two triangles appear to be parallel.)

28. A dilation.

29. (3, 4).

30. 2.

Set II (pages 410–413)

The story of the Washington Monument is told by former engineering professor Robert Banks in *Slicing Pizzas, Racing Turtles, and Further Adventures in Applied Mathematics* (Princeton University Press, 1999). A summary of the story follows: The monument, 555 feet 5 inches in height, is made of marble and granite. The cornerstone for its construction was laid on July 4, 1848, when James Polk was president. As originally planned, the tower was not tapered and would have had a star rather than a pyramid at its top. After 5 years, it had risen to a height of 152 feet, but then the money ran out and construction didn't start up again until 1880. It was completed in 1884. As Banks points out, if the monument were not capped with the pyramid but extended until the tapered sides of the main column met at a point, it would be more than twice as tall, taller than the Empire State Building.

The rest of the exercises in Set II consider the SAS and SSS similarity theorems, which are nice to know but have been omitted from the main sequence of theorems. In most applications of similar triangles, it is angles that we know to be equal rather than sides being proportional. It is unfortunate that the proofs of these theorems are so cumbersome with so many algebraic steps. As is often the case with proofs of this nature, after the stage has been set with the extra line forming the third triangle, it is probably easier to construct the proofs rather than to read them.

Washington Monument.

•31. 555.4 ft.

32. $\triangle AGF \sim \triangle EHF$, $\triangle BGF \sim \triangle CHF$, and $\triangle ABF \sim \triangle ECF$.

33. $\dfrac{FG}{FH} = \dfrac{AB}{EC}$, $\dfrac{x}{x-500.4} = \dfrac{55}{34.4}$,
$34.4x = 55x - 27{,}522$, $20.6x = 27{,}522$,
$x \approx 1{,}336$ ft.

•34. Corresponding altitudes of similar triangles have the same ratio as that of the corresponding sides.

Congruence and Similarity.

•35. ASA.

•36. AA.

37. AAS.

38. AA.

39. Yes. If $\dfrac{a}{b} = \dfrac{c}{d} = 1$, then $a = b$ and $c = d$.

The triangles are congruent by SAS.

SAS Similarity Theorem.

•40. Ruler Postulate.

•41. Through a point not on a line, there is exactly one line parallel to the given line.

42. A line parallel to one side of a triangle cuts off segments on the other two sides proportional to the sides.

43. Substitution.

44. Multiplication.

45. SAS.

46. Corresponding parts of congruent triangles are equal.

•47. Parallel lines form equal corresponding angles.

48. Substitution.

•49. AA.

50. Yes. If $\dfrac{a}{b} = \dfrac{c}{d} = \dfrac{e}{f} = 1$, then $a = b$, $c = d$, and $e = f$. The triangles are congruent by SSS.

SSS Similarity Theorem.

51. Ruler Postulate.

52. Two points determine a line.

•53. Reflexive.

54. If an angle of one triangle is equal to an angle of another triangle and the sides including these angles are proportional, then the triangles are similar.

•55. Corresponding sides of similar triangles are proportional.

56. Substitution.

57. Multiplication.

•58. SSS Congruence Theorem.

59. Corresponding parts of congruent triangles are equal.

60. AA Similarity Theorem.

61. Two triangles similar to a third triangle are similar to each other.

Set III (page 413)

Camera Experiment.

1. It gets larger.

2. If $\triangle AHB \sim \triangle CHD$, $\dfrac{x}{y} = \dfrac{AB}{CD}$. So $x\,CD = y\,AB$, and $CD = \dfrac{y}{x}\,AB$.

3. It gets bigger.

4. It gets smaller.

5. It gets bigger.

6. It gets bigger.

Chapter 10, Lesson 6

Set I (pages 416–417)

Judo, which means "gentle way" in Japanese, didn't become an Olympic sport until 1972. The three regions of its competition area, although square in shape, are covered with "tatami," rectangular mats 1 meter wide and 2 meters long. (Tatami are the same mats used by Japanese architects in creating floor plans.) A consequence of the use of these mats is that, when the judo competition area was first designed, its inner space, the "contest area," had to have sides that were an even number of meters. Interested students might enjoy exploring this requirement and explaining why.

Lewis Carroll included whimsical references to maps in two of his books. In *The Hunting of the Snark* (1876), he wrote: "He had brought a large map representing the sea, Without the least vestige of land: And the crew were much pleased when they found it to be A map they could all understand." Martin Gardner remarks in *Lewis Carroll's The Hunting of the Snark—Centennial Edition* (William Kaufman, Inc., 1981): "In contrast, a map in Carroll's *Sylvie and Bruno Concluded*, Chapter 11, has *everything* on it. The German Professor explains how his country's cartographers experimented with larger and larger maps until they finally made one with a scale of a mile to the mile."

Triangle Ratios.

•1. $\frac{1}{2}$.

•2. $\frac{1}{4}$.

3. $\frac{1}{3}$.

4. $\frac{1}{9}$.

5. $\frac{2}{3}$.

6. $\frac{4}{9}$.

7. $\frac{1}{2} \cdot (\frac{2}{4}.)$

8. $\frac{1}{4} \cdot (\frac{4}{16}.)$

9. The ratio of the areas of two similar polygons is equal to the square of the ratio of the corresponding sides.

Judo Mat.

•10. side of 1 = 8 m.

•11. $\rho 1$ = 32 m.

•12. $\alpha 1$ = 64 m².

13. side of 2 = 10 m.

14. $\rho 2$ = 40 m.

15. $\alpha 2$ = 100 m².

16. side of 3 = 16 m.

17. $\rho 3$ = 64 m.

18. $\alpha 3$ = 256 m².

•19. $\frac{4}{5} \cdot (\frac{\text{side of } 1}{\text{side of } 2} = \frac{8}{10}.)$

20. $\frac{4}{5} \cdot (\frac{\rho 1}{\rho 2} = \frac{32}{40}.)$

•21. $\frac{16}{25} \cdot (\frac{\alpha 1}{\alpha 2} = \frac{64}{100}.)$

22. $\frac{5}{8} \cdot (\frac{\text{side of } 2}{\text{side of } 3} = \frac{10}{16}.)$

23. $\frac{5}{8} \cdot (\frac{\rho 2}{\rho 3} = \frac{40}{64}.)$

24. $\frac{25}{64} \cdot (\frac{\alpha 2}{\alpha 3} = \frac{100}{256}.)$

Map Scaling.

•25. They must be proportional.

26. They must be equal.

27. They must be proportional to the square of the ratio of the corresponding distances.

•28. 440. (The scale is $\frac{1}{5,280}$, and so 1 inch represents 5,280 inches; $\frac{5,280 \text{ in}}{12}$ = 440 ft.)

29. 193,600. [(440 ft)² = 193,600 ft².]

•30. $\frac{1}{63,360} \cdot (\frac{1 \text{ inch}}{5,280 \times 12 \text{ inches}}.)$

31. $\frac{1}{1}$ or 1.

Set II (pages 417–419)

The question asked in the SAT problem of exercises 32 through 36 is the area of ΔADE, for which four numerical choices were listed, followed by the choice, "It cannot be determined from the information given." This choice, though incorrect, is somewhat tempting in that there is no way to know the lengths of ΔADE's sides.

Cranberries are grown commercially in level bog areas that can be drained. When the berries are ready to harvest, the bog is flooded and they are shaken loose from the vines by tractors. They then float to the surface where they are corralled as shown in the photograph before being transported.

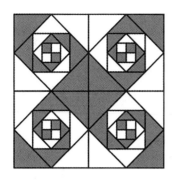

The pattern used by Barbara Dean in designing her quilt is called the "Virginia Reel." Copies of it are turned through 90°, 180°, and 270° to form the four-fold rotation-symmetric design shown above. Jinny Beyer presents this design and others in Chapter 9, "Creating Geometric Tessellations," of her beautiful book titled *Designing Tessellations* (Contemporary Books, 1998).

SAT Problem.

•32. Yes. $\triangle ADE \sim \triangle ABC$ by AA.

33. Yes. Corresponding sides of similar triangles are proportional.

•34. No.

35. No.

36. Yes. The ratio of the areas of two similar polygons is equal to the square of the ratio of the corresponding sides; so

$$\frac{\alpha\triangle ADE}{\alpha\triangle ABC} = (\frac{1}{3})^2.$$

$$\alpha\triangle ADE = \frac{1}{9}\alpha\triangle ABC = \frac{1}{9}(54) = 6.$$

Quilt Pattern.

•37. Six.

38. $\frac{1}{4}$.

39. $\frac{1}{2}$.

•40. $\frac{1}{2}$.

41. $\frac{2}{1}$ or 2.

42. $\sqrt{\frac{2}{1}} = \frac{\sqrt{2}}{1}$ or $\sqrt{2}$.

43. $\frac{\sqrt{2}}{1}$ or $\sqrt{2}$.

Cranberry Circles.

•44. Because all circles have the same shape.

45. 45 ft and 60 ft.

•46. $90\pi \approx 283$ ft and $120\pi \approx 377$ ft.

47. $2{,}025\pi \approx 6{,}362$ ft² and $3{,}600\pi \approx 11{,}310$ ft².

•48. $\frac{3}{4} = 0.75$. ($\frac{90}{120}$.)

49. $\frac{3}{4} = 0.75$. ($\frac{45}{60}$.)

50. $\frac{3}{4} = 0.75$. ($\frac{90\pi}{120\pi}$.)

51. $\frac{9}{16} = 0.5625$. [$\frac{2{,}025\pi}{3{,}600\pi}$ or $(\frac{3}{4})^2$.]

Pentagon Measurements.

52. 108°.

•53. 921 ft. (The scale is $\frac{1}{11{,}052}$; so 1 inch represents 11,052 in; $\frac{11{,}052 \text{ in}}{12} = 921$ ft.)

54. 4,605 ft. (5 × 921.)

55. Approximately 29 acres.
$$[\frac{1.5 \text{ in}^2}{x \text{ in}^2} = (\frac{1}{11{,}052})^2, x = 183{,}220{,}056 \text{ in}^2,$$
$$\frac{183{,}220{,}056 \text{ in}^2}{144} = 1{,}272{,}361.5 \text{ ft}^2,$$
$$\frac{1{,}272{,}361.5 \text{ ft}^2}{43{,}560} \approx 29 \text{ acres.}]$$

Set III (page 419)

This exercise is based on Proposition 31 of Book VI of the *Elements*. Euclid stated it as follows: "In right-angled triangles the figure on the side subtending the right angle is equal to the similar and similarly described figures on the sides containing the right angle."

Pythagoras on the Sides.

1. The ratio of the areas of two similar polygons is equal to the square of the ratio of the corresponding sides.

2. Adding $\dfrac{\alpha\,Py.a}{\alpha\,Py.c} = (\dfrac{a}{c})^2$ and $\dfrac{\alpha\,Py.b}{\alpha\,Py.c} = (\dfrac{b}{c})^2$

   gives $\dfrac{\alpha\,Py.a}{\alpha\,Py.c} + \dfrac{\alpha\,Py.b}{\alpha\,Py.c} = (\dfrac{a}{c})^2 + (\dfrac{b}{c})^2$, or

   $\dfrac{\alpha\,Py.a + \alpha\,Py.b}{\alpha\,Py.c} = \dfrac{a^2 + b^2}{c^2}$. Because a, b, and

   c are the lengths of the sides of the right

   triangle, $a^2 + b^2 = c^2$; so $\dfrac{\alpha\,Py.a + \alpha\,Py.b}{\alpha\,Py.c} = 1$.

   Multiplying, $\alpha Py.a + \alpha Py.b = \alpha Py.c$.

Chapter 10, Review

Set I (pages 421–423)

J. L. Heilbron describes Galileo's figure for objects in free fall in *Geometry Civilized* (Clarendon Press, 1998): "After many false starts, he [Galileo] discovered that the fundamental characteristic of free fall is constant acceleration, which means that the velocity with which the body moves at any time is proportional to the elapsed time. . . . Galileo derived his rule relating distance and time using geometry. . . . The angle EOF is drawn so that AC = gOA [where g is a constant number that measures the pull of gravity]. . . . In a moment of great insight, he identified the *areas* of the triangles AOC, BOD, with the *distances* covered by the falling body in the times t_A, t_B." Heilbron goes on to explain the rest of Galileo's argument. By associating distances covered with areas of triangles, Galileo reasoned that, because OA = t_A and AC = gOA = gt_A,

$d_A = \alpha\triangle AOC = \dfrac{1}{2}OA \cdot AC = \dfrac{1}{2}t_A(gt_A) = \dfrac{1}{2}gt_A^2.$

Because A is an arbitary constant, it follows that

$d = \dfrac{1}{2}gt^2.$

Body Ratios.

•1. $\dfrac{1}{2}. (\dfrac{2}{4}.)$

2. $\dfrac{1}{5}.$

3. $\dfrac{1}{3}.(\dfrac{2}{6}.)$

4. $\dfrac{5}{12}.(\dfrac{2.5}{6}.)$

Perspective Law.

5. If a line parallel to one side of a triangle intersects the other two sides in different points, it cuts off segments proportional to the sides.

6. AA. ($\angle A = \angle A$ and the corresponding angles formed by BC and DE with AD and AE are equal.)

7. AC = 2 cm, AE = 5 cm, DE = 4 cm.

8. $\dfrac{AC}{AE} = \dfrac{BC}{DE}$; so $\dfrac{2}{5} = \dfrac{BC}{4}$, 5BC = 8, BC = 1.6.

•9. 2.5. $(\dfrac{DE}{BC} = \dfrac{4 \text{ cm}}{1.6 \text{ cm}} = 2.5.)$

•10. It is 6.25 times as great. ($2.5^2 = 6.25$.)

Proportion Practice.

11. 8. $(\dfrac{x}{12} = \dfrac{10}{15}, 15x = 120, x = 8.)$

•12. 6.4. $(\dfrac{x}{4} = \dfrac{8}{5}, 5x = 32, x = 6.4.)$

13. 5.6. $(\dfrac{x}{8} = \dfrac{7}{10}, 10x = 56, x = 5.6.)$

•14. 12.8. $(\dfrac{x}{8} = \dfrac{8}{5}, 5x = 64, x = 12.8.)$

Similar Triangles.

15. Nine.

16.

Reptiles.

•17. Concave pentagons.

•18. ABHIG $\cong$ EKJGI.

19. AFEDC ~ GABHI.

•20. GFEI is a parallelogram. Both pairs of its opposite sides are equal because GFEKJ ≅ EIGJK.

21. αAFEDC = 4αABHIG.

22. ρAFEDC = 2ρABHIG.

23. $\frac{2}{1}$, or 2.

Free Fall.

24. In a plane, two lines perpendicular to a third line are parallel.

•25. It increases. (For example, BD > AC.)

26. AA. (∠O = ∠O and ∠OAC and ∠OBD are right angles.)

27. Corresponding sides of similar triangles are proportional.

28. The ratio of the areas of two similar triangles is equal to the square of the ratio of their corresponding sides.

Type Transformations.

29. Dilations.

•30. 2. ($\frac{48}{24}$.)

31. $\frac{3}{4}$ or 0.75. ($\frac{72}{96}$.)

•32. They are multiplied by 3. ($\frac{72}{24}$.)

33. They do not change.

Set II (pages 423–424)

The crossed-ladders problem on which exercises 45 through 50 are based is a famous puzzle in recreational mathematics. The puzzle concerns two crossed ladders of unequal length that lean against two buildings as shown in the figure: given the lengths of the ladders and the height of their crossing point, the problem is to find the distance between the two buildings. The puzzle is famous because the solution, based on similar triangles, leads to a difficult quartic equation. Different numbers appear in various versions of the puzzle and the solution is generally not an integer. The simplest set of numbers for which all

of the distances in the figure are integers was reported in the *American Mathematical Monthly* in 1941, and these numbers are used in the version of the puzzle presented in the exercises. More on the crossed-ladders problem can be found in Chapter 5 of Martin Gardner's *Mathematical Circus* (Knopf, 1979).

Trick Card.

•34. $\frac{10}{x} = \frac{x}{5}$, $x^2 = 50$, $x = \sqrt{50} = 5\sqrt{2} \approx 7.1$.

The width of the original card is approximately 7.1 cm.

35. $50\sqrt{2} \approx 71$ cm^2.

•36. $\frac{\sqrt{50}}{10} = \frac{5\sqrt{2}}{10} = \frac{\sqrt{2}}{2} \approx 0.71$.

37. $(\frac{\sqrt{50}}{10})^2 = \frac{50}{100} = \frac{1}{2}$.

38. Yes. The second card is formed by folding the first card in half.

Catapult.

39. AB = 108. (∆CFD ~ ∆CAB by AA; so $\frac{CF}{CA} = \frac{FD}{AB}$. $\frac{60}{162} = \frac{40}{AB}$, 60AB = 6,480, AB = 108.)

Leg Splitter.

40.

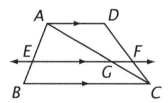

•41. In a plane, two lines parallel to a third line are parallel to each other. (EF ∥ BC and AD ∥ BC.)

42. If a line parallel to one side of a triangle intersects the other two sides in different points, it divides the sides in the same ratio.

43. Substitution.

44. They prove that the line divides the legs of the trapezoid in the same ratio.

Crossed Ladders.

45.

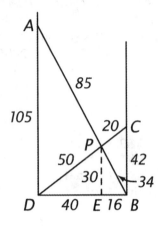

•46. 30 ft. ($\triangle PED \sim \triangle CBD$; so $\dfrac{PE}{42} = \dfrac{40}{56}$.)

47. 105 ft. ($\triangle ADB \sim \triangle PEB$; so $\dfrac{AD}{30} = \dfrac{56}{16}$.)

•48. 70 ft. ($CD^2 = 56^2 + 42^2$; so $CD = \sqrt{4{,}900} = 70$.)

49. 119 ft. ($AB^2 = 105^2 + 56^2$; so $AB = \sqrt{14{,}161} = 119$.)

50. $PA = 85$, $PB = 34$, $PC = 20$, $PD = 50$.
($PD^2 = 30^2 + 40^2$; so $PD = \sqrt{2{,}500} = 50$.

$PB^2 = 30^2 + 16^2$; so $PB = \sqrt{1{,}156} = 34$.

In $\triangle ADB$, $PE \parallel AD$; so $\dfrac{PA}{34} = \dfrac{40}{16}$,

$16PA = 1{,}360$, $PA = 85$.

In $\triangle CBD$, $PE \parallel CB$; so $\dfrac{PC}{50} = \dfrac{16}{40}$, $40PC = 800$,

$PC = 20$.)

Seven Triangles.

51. $\triangle ABD \sim \triangle DEF$, $\triangle ABC \sim \triangle DEC$,
$\triangle BDC \sim \triangle EFC$.

52.

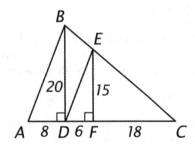

•53. 15. ($\triangle EFD \sim \triangle BDA$; so $\dfrac{EF}{20} = \dfrac{6}{8}$, $8EF = 120$,
EF = 15.)

54. 18. ($\triangle EFC \sim \triangle BDC$; so $\dfrac{FC}{6+FC} = \dfrac{15}{20}$,
$20FC = 90 + 15FC$, $5FC = 90$, $FC = 18$.)

•55. $\dfrac{4}{3}$. ($\triangle ABD \sim \triangle DEF$; so $\dfrac{20}{15} = \dfrac{8}{6} = \dfrac{4}{3}$.)

•56. 80. [$\alpha\triangle ABD = \dfrac{1}{2}(8 \cdot 20)$.]

57. 45. [$\alpha\triangle DEF = \dfrac{1}{2}(6 \cdot 15)$.]

•58. $\dfrac{16}{9}$. ($\dfrac{80}{45}$.)

59. $\dfrac{1}{3}$. (In $\triangle BCD$, EF $\parallel$ BD; so $\dfrac{BE}{EC} = \dfrac{6}{18} = \dfrac{1}{3}$.)

60. 60. [$\alpha\triangle BED = \dfrac{1}{2}(20 \cdot 6)$.]

61. 180. [$\alpha\triangle DEC = \dfrac{1}{2}(24 \cdot 15)$.]

62. $\dfrac{1}{3}$. ($\dfrac{60}{180}$.)

Chapter 10, Algebra Review 1 (page 425)

•1. $(x + 7)(x - 4) = 0$,
$x + 7 = 0$ or $x - 4 = 0$,
$x = -7$ or $x = 4$,
4 and −7.

•2. $(5x + 1)(x - 1) = 0$,
$5x + 1 = 0$ or $x - 1 = 0$,
$x = -\dfrac{1}{5}$ or $x = 1$,
1 and $-\dfrac{1}{5}$.

•3. $x^2 - 8x + 5 = 0$,
$x = \dfrac{-(-8) \pm \sqrt{(-8)^2 - 4(1)(5)}}{2(1)} = \dfrac{8 \pm \sqrt{64 - 20}}{2} =$
$\dfrac{8 \pm \sqrt{44}}{2} = \dfrac{8 \pm 2\sqrt{11}}{2} = 4 \pm \sqrt{11}$,
$4 + \sqrt{11}$ and $4 - \sqrt{11}$.

•4. $x^2 - 49 = 0$,
$(x + 7)(x - 7) = 0$,
$x + 7 = 0$ or $x - 7 = 0$,
$x = -7$ or $x = 7$.
7 and –7.

•5. $x^2 + 4x - 20 = 0$,
$$x = \frac{-4 \pm \sqrt{4^2 - 4(1)(-20)}}{2(1)} = \frac{-4 \pm \sqrt{16 - (-80)}}{2}$$
$$= \frac{-4 \pm \sqrt{96}}{2} = \frac{-4 \pm 4\sqrt{6}}{2} = -2 \pm 2\sqrt{6}.$$
$-2 + 2\sqrt{6}$ and $-2 - 2\sqrt{6}$.

•6. $(3x - 1)^2 = 0$,
$3x - 1 = 0$,
$$x = \frac{1}{3}.$$
$$\frac{1}{3}.$$

•7. $3x^2 + 5x - 12 = 0$,
$(3x - 4)(x + 3) = 0$,
$3x - 4 = 0$ or $x + 3 = 0$,
$$x = \frac{4}{3} \text{ or } x = -3.$$
$\frac{4}{3}$ and –3.

•8. $x^2 - 7x + 12 = 2$,
$x^2 - 7x + 10 = 0$,
$(x - 2)(x - 5) = 0$,
$x - 2 = 0$ or $x - 5 = 0$,
$x = 2$ or $x = 5$.
2 and 5.

•9. $x^2 + 10x + 25 = x + 17$,
$x^2 + 9x + 8 = 0$,
$(x + 1)(x + 8) = 0$,
$x + 1 = 0$ or $x + 8 = 0$,
$x = -1$ or $x = -8$.
–1 and –8.

•10. $5x^2 - 5 = 2x$,
$5x^2 - 2x - 5 = 0$,
$$x = \frac{-(-2) \pm \sqrt{(-2)^2 - 4(5)(-5)}}{2(5)} =$$
$$\frac{2 \pm \sqrt{4 - (-100)}}{10} = \frac{2 \pm \sqrt{104}}{10} = \frac{2 \pm 2\sqrt{26}}{10} =$$
$$\frac{1 \pm \sqrt{26}}{5}.$$
$$\frac{1 + \sqrt{26}}{5} \text{ and } \frac{1 - \sqrt{26}}{5}.$$

Chapter 10, Algebra Review 2 (page 426)

•1. $r = \dfrac{c}{2\pi}$.

•2. $p_1 = \dfrac{p_2 v_2}{v_1}$.

•3. $v_2 = \dfrac{p_1 v_1}{p_2}$.

•4. $c = \sqrt{a^2 + b^2}$.

•5. $c^2 - b^2 = a^2$,
$a = \sqrt{c^2 - b^2}$.

•6. $2A = bh$,
$$h = \frac{2A}{b}.$$

•7. $V^2 = 2gh$,
$$h = \frac{V^2}{2g}.$$

•8. $\dfrac{A}{4\pi} = r^2$,
$$r = \sqrt{\frac{A}{4\pi}} \quad \text{or} \quad r = \frac{1}{2}\sqrt{\frac{A}{\pi}}.$$

•9. $IR = E$,
$$R = \frac{E}{I}.$$

•10. $3V = \pi r^2 h$,
$$h = \frac{3V}{\pi r^2}.$$

• 11. $S(1 - r) = a$,
$a = S(1 - r)$ or $a = S - Sr$.

• 12. $S(1 - r) = a$,   or $S(1 - r) = a$,
$S - Sr = a$,
$S - a = Sr$, $1 - r = \dfrac{a}{S}$,
$r = \dfrac{S - a}{S}$ $r = 1 - \dfrac{a}{S}$.

• 13. $8M = wL^2$,
$w = \dfrac{8M}{L^2}$.

• 14. $W^2 = w_1 w_2$,
$w_1 = \dfrac{W^2}{w_2}$.

• 15. $r = \dfrac{R}{A - 150}$.

• 16. $R = rA - 150r$,   or $\dfrac{R}{r} = A - 150$,
$rA = R + 150r$,
$A = \dfrac{R + 150r}{r}$ $A = \dfrac{R}{r} + 150$.

• 17. $Ed^2 = I$,
$d^2 = \dfrac{I}{E}$,
$d = \sqrt{\dfrac{I}{E}}$.

• 18. $\dfrac{T}{2\pi} = \sqrt{\dfrac{l}{g}}$,
$\dfrac{l}{g} = \dfrac{T^2}{4\pi^2}$,
$l = \dfrac{gT^2}{4\pi^2}$.

• 19. $I_n - I_x = Ad^2$,
$A = \dfrac{I_n - I_x}{d^2}$.

• 20. $I_n - I_x = Ad^2$,
$\dfrac{I_n - I_x}{A} = d^2$,
$d = \sqrt{\dfrac{I_n - I_x}{A}}$.

• 21. $r_1 r_2 = Rr_2 + Rr_1$,   or
$R(r_1 + r_2) = r_1 r_2$,
$R = \dfrac{r_1 r_2}{r_1 + r_2}$

$R = \dfrac{1}{\dfrac{1}{r_1} + \dfrac{1}{r_2}}$,

$R = \dfrac{r_1 r_2}{r_1 + r_2}$.

• 22. $r_1 r_2 = Rr_2 + Rr_1$,   or
$r_1 r_2 - Rr_1 = Rr_2$,
$r_1(r_2 - R) = Rr_2$,
$r_1 = \dfrac{Rr_2}{r_2 - R}$

$\dfrac{1}{r_1} = \dfrac{1}{R} - \dfrac{1}{r_2}$,

$r_1 = \dfrac{1}{\dfrac{1}{R} - \dfrac{1}{r_2}}$,

$r_1 = \dfrac{Rr_2}{r_2 - R}$.

• 23. $nur + vr = vu(n - 1)$,
$nur + vr = vun - vu$,
$nur - vun = -vr - vu$,
$n(ur - vu) = -(vr + vu)$,
$n = -\dfrac{vr + vu}{ur - vu}$ or $n = -\dfrac{v(r + u)}{u(r - v)}$ or
$n = \dfrac{vr + vu}{vu - ur}$ or $n = \dfrac{v(r + u)}{u(v - r)}$.

• 24. $F - 32 = \dfrac{9}{5}C$,   or $F - 32 = \dfrac{9}{5}C$,
$C = \dfrac{5}{9}(F - 32)$ $9C = 5(F - 32)$,
$C = \dfrac{5}{9}(F - 32)$.

• 25. $\dfrac{f_0}{f_s} = \sqrt{\dfrac{c + v}{c - v}}$,
$\dfrac{f_0^2}{f_s^2} = \dfrac{c + v}{c - v}$,
$f_0^2(c - v) = f_s^2(c + v)$,
$f_0^2 c - f_0^2 v = f_s^2 c + f_s^2 v$,
$f_0^2 c - f_s^2 c = f_0^2 v + f_s^2 v$,
$c(f_0^2 - f_s^2) = v(f_0^2 + f_s^2)$,
$v = c\dfrac{f_0^2 - f_s^2}{f_0^2 + f_s^2}$.

Set I (pages 430–431)

The chameleon's coiled tail is a pretty good approximation of a logarithmic spiral different from that of the chambered nautilus. Its graph can be obtained from the polar equation, $r = 1.1167^\theta$, where θ is the measure of the rotation in radians and $-2\pi \le \theta \le 5\pi$.

Exercises 20 through 30 show a special case of the fact that, in any right triangle, the altitude to the hypotenuse divides the hypotenuse into two segments that have the same ratio as the square of the ratio of the legs. The general case will be considered in a later exercise.

Exercise 31 appears in the *Elements* as Proposition 17 of Book VI: "If three straight lines be proportional, the rectangle contained by the extremes is equal to the square on the mean." (Euclid includes its converse in the same proposition.)

Chameleon Tail.

1. A spiral.

•2. The altitude to the hypotenuse.

•3. AP and PC.

4. The altitude to the hypotenuse of a right triangle is the geometric mean between the segments into which it divides the hypotenuse.

•5. EP.

•6. EP and EG.

7. Each leg of a right triangle is the geometric mean between the hypotenuse and its projection on the hypotenuse.

Shadows.

8. AD.

9. DB.

10. The legs.

•11. Their projections on the hypotenuse.

•12. AC.

13. CD.

14. CB.

Find the Lengths.

•15. 4. ($\frac{8}{x} = \frac{x}{2}$, $x^2 = 16$, $x = 4$.)

16. 6. ($\frac{9}{x} = \frac{x}{4}$, $x^2 = 36$, $x = 6$.)

•17. $2\sqrt{10} \approx 6.32$. ($\frac{8}{x} = \frac{x}{5}$, $x^2 = 40$,
$x = \sqrt{40} = 2\sqrt{10} \approx 6.32$.)

•18. 4.05. ($\frac{20}{9} = \frac{9}{x}$, $20x = 81$, $x = 4.05$.)

19. 10. ($\frac{x+8}{12} = \frac{12}{8}$, $8x + 64 = 144$, $8x = 80$,
$x = 10$.)

With and Without Pythagoras.

20.

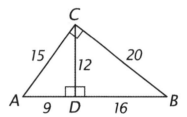

•21. 16. ($\frac{9}{12} = \frac{12}{\text{DB}}$, $9\text{DB} = 144$, $\text{DB} = 16$.)

22. 15. ($\frac{25}{\text{AC}} = \frac{\text{AC}}{9}$, $\text{AC}^2 = 225$, $\text{AC} = 15$.)

23. 15. ($\text{AC}^2 = 9^2 + 12^2 = 81 + 144 = 225$,
$\text{AC} = 15$.)

24. 20. ($\frac{25}{\text{BC}} = \frac{\text{BC}}{16}$, $\text{BC}^2 = 400$, $\text{BC} = 20$.)

25. 20. ($\text{BC}^2 = 12^2 + 16^2 = 144 + 256 = 400$,
$\text{BC} = 20$.)

•26. $\frac{9}{16}$.

27. $\frac{3}{4}$. ($\frac{15}{20}$.)

28. $\frac{9}{16}$.

•29. False.

30. True.

Rectangle and Square.

31. They are equal because $\frac{a}{b} = \frac{b}{c}$; so $ac = b^2$.

Set II (pages 431–432)

Exercises 38-45 illustrate in theory an easy way to construct, given a segment of unit length, the successive integral powers of a number greater than 1. The process can be reversed to construct the corresponding powers of a number less than 1.

Strongest Beam.

32.

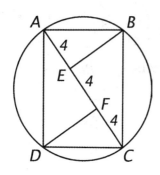

• 33. $4\sqrt{6}$. ($\frac{12}{BC} = \frac{BC}{8}$, $BC^2 = 96$,

$BC = \sqrt{96} = 4\sqrt{6}$.)

34. $4\sqrt{3}$. ($\frac{12}{AB} = \frac{AB}{4}$, $AB^2 = 48$,

$AB = \sqrt{48} = 4\sqrt{3}$.)

• 35. $\frac{\sqrt{2}}{1}$ (or $\sqrt{2}$.) ($\frac{4\sqrt{6}}{4\sqrt{3}}$.)

36. 68 in². [$(4\sqrt{6})(4\sqrt{3}) = 16\sqrt{18} \approx 68$.]

• 37. 113 in². ($\pi 6^2 = 36\pi \approx 113$.)

Zigzag.

• 38. $\frac{OA}{OB} = \frac{OB}{OC}$.

39. Each leg of a right triangle is the geometric mean between the hypotenuse and its projection on the hypotenuse.

40. $\frac{OB}{OC} = \frac{OC}{OD}$ and $\frac{OC}{OD} = \frac{OD}{OE}$.

• 41. r^2. ($\frac{1}{r} = \frac{r}{OC}$, $OC = r^2$.)

• 42. r^3. ($\frac{r}{r^2} = \frac{r^2}{OD}$, $r\,OD = r^4$, $OD = r^3$.)

43. r^4. ($\frac{r^2}{r^3} = \frac{r^3}{OE}$, $r^2\,OE = r^6$, $OE = r^4$.)

44. $OF = r^5$ and $OG = r^6$.

45. 1,000,000 inches (almost 16 miles!). (10^6.)

Plato's Proportions.

46. Yes. $\frac{OA}{OB} = \frac{OB}{OC}$ and $\frac{OB}{OC} = \frac{OC}{OD}$;

so $\frac{OA}{OB} = \frac{OC}{OD}$ by substitution.

47. Yes. $\triangle ABC \sim \triangle BCD$ by AA ($\angle A$ is complementary to $\angle ABO$ and $\angle ABO$ is complementary to $\angle DBC$; so $\angle A = \angle DBC$, and so $\angle ABC = \angle BCD$. $\frac{AB}{BC} = \frac{BC}{CD}$ because corresponding sides of similar triangles are proportional.)

SAT Problem.

• 48. 5. ($AC^2 = 4^2 + 3^2 = 25$, $AC = 5$.)

49. 3.2. ($\frac{5}{4} = \frac{4}{x}$, $5x = 16$, $x = 3.2$.)

50. 1.8.

51. 2.4. ($\frac{3.2}{z} = \frac{z}{1.8}$, $z^2 = 5.76$, $z = 2.4$.)

Set III (page 433)

Martin Gardner's column on the golden ratio can be found in *The 2nd Scientific American Book of Mathematical Puzzles and Diversions* (Simon & Schuster, 1961). As Gardner points out, a lot of material that has been written on the golden rectangle and golden ratio is crankish in nature.

The Golden Rectangle.

1.

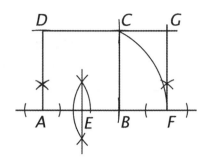

2. a) AD = 2.
   b) EB = 1.
   c) EC = $\sqrt{5}$.
   d) EF = $\sqrt{5}$.
   e) AF = $1 + \sqrt{5}$.
   f) $\dfrac{AF}{AD} = \dfrac{1+\sqrt{5}}{2}$.

3. 1.618033989 . . . .

4. a) GF = 2.
   b) BF = $\sqrt{5} - 1$. [BF = AF − AB = $(1 + \sqrt{5}) - 2$.]
   c) $\dfrac{GF}{BF} = \dfrac{2}{\sqrt{5}-1}$.

5. 1.618033989 . . . . [It is easy to show that

   $\dfrac{2}{\sqrt{5}-1} = \dfrac{1+\sqrt{5}}{2}$: $\dfrac{2}{\sqrt{5}-1} = \dfrac{2(\sqrt{5}+1)}{(\sqrt{5}-1)(\sqrt{5}+1)} =$

   $\dfrac{2(\sqrt{5}+1)}{5-1} = \dfrac{\sqrt{5}+1}{2}$.]

6. That it is also a golden rectangle (also that it is similar to rectangle AFGD.)

7. 0.618033989 . . . .

8. If r represents the golden ratio, it appears

   that $\dfrac{1}{r} = r - 1$. [This equality can be proved

   by solving the following equation for r:
   $1 = r(r - 1)$, $r^2 - r = 1$, $r^2 - r - 1 = 0$,

   $r = \dfrac{-(-1) \pm \sqrt{(-1)^2 - 4(1)(-1)}}{2(1)}$, $r = \dfrac{1 \pm \sqrt{1+4}}{2}$;

   $r > 0$, and so $r = \dfrac{1+\sqrt{5}}{2}$.]

Set I (pages 436–437)

Pythagorean triples have been a popular subject of recreational mathematics since ancient times. Martin Gardner includes many interesting facts about these numbers in the *Sixth Book of Mathematical Games from Scientific American* (W. H. Freeman and Company, 1971). Some of your students might enjoy trying to discover some of them for themselves. For example, one number must be a multiple of 3 and one of the other two numbers a multiple of 5. Another great source of information on this subject is the chapter titled "The Eternal Triangle" in *Recreations in the Theory of Numbers—The Queen of Mathematics Entertains,* by Albert H. Beiler (Dover, 1966). In *Geometry and Algebra in Ancient Civilizations* (Springer Verlag, 1983), B. L. van der Waerden points out that, although Pythagorean triples exist in the early mathematics of Egypt, India, Greece, and China, there is no mention of them in the work of Archimedes or Euclid.

Baseball Distances.

•1. 30 ft.

•2. About 95 ft. ($\sqrt{30^2 + 90^2} = \sqrt{9,000} \approx 95$.)

3. About 108 ft. ($\sqrt{60^2 + 90^2} = \sqrt{11,700} \approx 108$.)

4.

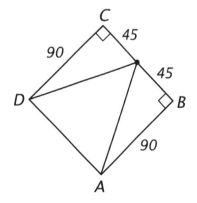

•5. About 101 ft. ($\sqrt{45^2 + 90^2} = \sqrt{10,125} \approx 101$.)

6. About 101 ft (by symmetry).

Pole Problem.

7.

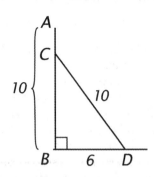

•8. 8 cubits. (BC = $\sqrt{10^2 - 6^2}$ = $\sqrt{64}$ = 8.)

9. 2 cubits.

Emergency Exits.

10.

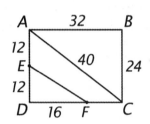

11. 40 ft. (AC = $\sqrt{32^2 + 24^2}$ = $\sqrt{1,600}$ = 40.)

•12. 20 ft. (EF $\geq \frac{1}{2}$AC.)

13. Yes. If F is midway between D and C,
DF = FC = 16. EF = $\sqrt{12^2 + 16^2}$ = $\sqrt{400}$ = 20.

14.

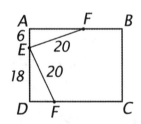

•15. About 8.7 ft. (DF = $\sqrt{20^2 - 18^2}$ = $\sqrt{76}$ ≈ 8.7.)

16. About 19 ft. (AF = $\sqrt{20^2 - 6^2}$ = $\sqrt{364}$ ≈ 19.)

17. Yes. It could be placed anywhere on wall
BC. (Every point on BC is more than 20 ft
from point E.)

Pythagorean Triples.

•18. $7^2 + 24^2 = 49 + 576 = 625 = 25^2$.

19. Yes. (The list contains many examples.)

20. No. The square of an odd number is odd.
The sum of two odd numbers is even.

•21. Each number in 14-48-50 is twice the
corresponding number in 7-24-25.

22. A dilation.

•23. 6-8-10 and 9-12-15.

24. 12-16-20; 15-20-25; 18-24-30; 21-28-35;
24-32-40; 27-36-45; 30-40-50.

25. 10-24-26 and 15-36-39.

Not Quite Right.

26. In ∆ABC, $15^2 + 10^2 = 325$ and $18^2 = 324$.
In ∆DEF, $19^2 + 9^2 = 442$ and $21^2 = 441$.

27. Acute. The longest side of each triangle is
slightly less than what it would be if the
triangle were a right triangle.

Distance Formula.

•28. AB.

•29. $y_2 - y_1$.

30. Because ∆ABC is a right triangle,
$AC^2 = AB^2 + BC^2$; so $d^2 = (x_2 - x_1)^2 + (y_2 - y_1)^2$
and
$$d = \sqrt{(x_2 - x_1)^2 + (y_2 - y_1)^2}.$$

Set II (pages 438–440)

The solution of the oil-well puzzle is based on a
fact whose generality is quite surprising. As is
demonstrated in exercises 47 through 49, given a
point inside a rectangle, the sum of the squares of
its distances from two opposite corners is equal to
the sum of the squares of its distances from the
other two corners. This is also true if the point is
outside the rectangle or even on it. In fact, the
point doesn't even have to lie in the plane of the
rectangle. It is true for *any* point whatsoever!

Matthys Levy and Mario Salvadori point out
in *Why Buildings Fall Down* (Norton, 1992) that
"the earliest suspension bridges in the world, in
China and South America, were 'vine bridges'
made of vegetable fiber ropes that served as
cables, hangers, and roadway."

In *Einstein's Legacy—The Unity of Space and Time* (Scientific American Library, 1986), Julian Schwinger reports that Einstein at age 12 "encountered and was deeply moved by the wonders of Euclidean geometry. An uncle had told him about the theorem of Pythagoras. . . . About this he later said, 'After much effort I succeeded in "proving" this theorem on the basis of the similarity of triangles.'" It is interesting that some of the basic equations of relativity come from the Pythagorean Theorem. Exercises 56 and 57 provide an example.

TV Screens.

•31. $d = 5x$. [$d^2 = (4x)^2 + (3x)^2$, $d^2 = 25x^2$, $d = 5x$.]

32. $x = \dfrac{1}{5}d$.

•33. 8.4 in by 11.2 in. [If $d = 14$, $x = \dfrac{1}{5}(14) = 2.8$; so $3x = 8.4$ and $4x = 11.2$.]

34. 16.2 in by 21.6 in. [If $d = 27$, $x = \dfrac{1}{5}(27) = 5.4$; so $3x = 16.2$ and $4x = 21.6$.]

•35. $\dfrac{27}{14}$ or about 1.9.

36. About 3.7 times as big. [$(\dfrac{27}{14})^2 = \dfrac{729}{196} \approx 3.7$.]

Grid Exercise.

37.

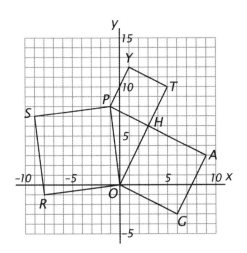

•38. OH $= \sqrt{(3-0)^2 + (6-0)^2} = \sqrt{9+36} = \sqrt{45}$ or $3\sqrt{5}$.

OP $= \sqrt{(8-0)^2 + (-1-0)^2} = \sqrt{64+1} = \sqrt{65}$.

PH $= \sqrt{(3-(-1))^2 + (6-8)^2} = \sqrt{16+4} = \sqrt{20}$ or $2\sqrt{5}$.

39. $\triangle$OHP is a right triangle because OH^2 + PH^2 = OP^2. [$(\sqrt{45})^2 + (\sqrt{20})^2 = (\sqrt{65})^2$, $45 + 20 = 65$.]

40. They appear to be squares.

41. 20, 45, and 65 square units.

Rope Surveying.

•42. The length of AP.

•43. $x^2 + 20^2 = (40 - x)^2$,
$x^2 + 400 = 1{,}600 - 80x + x^2$, $80x = 1{,}200$,
$x = 15$.

44. $32 - y$.

45. $y^2 + 16^2 = (32 - y)^2$,
$y^2 + 256 = 1{,}024 - 64y + y^2$, $64y = 768$, $y = 12$.

46. The converse of the Pythagorean Theorem.

Oil Well.

47. $b^2 = x^2 + y^2$, $c^2 = x^2 + z^2$, $d^2 = w^2 + z^2$.

48. $a^2 + c^2 = w^2 + y^2 + x^2 + z^2$,
$b^2 + d^2 = x^2 + y^2 + w^2 + z^2$.

49. That $a^2 + c^2 = b^2 + d^2$.

50. 27,000 ft. ($21{,}000^2 + 18{,}000^2 = 6{,}000^2 + d^2$, $d^2 = 729{,}000{,}000$, $d = 27{,}000$.)

Suspension Bridge.

51.

50   50
50.5   50.5

52. About 7 ft. ($\sqrt{50.5^2 - 50^2} = \sqrt{50.25} \approx 7.1$.)

Turning Radius.

•53. $D = R - \sqrt{R^2 - B^2}$.

54. About 2.8 ft.
$(27 - \sqrt{27^2 - 12^2} = 27 - \sqrt{585} \approx 2.8.)$

55. It becomes about half as large.
$(54 - \sqrt{54^2 - 12^2} = 54 - \sqrt{2{,}772} \approx 1.4.)$

Rocket Car.

56. Because $A^2 + B^2 = C^2$, $\dfrac{A^2}{C^2} + \dfrac{B^2}{C^2} = 1$; so

$\dfrac{A^2}{C^2} = 1 - \dfrac{B^2}{C^2}$. Because $\left(\dfrac{A}{C}\right)^2 = 1 - \left(\dfrac{B}{C}\right)^2$,

$\dfrac{A}{C} = \sqrt{1 - \left(\dfrac{B}{C}\right)^2}$.

57. About 0.44. $[\sqrt{1 - (0.90)^2} = \sqrt{0.19} \approx 0.44.]$

Set III (page 440)

Fermat's Last Theorem—Unlocking the Secret of an Ancient Mathematical Problem, by Amir D. Aczel (Four Walls Eight Windows, 1996) is an appropriate book for students who are curious about the subject and its geometric aspects. As *Kirkus Reviews* describes the book, it is "an excellent short history of mathematics, viewed through the lens of one of its great problems—and achievements."

Fermat's Last Theorem.

1. Infinitely many.

2. Infinitely many. There are infinitely many Pythagorean triples.

3. None.

4. The reporter observed that all positive integer powers of a number whose last digit is 4 end in either 4 or 6 and that all positive integer powers of a number whose last digit is 1 end in 1. The last digits of the three numbers in Krieger's equation, then, are either 4, 1, and 1 or 6, 1, and 1. But $4 + 1 = 5$, not 1, and $6 + 1 = 7$, not 1; so regardless of the value of n chosen, the equation cannot be true.

Set I (pages 443–445)

Isosceles Right Triangles.

•1. Yes. They are similar by AA.

•2. $\dfrac{a}{1} = \dfrac{c}{\sqrt{2}}$.

•3. $c = a\sqrt{2}$.

•4. $a = \dfrac{c}{\sqrt{2}}$.

•5. In an isosceles right triangle, the hypotenuse is $\sqrt{2}$ times the length of a leg.

30°-60° Right Triangles.

6. Yes.

7. $\dfrac{a}{1} = \dfrac{b}{\sqrt{3}} = \dfrac{c}{2}$.

8. $b = a\sqrt{3}$.

9. $c = 2a$.

10. $a = \dfrac{b}{\sqrt{3}}$.

•11. $c = \dfrac{2b}{\sqrt{3}}$.

12. $a = \dfrac{c}{2}$.

13. $b = \dfrac{c\sqrt{3}}{2}$.

14. In a 30°-60° right triangle, the hypotenuse is twice the shorter leg and the longer leg is $\sqrt{3}$ times the shorter leg.

Length Problems.

•15. $8\sqrt{2}$.

•16. $5\sqrt{2}$. $(x = \dfrac{10}{\sqrt{2}} = \dfrac{10\sqrt{2}}{2} = 5\sqrt{2}.)$

•17. $x = 12, y = 6\sqrt{3}$.

18. $x = 2, y = 2\sqrt{3}$.

•19. $x = 5\sqrt{3}$, $y = 10\sqrt{3}$. ($x = \dfrac{15}{\sqrt{3}} = \dfrac{15\sqrt{3}}{3} = 5\sqrt{3}$.)

20. $12\sqrt{2}$.

21. 14.

22. $10\sqrt{2}$. ($x = \dfrac{20}{\sqrt{2}} = \dfrac{20\sqrt{2}}{2} = 10\sqrt{2}$.)

Equilateral Triangles.

•23. $\sqrt{3}$. [$\dfrac{\sqrt{3}}{4}(2)^2$.]

•24. $4\sqrt{3}$. [$\dfrac{\sqrt{3}}{4}(4)^2$.]

25. $9\sqrt{3}$. [$\dfrac{\sqrt{3}}{4}(6)^2$.]

26. ΔADE contains 4 triangles congruent to ΔABC; ΔAFG contains 9 triangles congruent to ΔABC.

Line of Sight.

•27. $90\sqrt{2} \approx 127$ ft.

28. 150 ft. (A quick method: this triangle is a "3-4-5" right triangle with sides 3×30, 4×30, and 5×30.)

29. $120\sqrt{2} \approx 170$ ft.

Cathedral Design.

•30. $75\sqrt{3} \approx 130$ ft. ($\dfrac{\sqrt{3}}{2}150$.)

•31. About 23 in. ($20 = \dfrac{\sqrt{3}}{2}s$, $s = \dfrac{40}{\sqrt{3}} \approx 23$.)

Olympic Course.

32. About 2.3 nautical miles. ($\dfrac{3.25}{\sqrt{2}} \approx 2.3$.)

33. About 7.8 nautical miles.
[$3.25 + 2(\dfrac{3.25}{\sqrt{2}}) \approx 7.8$.]

Set II (pages 445–447)

Exercises 38 through 41 deal with a special case of the navigator's rule called "doubling the angle on the bow." Another special case is the one in which the angle is doubled from 22.5° to 45°. It is called the "$\dfrac{7}{10}$ rule," on the basis of the fact that

$$\dfrac{\sqrt{2}}{2} \approx 0.7 = \dfrac{7}{10}.$$

Exercises 42 through 47 provide a preview to students who continue their study of mathematics of the prominence of the 30°-60° and isosceles right triangle theorems in quick-calculation trigonometry.

Exercises 48 through 52 are based on a special case of "rep-tiles," polygons that can be put together or subdivided to make larger or smaller copies of themselves. According to Solomon Golomb, Professor of Electrical Engineering at the University of Southern California, a rep-tile of order k is one that can be divided into k copies similar to it and congruent to one another. As Martin Gardner explains in the chapter on rep-tiles in *The Unexpecting Hanging and Other Mathematical Diversions* (Simon & Schuster, 1969), *every* triangle is of order 4. The 30°-60° right triangle is unusual (unique?) in that it is also of order 3, as the construction demonstrates.

Not Quite Right.

•34. 21 square units. [$\dfrac{3}{7}(7)^2 = 3(7) = 21$.]

35. $\dfrac{\sqrt{3}}{4}s^2$.

36. 21.2 square units. [$\dfrac{\sqrt{3}}{4}(7)^2 = \dfrac{49\sqrt{3}}{4} \approx 21.2$.]

37. If $\sqrt{3} = \dfrac{12}{7}$, $\dfrac{\sqrt{3}}{4}s^2 = \dfrac{\frac{12}{7}}{4}s^2 = \dfrac{3}{7}s^2$.

Navigator's Rule.

38. $AB = BD$. $\angle ABD = 120°$; so $\angle ADB = 30° = \angle A$. Because two angles of ΔABD are equal, the sides opposite them are equal.

•39. $\dfrac{x}{2}$. (Because $AB = BD = x$, in 30°-60° right ΔBCD, $BC = \dfrac{x}{2}$.)

40. $\dfrac{x}{2}\sqrt{3}$.

41. AB = x and CD = $\frac{x}{2}\sqrt{3} = \frac{\sqrt{3}}{2}x$.

$\frac{\sqrt{3}}{2} \approx 0.866 \approx 0.875 = \frac{7}{8}$.

Trigonometric Angles.

42. $\angle A = 90°$, $\angle O = 60°$, and $\angle P = 30°$.

•43. OA = $\frac{1}{2}$, AP = $\frac{1}{2}\sqrt{3}$.

•44. $(-\frac{1}{2}, \frac{1}{2}\sqrt{3})$.

45. $\angle A = 90°$, $\angle O = 45°$, and $\angle P = 45°$.

46. OA = AP = $\frac{1}{\sqrt{2}} = \frac{\sqrt{2}}{2}$ or $\frac{1}{2}\sqrt{2}$.

47. $(\frac{1}{2}\sqrt{2}, -\frac{1}{2}\sqrt{2})$.

Construction Exercise.

48.

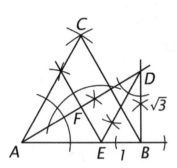

49. $\triangle ABD$, $\triangle AEF$, $\triangle DEF$, and $\triangle BDE$.

50. The following lengths are in inches: DE = 2, DB = $\sqrt{3}$, DF = $\sqrt{3}$, EF = 1, AE = 2, AF = $\sqrt{3}$.

51. *(Student answer.)* (Measurements of DB, DF, and AF suggest that they are $1\frac{3}{4}$ inches long, and $\sqrt{3} \approx 1\frac{3}{4}$.)

52. Yes. Because DB = $\sqrt{3}$, it follows that AD = $2\sqrt{3} = \sqrt{3} + \sqrt{3}$ = AF + FD, and AB = $\sqrt{3}\sqrt{3}$ = 3.

Area Problem.

•53. $\frac{1}{2}$ or 0.5. $(\frac{1}{2}1^2 = \frac{1}{2}.)$

•54. $\frac{\sqrt{3}}{4}$. $(\frac{\sqrt{3}}{4}1^2 = \frac{\sqrt{3}}{4}.)$

55. $6 + 3\sqrt{3}$. $[12(\frac{1}{2}) + 12(\frac{\sqrt{3}}{4}).]$

56. 11.2 square units.

Moscow Challenge.

57.

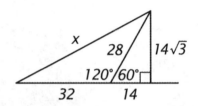

$x = 52$. [Drawing the altitude to the side labeled 32 produces a 30°-60° right triangle with hypotenuse 28 and legs 14 and $14\sqrt{3}$. By the Pythagorean Theorem, $x^2 = 46^2 + (14\sqrt{3})^2 = 2{,}116 + 588 = 2{,}704$; so $x = 52$.]

Set III (page 447)

A Big Mistake.

1. A 3-4-5 triangle is a right triangle because $3^2 + 4^2 = 5^2$. If one acute angle of a right triangle has a measure of 45°, so must the other. But, if two angles of a triangle are equal, the sides opposite them are equal. This proves that a 3-4-5 right triangle cannot have a 45° angle.

2. If the object is to find how far the tree will reach when it falls, an isosceles right triangle should be used.

Chapter 11, Lesson 4

Set I (pages 450–451)

Students with graphing calculators will find it of interest to look at the graph of $y = \tan \theta$ over the interval $0° \le \theta < 90°$, especially in relation to exercises 10 through 16. The graph of this function climbs from 0 to 1 in the interval 0° to 45° and then from 1 to ∞ in the interval 45° to 90°. Exercises 27 and 28 suggest that, if two angles are complementary, their tangents are reciprocals of each other. This is nicely illustrated by comparing the graphs of $y = \tan \theta$ and $y = \dfrac{1}{\tan(90-\theta)}$.

Tangent Practice.

•1. $\dfrac{c}{b}$.

2. $\dfrac{d}{f}$.

•3. $\dfrac{f}{d}$.

•4. 0.445.

5. 0.017.

•6. 57.290.

•7. 5.711°.

8. 87.614°.

9. 89.943°.

Isosceles Right Triangle.

10. $\angle A = 45°$.

•11. $\tan A = \dfrac{a}{b} = \dfrac{a}{a} = 1$.

12. $\tan 45° = 1$.

•13. $\angle A > 45°$.

•14. $\tan A > 1$.

15. $\angle B < 45°$.

16. $\tan B < 1$.

30°-60° Right Triangle.

17. $AB = 2b$, $BC = b\sqrt{3}$.

•18. $\tan 60° = \dfrac{BC}{AC} = \dfrac{b\sqrt{3}}{b} = \sqrt{3}$.

19. $\tan 60° \approx 1.7320508$
and $\sqrt{3} \approx 1.7320508$.

20. $\tan 30° = \dfrac{AC}{BC} = \dfrac{b}{b\sqrt{3}} = \dfrac{1}{\sqrt{3}}$ or $\dfrac{\sqrt{3}}{3}$.

21. $\tan 30° \approx 0.5773502$
and $\dfrac{\sqrt{3}}{3} \approx 0.5773502$.

3-4-5 Triangle.

•22. A right triangle because $3^2 + 4^2 = 5^2$.

23. $\tan A = \dfrac{3}{4}$.

•24. $\tan B = \dfrac{4}{3}$.

•25. $\angle A \approx 37°$.

26. $\angle B \approx 53°$.

•27. $\angle A < \angle B$; $\angle A$ and $\angle B$ are complementary.

28. $\tan A < \tan B$; $\tan A$ is the reciprocal of $\tan B$.

Finding Lengths.

•29. 16.5. ($\tan 70° = \dfrac{x}{6}$, $x = 6 \tan 70° \approx 16.5$.)

30. 10.4. ($\tan 46° = \dfrac{x}{10}$, $x = 10 \tan 46° \approx 10.4$.)

31. 20.0. ($\tan 55° = \dfrac{x}{14}$, $x = 14 \tan 55° \approx 20.0$.)

Finding Angles.

•32. 20°. ($\tan x = \dfrac{4}{11}$, $x \approx 20°$.)

33. 36°. ($\tan x = \dfrac{18}{25}$, $x \approx 36°$.)

34. 57°. ($\tan x = \dfrac{20}{13}$, $x \approx 57°$.)

Set II (pages 451–453)

The lighthouse at Alexandria stood at the entrance to the most famous harbor of the ancient world. It was begun by the first Ptolemy and completed by his son in about 285 B.C. It survived storms and earthquakes until being destroyed by a particularly violent earthquake in 1303. Estimates of its height vary. The drawing of the lighthouse and the measurements of exercises 39 through 43 are based on material in Herman Thiersch's book *Der Pharos von Alexandrien* (1909), included in B. L. van der Waerden's *Science Awakening* (Oxford University Press, 1961).

In an article for the *Encyclopedia Britannica* 1970 Yearbook of *Science and the Future*, Jay Holmes wrote: "Because of the immense distances to be traversed, all present and contemplated space flights follow curved paths. . . . The distances proportionately magnify the effect of the slightest navigation error. A miscalculation of only one degree in the course from the moon back to the earth could result in missing the earth by more than 4,000 miles. And unlike sailors of old, modern astronauts could not turn around if they should find themselves so far off course."

Fly Ball.

•35. 58°. $(\tan e = \dfrac{80}{50}, e \approx 58.)$

36. 67°. $(\tan e = \dfrac{70}{30}, e \approx 67.)$

37. 72°. $(\tan e = \dfrac{60}{20}, e \approx 72.)$

38. It increases.

Lighthouse.

•39. 197 ft. $(\tan 33.3° = \dfrac{AB}{300},$
AB $= 300 \tan 33.3° \approx 197.)$

40. 295 ft. $(\tan 44.5° = \dfrac{AC}{300},$
AC $= 300 \tan 44.5° \approx 295.)$

41. 370 ft. $(\tan 51.0° = \dfrac{AD}{300},$
AD $= 300 \tan 51.0° \approx 370.)$

42. 392 ft. $(\tan 52.6° = \dfrac{AE}{300},$
AE $= 300 \tan 52.6° \approx 392.)$

•43. About 22 ft.
(DE $=$ AE $-$ AD $\approx 392 - 370 = 22.)$

Binoculars.

44. 0.13. $(\dfrac{AC}{BD} = \dfrac{130}{1,000}.)$

45. 65 m. $(\dfrac{130}{2}.)$

•46. 3.7°. $(\tan \angle ABD = \dfrac{AD}{BD} = \dfrac{65}{1,000},$
$\angle ABD \approx 3.72.)$

47. 7.4°. $(\angle ABC = 2\angle ABD.)$

Grid Triangle.

•48. 26.565051°. $(\tan \angle ABD = \dfrac{1}{2}.)$

49. 18.434949°. $(\tan \angle EBC = \dfrac{1}{3}.)$

50. 45°. $(\angle ABC = \angle ABD + \angle EBC.)$

•51. AB $= \sqrt{5}$, BC $= \sqrt{10}$, and AC $= \sqrt{5}$.
(AB² $= 1^2 + 2^2$, AB $= \sqrt{5}$; BC² $= 1^2 + 3^2$,
BC $= \sqrt{10}$; AC² $= 1^2 + 2^2$, AC $= \sqrt{5}$.)

52. ΔABC is an isosceles right triangle because
it has two equal sides and the sum of the
squares of these sides is equal to the square
of the third side.

Diamond Cut.

•53. 40.8°. $(\tan \angle 1 = \dfrac{0.431x}{0.5x} = 0.862; \angle 1 \approx 40.8°.)$

54. 34.6°. $[\tan \angle 2 = \dfrac{0.162x}{0.5(x - 0.53x)} = \dfrac{0.162}{0.235} \approx$
$0.689; \angle 2 \approx 34.6°.]$

Space-Flight Errors.

•55. About 4,172 mi. $(\tan 1° = \dfrac{PE}{239,000},$
PE $= 239,000 \tan 1° \approx 4,172.)$

56. About 14,254 mi. $(\tan \dfrac{1}{60}° = \dfrac{PE}{49,000,000},$
PE $= 49,000,000 \tan \dfrac{1}{60}° \approx 14,254.)$

Exam Problem.

57.

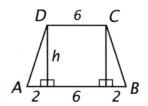

$\alpha ABCD = \dfrac{1}{2}h(10 + 6) = 8h.$ $\tan A = \dfrac{h}{2};$
so $h = 2 \tan A.$
So $\alpha ABCD = 8(2 \tan A) = 16 \tan A.$

58. About 68°. $(40 = 16 \tan A, \tan A = \dfrac{40}{16} = 2.5,$
$\angle A \approx 68°.)$

Set III (page 453)

The overhead photograph of the camel caravan
was taken in the Kerman Province of Iran by
aerial photographer Georg Gerster.

What Time is It?

To find out what time it is, we can find the angle of elevation of the sun as shown in this figure:

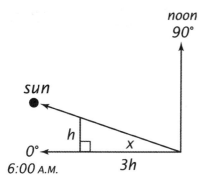

noon
90°

sun

h

0°
6:00 A.M.   x   3h

$\tan x = \dfrac{h}{3h} = \dfrac{1}{3}$; so $x \approx 18.435°$.

The sun's angle of elevation changes at a constant rate because the earth rotates at a constant rate. Let t = the number of hours elapsed since 6:00 A.M.: $\dfrac{18.435°}{90°} = \dfrac{t}{6}$,

$t \approx 1.229$ hours ≈ 1 hour, 14 minutes.

The time was about 7:14 A.M.

Chapter 11, Lesson 5

Set I (pages 456–457)

In addition to providing practice in using a calculator to find sines and cosines and their inverses, exercises 7 through 12 and the two examples illustrate the fact that a function of an angle is the cofunction of its complement. Exercises 13 through 22 reinforce this idea. The notion of the cosine of an angle developed long after the sine, and the name "cosine" was not introduced until the seventeenth century, the prefix "co-" referring to the word "complement."

More information about Bhaskara's calculation of sines can be found in B. L. van der Waerden's *Geometry and Algebra in Ancient Civilizations* (Springer Verlag, 1983). According to van der Waerden, Bhaskara divided an arc of 90° in a circle first into 6, then 12, and finally 24 equal parts. He computed the values of the sines rather than finding them by measurement.

Sine and Cosine Practice.

•1. $\dfrac{h}{a}$ or $\dfrac{b}{x+y}$.

•2. $\dfrac{h}{y}$ or $\dfrac{a}{b}$.

•3. $\dfrac{y}{b}$ or $\dfrac{b}{x+y}$.

4. $\dfrac{x}{a}$ or $\dfrac{a}{x+y}$.

5. $\dfrac{x}{a}$.

6. $\dfrac{h}{b}$.

•7. 0.996.

8. 0.719.

9. 0.719.

•10. 82.935°.

•11. 45.000°.

12. 45.000°.

Pythagorean Triangle.

13. A right triangle because $5^2 + 12^2 = 13^2$.

•14. $\sin A = \dfrac{12}{13}$.

•15. $\angle A \approx 67°$.

16. $\cos B = \dfrac{12}{13}$.

17. $\angle B \approx 23°$.

18. $\angle A > \angle B$; $\angle A$ and $\angle B$ are complementary.

19. $\sin A = \cos B$.

•20. $\cos A = \dfrac{5}{13}$.

21. $\sin B = \dfrac{5}{13}$.

22. $\cos A = \sin B$.

Finding Lengths and Angles.

•23. 5.3. $(\sin 32° = \dfrac{x}{10}, x = 10 \sin 32° \approx 5.3.)$

24. 18.3. $(\cos 66° = \dfrac{x}{45}, x = 45 \cos 66° \approx 18.3.)$

•25. 61°. $(\tan x = \dfrac{16}{9}, x \approx 61°.)$

•26. 36.5. $(\cos 70° = \dfrac{12.5}{x}$, $x \cos 70° = 12.5$,

$x = \dfrac{12.5}{\cos 70°} \approx 36.5$.)

27. 155.6. $(\sin 13° = \dfrac{35}{x}$, $x \sin 13° = 35$,

$x = \dfrac{35}{\sin 13°} \approx 155.6$.)

28. 60°. $(\cos x = \dfrac{27}{54} = 0.5$, $x = 60°$.)

Sines from a Circle.

•29. BH = 2.6 cm, CI = 5.0 cm, DJ = 7.1 cm,
EK = 8.7 cm, and FL = 9.7 cm.

30. $\sin 30° \approx \dfrac{5.0}{10} = 0.50$,

$\sin 45° \approx \dfrac{7.1}{10} = 0.71$,

$\sin 60° \approx \dfrac{8.7}{10} = 0.87$,

$\sin 75° \approx \dfrac{9.7}{10} = 0.97$.

31. $\sin 30° = 0.5$,
$\sin 45° \approx 0.7071067 \approx 0.71$,
$\sin 60° \approx 0.8660254 \approx 0.87$,
$\sin 75° \approx 0.9659258 \approx 0.97$.

32. It gets larger.

33. A number very close to zero.

34. A number very close to one.

35. According to a calculator, $\sin 0° = 0$ and $\sin 90° = 1$.

Set II (pages 458–460)

A wealth of information on flying kites to great altitudes can be found in Richard P. Synergy's book *Kiting To Record Altitudes* (Fly Right Publications, 1994). The line for the kite shown in the photograph was more than 4 miles long and was spooled from a winch. At times it exerted a force of 100 pounds! The calculation of the kite's altitude was more complicated than that implied in the exercises because the line for the kite was not actually straight; its weight made it sag considerably. The altitude of the kite was established through altimeter readings.

As the title of Robert B. Banks' *Towing Icebergs, Falling Dominoes, and Other Adventures in Applied Mathematics* (Princeton University Press, 1998) implies, he includes a good discussion of the mathematics and physics of toppling dominoes. The *Guinness Book of World Records* always has an entry on the latest record in this sport. In 1988, some students from three universities in the Netherlands spent a month setting up 1,500,000 dominoes. A single push caused 1,382,101 of them to fall!

Kite Flying.

•36. The sine of an acute angle of a right triangle is the ratio of the length of the opposite leg to the length of the hypotenuse.

37. Multiplication.

•38. In a 30°-60° right triangle, the side opposite the 30° angle is half the hypotenuse.

39. In $\triangle ABC$, $\sin 30° = \dfrac{h}{l}$. Because $\sin 30° = 0.5$,
$\dfrac{h}{l} = 0.5$, so $h = 0.5l$.

•40. Approximately 14,500 ft, or more than
2.7 mi. $(\sin 42° = \dfrac{h}{21,600}$,
$h = 21,600 \sin 42° \approx 14,453$; $\dfrac{14,453}{5,280} \approx 2.7$.)

Force Components.

41. $h = \dfrac{1}{2}(27,000) = 13,500$ pounds,
$v = 13,500\sqrt{3} \approx 23,383$ pounds.

42. $\sin 60° = \dfrac{v}{27,000}$,
$v = 27,000 \sin 60° \approx 23,383$ pounds.
$\cos 60° = \dfrac{h}{27,000}$,
$h = 27,000 \cos 60° = 13,500$ pounds.

Flight Path.

•43. About 806 ft. $(\sin 3.3° = \dfrac{CB}{14,000}$,
$CB = 14,000 \sin 3.3° \approx 806$.)

44. About 13.6°. $(\sin A = \dfrac{3,300}{14,000}$, $\angle A \approx 13.6°$.)

Leaning Tower.

45. About 10 ft. ($\sin 3° = \dfrac{x}{193}$,

  $x = 193 \sin 3° \approx 10.1$.)

•46. About 4°. ($\sin A = \dfrac{13.5}{193}$, $\angle A \approx 4°$.)

47. About 18.5 ft. ($\sin 5.5° = \dfrac{x}{193}$,

  $x = 193 \sin 5.5° \approx 18.5$.)

Fiber Optics.

•48. Parallel lines form equal alternate interior angles.

•49. AAS.

50. Corresponding parts of congruent triangles are equal.

•51. 5a.

52. 5d.

53. Because $\dfrac{5a}{5d} = \dfrac{a}{d}$ and the cosine of each

  numbered angle is $\dfrac{a}{d}$.

SAS Area.

54.

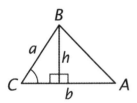

55. $\sin C = \dfrac{h}{a}$; so $h = a \sin C$.

  $\alpha \triangle ABC = \dfrac{1}{2}bh = \dfrac{1}{2}b(a \sin C) = \dfrac{1}{2}ab \sin C$.

Dominoes.

56. Parallel lines form equal alternate interior angles.

•57. $\sin \angle 2 = \dfrac{d-w}{h}$.

58. 30°. (If $d - w = \dfrac{1}{2}h$, $\sin \angle 2 = \dfrac{1}{2}$;

  so $\angle 2 = 30° = \angle 1$.)

59. 90°. (If $d - w = h$, the domino has room to turn 90° without hitting the next one. Also, if $d - w = h$, $\sin \angle 2 = 1$; so $\angle 2 = 90° = \angle 1$.)

60. The next domino might not fall over.

Set III (page 460)

The first distance to a star beyond the sun was determined by the German astronomer Friedrich Bessel in 1838. The star, 61 Cygni, was chosen because its position against the distant sky changes more than those of most stars. A Scottish astronomer, Thomas Henderson, found the distance to Alpha Centauri (and its faint companion Proxima Centauri) a year later. The measure of the angle ∠ACS in the figure is called the "parallax" of the star. The unit now used to express distances in space is the "parsec," defined as the distance of a star whose parallax is 1 second. More details on the process of determining astronomical distances by parallax can be found in Eli Maor's book titled *Trigonometric Delights* (Princeton University Press, 1998). Another excellent reference source is *Parallax—The Race to Measure the Cosmos*, by Alan W. Hirshfeld (W. H. Freeman and Company, 2001).

Star Distance.

1. About 25,000,000,000,000 mi.

  $(0.76" = \dfrac{0.76°}{3{,}600}$, $\sin \dfrac{0.76°}{3{,}600} = \dfrac{93{,}000{,}000}{AC}$,

  $AC = \dfrac{93{,}000{,}000}{\sin \dfrac{0.76°}{3{,}600}} \approx 2.5 \times 10^{13}$.)

2. About 4.3 light-years. $\left(\dfrac{2.5 \times 10^{13}}{5.88 \times 10^{12}} \approx 4.3.\right)$

Chapter 11, Lesson 6

Set I (pages 463–465)

About 90% of all avalanches occur on slopes between 30° and 45°, although very wet snow can avalanche on slopes of only 10° to 25°. Skiers sometimes measure the slope angle with a clinometer and sometimes merely estimate it with a ski pole.

  According to Cleo Baldon and Ib Melchior in *Steps and Stairways* (Rizzoli, 1989), the eighteenth-century French architect Jacques-François Blondel, "concluded that the stride or pace of walking was 24 inches and that when negotiating a stairway this pace should be decreased in proportion to the height of the riser. His theory was that two times the height of the riser plus the depth of the tread should equal the 24 inches of the pace." Blondel's formula is still used today, but the pace number has been increased to 26.

Slope Practice.

•1. h.

2. c.

•3. The x-axis.

4. The y-axis.

•5. e.

•6. $j, \frac{1}{4}; e, -4$.

7. g.

8. $g, 2; b, -\frac{1}{2}$.

9. Two nonvertical lines are perpendicular iff the product of their slopes is –1.

10. No. Two lines with the same slope would be parallel. All of the lines named with letters intersect at the origin.

Snow Avalanches.

•11. The slope is tan 30° ≈ 0.58.

12. tan 45° = 1.

Hydraulic Gradient.

•13. –60 m.

14. 1,000 m.

15. –0.06. ($\frac{-60}{1,000}$.)

•16. About 3.4°. (tan ∠1 = 0.06; ∠1 ≈ 3.4°.)

Stair Design.

17.

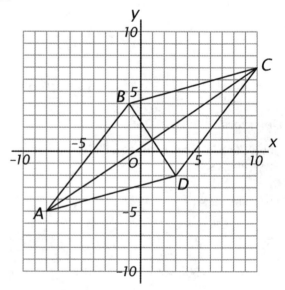

•18. $\frac{5}{9}$ (or about 0.56). ($\frac{10}{18}$.)

19. About 29°. (tan A = $\frac{5}{9}$, ∠A ≈ 29°.)

20.

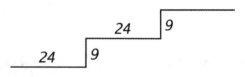

21. $\frac{3}{8}$ (or 0.375). ($\frac{9}{24}$.)

22. About 21°. (tan A = $\frac{3}{8}$, ∠A ≈ 20.6°.)

Quadrilateral Problem.

•23. $\frac{3}{11}$.

24. $\frac{9}{7}$.

25. $\frac{3}{11}$.

26. $\frac{2}{3}$.

•27. $-\frac{3}{2}$.

28. $\sqrt{130}$.

29. $\sqrt{130}$.

30. $\sqrt{130}$.

31.

32. It is a rhombus because all of its sides are equal. (Or, it is a parallelogram because both pairs of opposite sides are parallel.)

33. They are perpendicular because the diagonals of a rhombus are perpendicular.

34. They are equal.

35. Their product is –1. (Or, one slope is the opposite of the reciprocal of the other.)

Set II (pages 465–467)

The two theorems of this lesson about the slopes of parallel and perpendicular lines are only partly proved in exercises 52 through 65. Exercises 52 through 59 establish that, if two nonvertical lines are parallel, their slopes are equal. The converse of this theorem can be proved by showing that $\dfrac{AC}{BC} = \dfrac{DF}{EF}$; so $\triangle ABC \sim \triangle DEF$ by SAS similarity, from which it follows that $\angle 1 = \angle 2$, and so $l_1 \parallel l_2$.

Exercises 60 through 65 establish that, if two nonvertical lines are perpendicular, the product of their slopes is –1. The converse theorem can be proved by showing that, if $\dfrac{AD}{BD} \cdot \dfrac{-DC}{BD} = -1$, then $\dfrac{AD}{BD} = \dfrac{BD}{DC}$. From this it follows that $\triangle ADB \sim \triangle BDC$ by SAS similarity. Because $\triangle ADB$ and $\triangle BDC$ are right triangles, the facts that corresponding angles of similar triangles are equal and the acute angles of a right triangle are complementary can be used to show that $\angle ABC$ is a right angle; so $l_1 \perp l_2$.

Slope and x-Intercept.

•36. Its run is 550 m and its rise is –10 m.

37. 290 m above the ground. (300 – 10.)

38. Its run is x m and its rise is $-\dfrac{1}{55}x$ (or $-\dfrac{x}{55}$) m.

•39. $(300 - \dfrac{x}{55})$ m above the ground.

•40. 16,500 m. $(300 - \dfrac{1}{55}x = 0, \dfrac{1}{55}x = 300, x = 16,500.)$

41. Yes. (16,500 m = 16.5 km > 15 km.)

Point–Slope Equation of a Line.

•42. Run = $x - x_1$; rise = $y - y_1$.

•43. $y = m(x - x_1) + y_1$. $[m = \dfrac{y - y_1}{x - x_1}$, $m(x - x_1) = y - y_1$, $y = m(x - x_1) + y_1.]$

44. $y = -\dfrac{1}{55}x + 300.$

45. *(Student answer.)* (It is hoped, yes.)

46. The points (5,500, 200) and (1,100, 280) are on the path. The point (2,200, 250) is not.

$[y = -\dfrac{1}{55}(5,500) + 300 = 200;$

$y = -\dfrac{1}{55}(1,100) + 300 = 280;$

$y = -\dfrac{1}{55}(2,200) + 300 = 260 \neq 250.]$

100-Meter Dash.

•47. From the start to about 2.5 seconds.

48. From about 5 to 8 seconds.

49. From about 2.5 to 5 seconds.

50. The runner's speed is decreasing.

51. The runner's speed is not changing.

Theorem 52.

•52. $\dfrac{AC}{BC}.$

53. $\dfrac{DF}{EF}.$

54. Parallel lines form equal corresponding angles.

55. AA.

•56. Corresponding sides of similar triangles are proportional.

57. Multiplication.

•58. Division.

59. Substitution.

Theorem 53.

•60. $l_2.$

61. $\dfrac{AD}{BD}.$

•62. The altitude to the hypotenuse of a right triangle is the geometric mean between the segments of the hypotenuse.

63. $-\dfrac{1}{m_2}$.

64. Substitution.

65. Multiplication.

Fire Speed.

66. First figure, $\tan 25° \approx 0.47$; second figure, $\tan 50° \approx 1.19$.

67. About 6.5 times as fast. The squares of the two slopes are approximately $(0.47)^2 \approx 0.22$ and $(1.19)^2 \approx 1.42$; $\dfrac{1.42}{0.22} \approx 6.5$.

Set III (page 467)

The second-century Greek physician Galen first suggested a temperature scale based on ice and boiling water, but not until the seventeenth century were thermometers invented. In 1714, German physicist Gabriel Fahrenheit chose the temperature of a freezing mixture of ammonium chloride, ice, and water for the zero on his scale and the temperature of the human body as 96. The number 96 has many more divisors than 100 does, and Fahrenheit thought it would be more convenient for calculations! Swedish astronomer Anders Celsius introduced his scale, originally called the centigrade scale, in 1742. Strangely, Celsius chose 100° for the *freezing* point of water and 0° for its *boiling* point.

Celsius and Fahrenheit.

1. $\dfrac{9}{5}$ or 1.8.

   ($m = \dfrac{\text{rise}}{\text{run}} = \dfrac{212-32}{100-0} = \dfrac{180}{100} = \dfrac{9}{5} = 1.8$.)

2. $\dfrac{F-32}{C}$.

3. $\dfrac{F-32}{C} = \dfrac{9}{5}$.

4. $C = \dfrac{5}{9}(F-32)$.

   [$9C = 5(F-32)$, $C = \dfrac{5}{9}(F-32)$.]

5. $F = \dfrac{9}{5}C + 32$. ($F - 32 = \dfrac{9}{5}C$, $F = \dfrac{9}{5}C + 32$.)

6. Yes. The temperature is –40°. If F = C, then $C = \dfrac{9}{5}C + 32$; so $5C = 9C + 160$, $-4C = 160$, $C = -40$.

Chapter 11, Lesson 7

Set I (pages 471–472)

According to David Eugene Smith in his *History of Mathematics* (Ginn, 1925; Dover, 1958), the Law of Sines "was known to Ptolemy (c.150) in substance, although he expressed it by means of chords." About the Law of Cosines, Smith wrote that it is "essentially a geometric theorem of Euclid" (Book II, Propositions 12 and 13) and that "in that form it was known to all medieval mathematicians."

Greek Equations.

1. Sine.

•2. They are opposite each other.

3. $\dfrac{\gamma}{\eta\mu\Gamma}$.

•4. The Law of Sines.

5. Cosine.

6. $\angle A$ is opposite side α.

•7. They include $\angle A$.

8. $\angle B$ is opposite side β.

9. They include $\angle B$.

10. $\gamma^2 = \alpha^2 + \beta^2 - 2\alpha\beta \, \text{συν}\Gamma$.

11. The Law of Cosines.

Law of Sines.

•12. 40°. ($\dfrac{\sin x}{50} = \dfrac{\sin 75°}{75}$,

   $\sin x = \dfrac{50 \sin 75°}{75} \approx 0.644$, $x \approx 40°$.)

•13. 17.0. ($\dfrac{x}{\sin 50°} = \dfrac{17}{\sin 50°}$, $x = 17$. This is consistent with the fact that the triangle is isosceles.)

14. $75°$. ($\dfrac{\sin x}{4} = \dfrac{\sin 14°}{1}$,

$\sin x = 4 \sin 14° \approx 0.968$, $x \approx 75°$.)

Law of Cosines.

•15. $60°$. [$19^2 = 16^2 + 21^2 - 2(16)(21) \cos x$,
   $361 = 256 + 441 - 672 \cos x$, $672 \cos x = 336$,
   $\cos x = \dfrac{336}{672} = 0.5$, $x = 60°$.]

•16. 32.0. [$x^2 = 30^2 + 40^2 - 2(30)(40) \cos 52°$,
   $x^2 = 900 + 1600 - 2400 \cos 52°$, $x^2 \approx 1022$,
   $x \approx 32.0$.]

17. 15.0. [$x^2 = 26^2 + 30^2 - 2(26)(30)\cos 30°$,
   $x^2 = 676 + 900 - 1560 \cos 30°$, $x^2 \approx 225$,
   $x \approx 15.0$.]

Sines of Supplementary Angles.

•18. $\dfrac{\sin \angle 1}{y} = \dfrac{\sin \angle 4}{x}$.

19. $\dfrac{\sin \angle 4}{x} = \dfrac{\sin \angle 3}{y}$.

20. They imply that $\sin \angle 1 = \sin \angle 3$. Substituting
   for $\dfrac{\sin \angle 4}{x}$, we get $\dfrac{\sin \angle 1}{y} = \dfrac{\sin \angle 3}{y}$;
   multiplying by y gives $\sin \angle 1 = \sin \angle 3$.

•21. $108°$. ($\angle 3 = 180° - \angle 2 = 180° - \angle 1$.)

22. $\sin \angle 1 = \sin 72° \approx 0.9510565 \ldots$

23. $\sin \angle 3 = \sin 108° \approx 0.9510565 \ldots$

24. They suggest that the sines of
   supplementary angles are equal.

Set II *(pages 472–474)*

In a caption to a figure on which the figure for
exercises 25 through 30 is based, John Keay, the
author of *The Great Arc—The Dramatic Tale of
How India Was Mapped and Everest Was Named*
(Harper Collins, 2000), describes the triangulation
process: "Given the distance between points A and
B (the base-line), the distance from each to point
C is calculated by trigonometry using the AB
base-line measurement plus the angles CAB and
ABC as measured by sighting with a theodolite.
The distance between C and B having now been
established, it may be used as the base-line for
another triangle to plot the position of D.

CD may then be used as the base for a third
triangle, and so on." Keay points out that the
accuracy of the Survey of India was important
not only in measuring a subcontinent but also in
determining the curvature of the earth.

As it travels in its orbit around the earth, the
moon's distance from the earth varies between
221,463 miles and 252,710 miles. In his fascinating
book titled *What If the Moon Didn't Exist?*
(Harper Collins, 1993), astronomer Neil F. Comins
observes that, ever since the Apollo 12 astronauts
left the laser reflector on the moon in 1969, the
distance at any given time can be determined to
within a few inches. These measurements reveal
that the moon is moving away from the earth at
a rate of about 2 inches per year.

Triangulation.

•25. $\angle 3 = 56°$. ($\angle 3 = 180° - \angle A - \angle 1 =$
   $180° - 73° - 51° = 56°$.)

•26. $CB \approx 27.7$ mi. ($\dfrac{CB}{\sin 73°} = \dfrac{24.0}{\sin 56°}$,

   $CB = \dfrac{24.0 \sin 73°}{\sin 56°} \approx 27.7$.)

27. $\angle D = 53°$. ($\angle D = 180° - \angle 2 - \angle 4 =$
   $180° - 42° - 85° = 53°$.)

•28. $DB \approx 34.6$ mi. ($\dfrac{DB}{\sin 85°} = \dfrac{27.7}{\sin 53°}$,

   $DB = \dfrac{27.7 \sin 85°}{\sin 53°} \approx 34.6$.)

29. $CD \approx 23.2$ mi. ($\dfrac{CD}{\sin 42°} = \dfrac{27.7}{\sin 53°}$,

   $CD = \dfrac{27.7 \sin 42°}{\sin 53°} \approx 23.2$.)

30. $CD^2 = CB^2 + DB^2 - 2(CB)(DB)\cos 42°$,
   $CD^2 = 27.7^2 + 34.6^2 - 2(27.7)(34.6) \cos 42°$,
   $CD^2 \approx 539.96$, $CD \approx 23.2$.

Distance to the Moon.

31. $\angle M = 1.4°$.

•32. $238,000$ mi. ($\dfrac{AM}{\sin 89.2°} = \dfrac{5,820}{\sin 1.4°}$,

   $AM = \dfrac{5,820 \sin 89.2°}{\sin 1.4°} \approx 238,187$.)

33. 238,000 mi. ($\frac{BM}{\sin 89.4°} = \frac{5,820}{\sin 1.4°}$,

$BM = \frac{5,820 \sin 89.4°}{\sin 1.4°} \approx 238,197$.)

The Case of the Equilateral Triangle.

•34. $a^2 = b^2 + c^2 - 2bc \cos A$.

•35. $a^2 = a^2 + a^2 - 2aa \cos A$, $a^2 = 2a^2 - 2a^2 \cos A$,

$2a^2 \cos A = a^2$, $\cos A = \frac{1}{2}$.

36. *Example figure:*

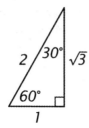

$\cos 60° = \frac{1}{2}$.

The Case of the Right Triangle.

37. $\frac{\sin A}{a} = \frac{\sin B}{b} = \frac{\sin C}{c}$.

38. 1.

39. $\frac{\sin A}{a} = \frac{\sin B}{b} = \frac{1}{c}$.

•40. By multiplication.

41. The definitions of the sine and cosine based on a right triangle.

42. $c^2 = a^2 + b^2 - 2ab \cos C$.

43. 0.

•44. $c^2 = a^2 + b^2$.

45. The Pythagorean Theorem.

The Case of the "Flattened" Triangle.

46. $c^2 = a^2 + b^2 - 2ab \cos C$.

47. −1.

48. $c^2 = a^2 + b^2 + 2ab$.

49. Yes. Because $c^2 = a^2 + 2ab + b^2 = (a + b)^2$, $c = a + b$.

Euclid's Law of Cosines.

50. The area of the rectangle is cx. In right $\triangle ADC$, $\cos A = \frac{x}{b}$; so $x = b \cos A$.

Substituting for x, the area of the rectangle is $c(b \cos A) = bc \cos A$.

Set III (page 474)

The idea for this exercise came from Carl Eckart's book titled *Our Modern Idol — Mathematical Science* (Scripps Institution of Oceanography, University of California, 1984). The 780-1351-1560 triangle is a remarkably good approximation of a 30°-60° right triangle because $\frac{1351}{780}$ matches $\sqrt{3}$ to five decimal places ($\frac{1351}{780}$ = 1.7320512 . . . and $\sqrt{3}$ = 1.7320508 . . .) and $\frac{1560}{780}$ = 2. Using *Mathematica* reveals that, to four decimal places, $\triangle ABC$ is a 30.0000°-60.0000°-90.0000° triangle. To five decimal places, it is a 30.00000°-60.00003°-89.99997° triangle. To eleven decimal places, $\angle A = 30.00000000000°$ but, to twelve decimal places, $\angle A = 29.999999999996°$! Eckart remarks that "such a triangle differs from a right triangle by less than one-tenth of a second" and observes that it cannot be drawn on paper, because "paper is too flexible, and even printed straight lines are too coarse."

A Possible Case of Mistaken Identity.

1. About 30.000000°.
   [$780^2 = 1351^2 + 1560^2 - 2(1351)(1560) \cos A$, $\angle A \approx 30.000000°$.]

2. About 60.000027°. [By the Law of Sines,
   $\frac{\sin B}{1351} = \frac{\sin 30°}{780}$, $\sin B = \frac{1351 \sin 30°}{780}$,
   $\angle B \approx 60.000027°$ or, by the Law of Cosines,
   $1351^2 = 780^2 + 1560^2 - 2(780)(1560) \cos B$,
   $\angle B \approx 60.000027°$.]

3. $\triangle ABC$ is an acute triangle. From the measures of $\angle A$ and $\angle B$, it follows that $\angle C \approx 89.999973°$. Also, $1351^2 + 780^2 = 2,433,601$, but $1560^2 = 2,433,600$; so $a^2 + b^2 > c^2$. (The triangle is a very good imitation, however, of a 30°-60° right triangle!)

Chapter 11, Review

Set I (pages 476–478)

Exercises 23 through 28 provide another example of a pair of triangles that have five pairs of equal parts and yet are not congruent. Letting AD = 1 and AC = x and using the fact that each leg of a right triangle is the geometric mean between the hypotenuse and its projection on the hypotenuse, we have $\frac{x+1}{x} = \frac{x}{1}$. Solving for x, $x^2 = x + 1$,

$x^2 - x - 1 = 0$, $x = \frac{1+\sqrt{5}}{2}$, the golden ratio! In

$\triangle$ACD, cos A = $\frac{1}{x}$; so cos A = $\frac{2}{1+\sqrt{5}}$,

$\angle$A = 51.827292 . . .° and $\angle$B = 38.172707 . . .°

According to J. V. Field in *The Invention of Infinity—Mathematics and Art in the Renaissance* (Oxford University Press, 1997), "the great invention of the military engineers [of the sixteenth century] is the polygonal fort. . . . The commonest choice is the pentagon. A construction for a regular pentagon can be found in *Elements*, Book 4, but it is hardly the sort of construction one would wish to use in laying out the ground plan of a fortress. . . . The practical method used by the Venetians, preserved in a file in the State Archives, is shown in [the figure used in the exercises]. We are given the sides of a right-angled triangle. The triangle can be constructed from rope, or by linking together the chains used by surveyors, and repetitions of it will then allow the complete pentagon to be laid out. The person who actually lays out the ground plan using this method does not need to be very good at mathematics, but it is clear that some mathematical expertise was needed to devise the method, and to calculate the lengths of the sides of the triangle."

Greek Cross.

- •1. 2.

2. $-\frac{1}{2}$.

3. 2.

4. $\frac{3}{4}$.

- •5. $-\frac{4}{3}$.

6. AC $\perp$ CD.

7. AC $\parallel$ DE.

- •8. HF $\perp$ BG.

9. They are equal.

10. Their product is –1. (Or, one slope is the opposite of the reciprocal of the other.)

Doubled Square.

- •11. $d = s\sqrt{2}$.

12. s^2.

13. d^2.

14. $2s^2$. [$d^2 = (s\sqrt{2})^2 = 2s^2$.]

Escalator Design.

- •15. 28 ft. [$c = 2a = 2(14)$.]

- •16. About 24.2 ft. ($b = a\sqrt{3} = 14\sqrt{3} \approx 24.2$.)

17. 20 ft. ($40 = 2a$.)

18. About 34.6 ft. ($b = a\sqrt{3} = 20\sqrt{3} \approx 34.6$.)

- •19. About 195 ft. ($b = 338$; so $a = \frac{338}{\sqrt{3}} \approx 195$.)

20. About 390 ft. [$c = 2a \approx 2(195)$.]

Basketball Angles.

- •21. About 5.7°.
  (tan B = $\frac{h}{D}$ = $\frac{10-7.5}{25}$ = $\frac{2.5}{25}$ = 0.1, $\angle$B $\approx$ 5.7°.)

22. About 14°. (tan B = $\frac{2.5}{10}$ = 0.25, $\angle$B $\approx$ 14°.)

Equal Parts.

23. Three.

24. The altitude to the hypotenuse of a right triangle forms two triangles similar to it and to each other.

25. Three.

- •26. Two.

27. Five.

28. No. The longest sides of the two triangles are not equal.

Pentagon.

29. $\angle BOC = 36°$. ($\frac{360°}{10}$.)

•30. BC = 100 ft. ($\frac{1,000}{10}$.)

•31. OB ≈ 137.6 ft. (tan 36° = $\frac{100}{OB}$,

OB = $\frac{100}{\tan 36°}$ ≈ 137.6.)

32. OC ≈ 170.1 ft. (sin 36° = $\frac{100}{OC}$,

OC = $\frac{100}{\sin 36°}$ ≈ 170.1.)

33. $OB^2 + BC^2$ ≈ $137.6^2 + 100^2$ ≈ 28,934 and OC^2 ≈ 170.1^2 ≈ 28,934.

•34. About 408 feet.
($\rho\triangle OBC$ ≈ 137.6 + 100 + 170.1 ≈ 408.)

Three Ratios.

•35. 0.87. (Letting the sides of the triangle and

square be 1 unit, CF = $\frac{\sqrt{3}}{2}$ and EA = 1; so

$\frac{CF}{EA}$ = $\frac{\sqrt{3}}{2}$ ≈ 0.87.)

36. 0.75. ($\rho\triangle ABC$ = 3 and $\rho ABDE$ = 4; so

$\frac{\rho\triangle ABC}{\rho ABDE}$ = $\frac{3}{4}$ = 0.75.)

37. 0.43. ($\alpha\triangle ABC$ = $\frac{\sqrt{3}}{4}(1)^2$ and $\alpha ABDE$ = 1^2; so

$\frac{\alpha\triangle ABC}{\alpha ABDE}$ = $\frac{\sqrt{3}}{4}$ ≈ 0.43.)

Set II (pages 478–480)

A discussion of various Steiner problems can be found in chapter 3, "Shortest and Quickest Connections," of *The Parsimonious Universe— Shape and Form in the Natural World*, by Stefan Hildebrandt and Anthony Tromba (Copernicus, 1996). Given a general set of points, the shortest network of segments connecting them is called their "minimal Steiner tree." These trees are important in the design of computer chips and have even been used by biologists studying the relationships of species. A rather amazing natural solution to finding Steiner trees comes from

dipping models made from glass or plastic plates into soap solutions!

The Babylonian clay tablet in exercises 65 and 66 was deciphered in 1945. David M. Burton remarks in his *History of Mathematics* (Allyn & Bacon, 1985): "This tablet is written in Old Babylonian script, which dates it somewhere between 1900 B.C. and 1600 B.C. The analysis of this extraordinary group of figures establishes beyond any doubt that the so-called Pythagorean theorem was known to Babylonian mathematicians more than a thousand years before Pythagoras was born." Burton observes that the 541–481 error was probably that of a scribe because, in the sexagesimal notation of the Babylonians, 541 is written as 9,1 and 481 as 8,1. Eleanor Robson shows in "Words and Pictures—New Light on Plimpton 322" (*American Mathematical Monthly*, February 2002) that it is likely that these Pythagorean triples actually arose from computations concerning reciprocals. Even so, "Pythagorean triples" were of interest long before Pythagoras.

Isosceles Right Triangles.

38.

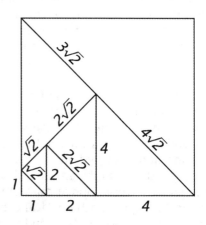

39. $\sqrt{2}$, 2, $2\sqrt{2}$, 4, $3\sqrt{2}$, and 7 units.

•40. 7 units.

41. $\frac{1}{2} + 1 + 2 + 4 + 8 + 9 + 24\frac{1}{2} = 49$.

Two Birds.

42.

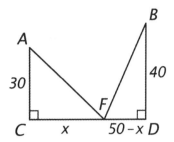

•43. AF = BF.

44. $50 - x$.

45. $30^2 + x^2 = 40^2 + (50 - x)^2$,
$900 + x^2 = 1{,}600 + 2{,}500 - 100x + x^2$,
$100x = 3{,}200$, $x = 32$.

46. 32 ft and 18 ft.

•47. About 44 ft. ($AF^2 = 30^2 + 32^2 = 1{,}924$,
$AF \approx 43.9$, $BF^2 = 40^2 + 18^2 = 1{,}924$, $BF \approx 43.9$.)

Road Systems.

48. 3 mi.

49.

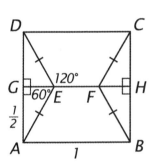

Approximately 2.7 mi.
In 30°-60° right △AGE, the longer leg

$AG = \dfrac{1}{2}$; so the shorter leg $GE = \dfrac{\frac{1}{2}}{\sqrt{3}} = \dfrac{1}{2\sqrt{3}}$

and the hypotenuse $AE = 2GE = \dfrac{1}{\sqrt{3}}$.

$EF = 1 - 2(\dfrac{1}{2\sqrt{3}}) = 1 - \dfrac{1}{\sqrt{3}}$. The sum of the

five segments is

$(1 - \dfrac{1}{\sqrt{3}}) + 4(\dfrac{1}{\sqrt{3}}) = 1 + \dfrac{3}{\sqrt{3}} = 1 + \sqrt{3} \approx 2.732$.

50. The second solution is better.

Sun and Moon.

51. In right △OMS, $\cos \angle O = \dfrac{m}{s}$. Because

$\cos 87° \approx 0.052 \approx \dfrac{1}{19}$, $\dfrac{m}{s} = \dfrac{1}{19}$.

52. Because the product of the means is equal
to the product of the extremes (or
multiplication), $s = 19m$.

•53. About 380 times as far. ($\cos 89.85° = \dfrac{m}{s}$; so

$\dfrac{s}{m} = \dfrac{1}{\cos 89.85°} \approx 382$.)

Eye Chart.

54. 20 ft = 240 in. $\tan P = \dfrac{3.75}{240} = 0.015625$,

$\angle P \approx 0.895°$; so $\angle P \approx 0.895° \times 60$ minutes per
degree ≈ 54 minutes.

•55. Approximately $\dfrac{1}{3}$ in.

[4.8 minutes = $(\dfrac{4.8}{60})° = 0.08°$.

$\tan 0.08° = \dfrac{x}{240}$, $x = 240 \tan 0.08° \approx 0.34$ in.]

Hot Air Balloon Altitude.

56. 75°. ($180° - 60° - 45° = 75°$.)

•57. About 600 ft. ($\dfrac{AC}{\sin 45°} = \dfrac{820}{\sin 75°}$,

$AC = \dfrac{820 \sin 45°}{\sin 75°} \approx 600$.)

58. About 520 ft. ($\sin 60° \approx \dfrac{CD}{600}$,
$CD \approx 600 \sin 60° \approx 520$.)

59. About 735 ft. ($\dfrac{BC}{\sin 60°} = \dfrac{820}{\sin 75°}$,

$BC = \dfrac{820 \sin 60°}{\sin 75°} \approx 735$.)

60. About 520 ft; so it checks. ($\sin 45° \approx \dfrac{CD}{735}$,
$CD \approx 735 \sin 45° \approx 520$.)

Outdoor Lighting.

61.

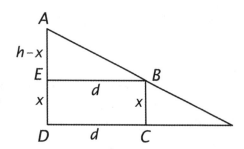

•62. $\tan \angle ABE = \dfrac{h-x}{d}$.

63. $x = h - (\tan \angle ABE)d$. ($d \tan \angle ABE = h - x$.)

64. They came from the equation for exercise 63 (and the fact that $\angle ABE = 90° - \angle A$.)

Number Mystery.

65. $\sqrt{75^2 - 45^2} = 60$

$\sqrt{97^2 - 65^2} = 72$

$\sqrt{481^2 - 319^2} = 360$

$\sqrt{769^2 - 541^2} = 546.516\ldots$

$\sqrt{1249^2 - 799^2} = 960$

$\sqrt{2929^2 - 1679^2} = 2400$

With one exception, each pair of numbers is part of a Pythagorean triple; specifically, the lengths of one leg and the hypotenuse.

66. As the list for exercise 65 shows, the mistake is in the pair of numbers 541, 769. (541 should be 481.)

Chapter 11, Algebra Review (page 482)

•1. $3x + 4y = 24$,
$4y = -3x + 24$,
$y = -\dfrac{3}{4}x + 6$.

•2. $m = -\dfrac{3}{4}$; $b = 6$.

•3.

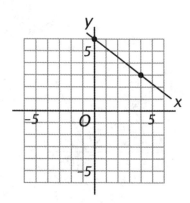

•4.

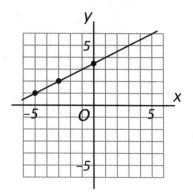

•5. $y - 1 = \dfrac{1}{2}(x - (-5))$,

$y - 1 = \dfrac{1}{2}(x + 5)$.

•6. $y = \dfrac{1}{2}x + \dfrac{5}{2} + 1$,

$y = \dfrac{1}{2}x + \dfrac{7}{2}$.

•7. $b = \dfrac{7}{2}$ (or 3.5).

•8.

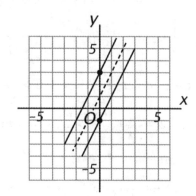

•9. They have the same slope, 2.

•10. $y = 2x + 1$.

• 11. $x + y = 5$,
    $y = -x + 5$;
    $5x - 5y = 15$,
    $-5y = -5x + 15$,
    $y = x - 3$.

• 12.

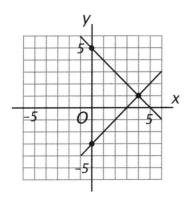

• 13. -1 and 1.

• 14. The product of their slopes is -1.

• 15. $-\dfrac{1}{3}$.

• 16. -6. $[4 - 4 = -\dfrac{1}{3}(x + 6), 0 = -\dfrac{1}{3}(x + 6),$
    $x + 6 = 0, x = -6.]$

• 17.

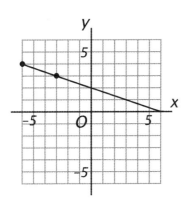

• 18. $y - 4 = -\dfrac{1}{3}(x + 6)$,
    $y = -\dfrac{1}{3}x - 2 + 4$,
    $y = -\dfrac{1}{3}x + 2$.

• 19. x-intercept, 6; y-intercept, 2.

• 20. $y = -\dfrac{1}{3}x + 2$,
    $\dfrac{1}{3}x + y = 2$,
    $x + 3y = 6$.

• 21.

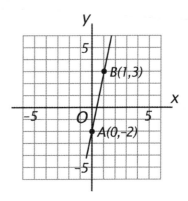

• 22. 5.

• 23. $y - 3 = 5(x - 1)$.

• 24. $y = 5x - 5 + 3$,
    $y = 5x - 2$.

• 25. $-5x + y = -2$,
    $5x - y = 2$.

• 26. $4x - 3y = 0$,
    $-3y = -4x$,
    $y = \dfrac{4}{3}x$ (or $y = \dfrac{4}{3}x + 0$).

• 27. $m = \dfrac{4}{3}$; $b = 0$.

• 28.

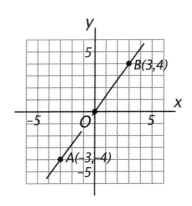

• 29. 0.

• 30. AB $= \sqrt{(3 - (-3))^2 + (4 - (-4))^2} = \sqrt{6^2 + 8^2} = \sqrt{100} = 10$.

Set I (pages 486–487)

According to Matthys Levy and Mario Salvadori in their book titled *Why Buildings Fall Down* (Norton, 1992), the Romans built semicircular arches of the type illustrated in exercises 1 through 5 that spanned as much as 100 feet. These arches were used in the bridges of 50,000 miles of roads that the Romans built all the way from Baghdad to London!

The conclusions of exercises 13 through 15 are supported by the fact that a diameter of a circle is its longest chord and by symmetry considerations. As the second chord rotates away from the diameter in either direction about the point of intersection, it becomes progressively shorter.

Roman Arch.

1. Radii.

•2. Chords.

•3. A diameter.

4. Isosceles. Each has two equal sides because the radii of a circle are equal.

5. They are congruent by SSS.

Old Figure.

•6. Five.

7. One.

8. Four.

•9. Six.

Butterfly Spots.

•10. Concentric.

11. It means that a set of circles lie in the same plane and have the same center.

12. *(Student answer.)* (The spots look like big eyes; so they might scare some predators away.)

Chord Problem.

13.

14. They are perpendicular.

15. The longest chord is always a diameter; so its length doesn't depend on the location of the point. The length of the shortest chord depends on how far away from the center the point is; the greater the distance, the shorter the chord.

Theorem 56.

16. Perpendicular lines form right angles.

•17. Two points determine a line.

•18. All radii of a circle are equal.

19. Reflexive.

20. HL.

21. Corresponding parts of congruent triangles are equal.

•22. If a line divides a line segment into two equal parts, it bisects the segment.

Theorem 57.

23. If a line bisects a line segment, it divides the segment into two equal parts.

24. Two points determine a line.

25. All radii of a circle are equal.

•26. In a plane, two points each equidistant from the endpoints of a line segment determine the perpendicular bisector of the line segment.

Theorem 58.

•27. If a line through the center of a circle bisects a chord that is not a diameter, it is also perpendicular to the chord.

•28. Through a point on a line, there is exactly one line perpendicular to the line.

29. CD contains O.

Set II (pages 488–490)

The SAT problem of exercises 30 and 31 is something of an oddity in that it is a rather well known "trick question." Martin Gardner included it in his "Mathematical Games" column in the November 1957 issue of *Scientific American*—many years before it appeared on the SAT!

The CBS eye, a creation of designer William Golden, first appeared on television in 1951. Hal

Morgan observes in his book titled *Symbols of America* (Viking, 1986) that, "because it is essentially a picture of an eye, it has proved a challenge to the police efforts of the CBS trademark attorneys. As one company executive admitted, 'Bill Golden, after all, didn't invent the eye. God thought of it first.'"

The story of the use of Jupiter to determine the speed of light is told by William Dunham in *The Mathematical Universe* (Wiley, 1994). After the invention of the telescope in the seventeenth century and the discovery of the moons of Jupiter, astronomers were puzzled by the delay in the eclipses of the moon Io when the earth and Jupiter were farthest apart. Ole Roemer reasoned that the time discrepancy was due to the extra distance that the light from Jupiter and its moon had to travel to get to the earth. Surprisingly, it was not the speed of light that especially interested Roemer but rather the conclusion that light does not travel instantaneously from one place to another, something that had not been known previously.

The definition of the roundness of a grain of sand was devised by Haakon Wadell in about 1940. He defined it to be (for cross sections of the grain) the "ratio of average radius of curvature of the corners to radius of maximum inscribed circle." From this definition, it follows that roundness can vary from 0 to 1, with perfect roundness being 1.

Exercises 43 through 47 are based on a simple theorem discovered in 1916 by the American geometer Roger Johnson: "If three circles having the same radius pass through a point, the circle through their other three points of intersection also has the same radius." The appearance of the cube in the explanation and the role played by two of its opposite vertices is quite unexpected. For more information on this elegant theorem, see chapter 10 of George Polya's *Mathematical Discovery*, vol. 2 (Wiley, 1965), and the fourth of the "Four Minor Gems from Geometry" in Ross Honsberger's *Mathematical Gems II* (Mathematical Association of America, 1976).

SAT Problem.

30. 10.

31. The diagonals of a rectangle are equal. The other diagonal of the rectangle is a radius of the circle.

CBS Eye.

32. *Example answer:*

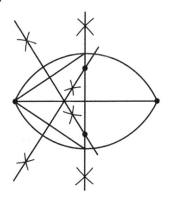

33. The perpendicular bisector of a chord of a circle contains the center of the circle.

Red-Spot Puzzle.

•34. An isosceles right triangle.

•35. About 46 mm. ($\frac{65}{\sqrt{2}} \approx 45.96$.)

36. Very difficult because the disks are barely big enough to cover the spot.

Speed of Light.

37. 186,000,000 mi. (2 × 93,000,000.)

•38. 1,000 seconds. (16 × 60 + 40.)

39. 186,000 miles per second. ($\frac{186,000,000}{1,000}$.)

Roundness.

•40. $\frac{10}{27}$, or about 0.37. [$\frac{2+3+5}{3(9)}$.]

41. $\frac{20}{54} = \frac{10}{27}$, or about 0.37. [$\frac{4+6+10}{3(18)}$.]

42. No. The figure would be larger but still have the same shape.

Three Circles.

43.

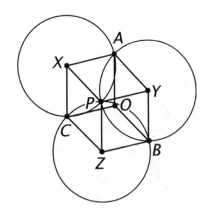

44. They are each 1 unit.

•45. They are rhombuses because all of their sides are equal.

46. They are equal. The opposite sides of a parallelogram are equal and the other segments all equal 1 unit.

47. The circle would have its center at O because OA = OB = OC = 1.

Steam Engine.

48.

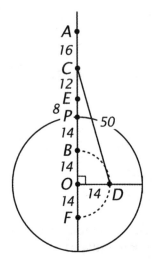

49. EF = 50 in. [EP + PF = 8 + 3(14).]

50. CD = 50 in. (CD = EF.)

51. CO = 48 in. (CO² + 14² = 50²; so
$CO = \sqrt{50^2 - 14^2} = \sqrt{2,304} = 48.$)

52. CE = 12 in. (CE = CO − EO = 48 − 36.)

53. AC = 16 in. (AC = AB − CB = 50 − 34.)

Set III (page 490)

When this puzzle appeared in Martin Gardner's "Mathematical Games" column in *Scientific American*, William B. Friedman suggested tying the rope to two places high on the tree at B so that the person could walk along the lower part while holding onto the upper part and not get wet.

Island Puzzle.

Tie one end of the rope to the tree at B. Holding the other end of the rope, walk around the lake so that the rope wraps around the tree at A. Pull the rope tight and tie it at the other end to the tree at B. Hold onto the rope to get to the island.

Chapter 12, Lesson 2

Set I (pages 493–494)

Exercises 15 through 18 reveal how the tangent ratio got its name. As Howard Eves remarks in *An Introduction to the History of Mathematics* (Saunders, 1990), "the meanings of the present names of the trigonometric functions, with the exception of sine, are clear from their geometrical interpretations when the angle is placed at the center of a circle of unit radius. . . . The functions tangent, cotangent, secant, and cosecant have been known by various other names, but these particular names appeared as late as the end of the 16th century."

Band Saw.

1. If a line is tangent to a circle, it is perpendicular to the radius drawn to the point of contact.

•2. In a plane, two lines perpendicular to a third line are parallel to each other.

3. The opposite sides of a parallelogram are equal.

Concentric Circles.

•4. OB ⊥ AC.

•5. If a line through the center of a circle bisects a chord that is not a diameter, it is also perpendicular to the chord.

6. AC is tangent to the smaller circle.

7. If a line is perpendicular to a radius at its outer endpoint, it is tangent to the circle.

8. 5.

•9. $\dfrac{4}{3}$.

•10. $-\dfrac{3}{4}$.

11. $-\dfrac{3}{4}$.

12. Lines *l* and *m* are perpendicular to line AB.

13. Lines *l* and *m* are parallel.

14. The slopes of *l* and *m* are equal; the slope of AB is the opposite of the reciprocal of their slope.

Double Meaning.

•15. AB is tangent to circle O.

•16. If a line is perpendicular to a radius at its outer endpoint, it is tangent to the circle.

17. The tangent of an acute angle of a right triangle is the ratio of the length of the opposite leg to the length of the adjacent leg.

•18. $\dfrac{AB}{1}$, or AB.

Railroad Curve.

19.

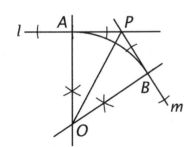

20. △APO ≅ △BPO by HL (they are right triangles with PO = PO and PA = PB).

21. Corresponding parts of congruent triangles are equal.

22. If a line is perpendicular to a radius at its outer endpoint, it is tangent to the circle.

23. No.

Tangents to Two Circles.

24.

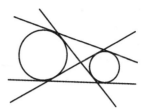

25.

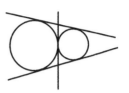

26. *Example figure:*

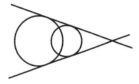

27. *Example figure:*

28. Yes. No lines can be drawn tangent to both circles if one circle is inside the other and the circles have no point in common.

Set II (pages 494–496)

Euclid's Book III, Proposition 17 is the following construction: "From a given point to draw a straight line touching a given circle." A construction based on the fact that an angle inscribed in a semicircle is a right angle is introduced in Lesson 6. In *Euclid—The Creation of Mathematics* (Springer, 1999), Benno Artmann says of Euclid's construction: "The unexpected way in which the tangent *l*, which at first sight has nothing to do with the problem, is used to produce the wanted tangent AB is a nice surprise and makes the construction very elegant."

Although the angle trisector called the "tomahawk" has been included in many twentieth-century textbooks, no one knows who invented it. The tomahawk is one of many methods for solving the trisection problem by going beyond the restrictions of classical constructions. According to Robert C. Yates in *The Trisection Problem* (Franklin Press, 1942; N.C.T.M., 1971), the tomahawk was mentioned in *Geométrie Appliquée a l'Industrie* published in 1835.

J. L. Heilbron in his *Geometry Civilized— History, Culture, and Technique* (Clarendon Press, 1998) writes that the Gothic arch of exercises 39 through 45 can be seen everywhere in churches built in the later middle ages and that it was often filled with other geometric shapes such as circles, squares, and smaller arches. The exercises reveal the surprising appearance of the 3-4-5 right triangle in the construction of an inscribed circle in a Gothic arch.

Construction Problem.

29.

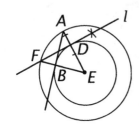

30. If a line is perpendicular to a radius at its outer endpoint, it is tangent to the circle.

31. SAS. (EF = EA, ∠E = ∠E, and ED = EB.)

•32. Because ∠FDE is a right angle and corresponding parts of congruent triangles are equal.

33. It is tangent to the smaller circle.

•34. Line AB.

Tomahawk.

35. SAS. (AB = BC, ∠ABY = ∠CBY, and BY = BY.)

•36. If a line is tangent to a circle, it is perpendicular to the radius drawn to the point of contact.

37. HL. (ΔCBY and ΔCPY are right triangles with CY = CY and BC = CD.)

38. ∠AYB = ∠BYC and ∠BYC = ∠CYP because corresponding parts of congruent triangles are equal. Because ∠XYZ has been divided into three equal angles, it is trisected.

Gothic Window.

39.

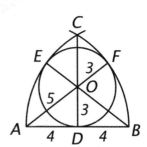

40. In a plane, two points each equidistant from the endpoints of a line segment determine the perpendicular bisector of the line segment.

•41. AD = 4 cm and AF = AB = 8 cm.

•42. $(8 - r)$ cm. (AO = AF − OF = 8 − r.)

43. $(8 - r)^2 = 4^2 + r^2$ by the Pythagorean Theorem. $64 - 16r + r^2 = 16 + r^2$, $16r = 48$, $r = 3$.

•44. A 3-4-5 right triangle.

45. $4\sqrt{3} \approx 6.9$ cm. [If AC and BC were drawn, CD would be an altitude of equilateral ΔABC; so CD = $\frac{\sqrt{3}}{2}$(8).]

Tangent Puzzle.

46. DC is perpendicular to AD and BC. If a line is tangent to a circle, it is perpendicular to the radius drawn to the point of contact.

47. They are parallel. In a plane, two lines perpendicular to a third line are parallel to each other.

•48. A parallelogram (also, a rectangle).

49. A right triangle.

50. By the Pythagorean Theorem: In a right triangle, the square of the hypotenuse is equal to the sum of the squares of the legs.

51. AB = $y + z$, AE = $y − z$, EB = x.

52. $(y + z)^2 = (y − z)^2 + x^2$.

53. $y^2 + 2yz + z^2 = y^2 − 2yz + z^2 + x^2$, $4yz = x^2$, $x^2 = 4yz$.

Set III (page 496)

On a trip from the earth to the moon, it is obvious that at the halfway point the earth would look bigger because it *is* bigger. What makes the earth and moon appear to have the same size as seen from point P is that the vertical angles formed by their common internal tangents are equal. When we observe an object in space, it is this angle, called its "angular size," rather than its true size, by which we judge it.

Trip to the Moon.

EP is approximately 184,000 miles. $\triangle AEP \sim \triangle BMP$ by AA ($\angle A$ and $\angle B$ are equal right angles and $\angle APE$ and $\angle BPM$ are equal vertical angles).
$\dfrac{EP}{MP} = \dfrac{EA}{MB}$ (corresponding sides of similar triangles are proportional). Letting EP = x,
$\dfrac{x}{234,000 - x} = \dfrac{3,960}{1,080}$, $1,080x = 926,640,000 - 3,960x$,
$5,040x = 926,640,000$, $x \approx 183,857$.

Chapter 12, Lesson 3

Set I (pages 500–501)

In a section of Chapter 2 of *Geometry Civilized—History, Culture, and Technique* (Clarendon Press, 1998), J. L. Heilbron treats the subject of the size of the earth. Columbus used a number for the circumference of the earth that was about 75% of what it actually is. Heilbron says that this mistake "had more profound consequences than any other geometrical error ever committed." To try to convince others of the practicality of his plan, Columbus took several steps to minimize the length of the trip. Fortunately for him, he encountered the New World almost exactly where he had expected to find Japan.

The "three-body problem," to understand the mutual effects of gravity on three objects in space, frustrated scientists from the time of Newton to the end of the nineteenth century, when it was proved that mathematical solutions of the type that they had been seeking do not exist. The only situation for which there is an exact and stable solution is the case in which the three objects lie at the vertices of an equilateral triangle: this case exists in our solar system with the two equilateral triangles formed by the sun, Jupiter, and the two sets of asteroids called the Trojans. It has been suggested that this arrangement might also be appropriate for establishing the locations of the earth's first colonies in space. The two triangles would be formed by the earth, the moon, and the two colonies.

Routes to Japan.

- •1. Central angles.
- •2. $\overset{\frown}{PJ}$.
- •3. $\overset{\frown}{PCJ}$.
- 4. 30°.
- •5. 283°. (253 + 30.)
- 6. 77°. (360 − 283.)
- •7. 77°.

Theorem 62.

- •8. Two points determine a line.
- 9. All radii of a circle are equal.
- •10. An arc is equal in measure to its central angle.
- 11. Substitution.
- 12. SAS.
- 13. Corresponding parts of congruent triangles are equal.

Jupiter's Asteroids.

14.

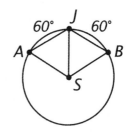

15. An arc is equal in measure to its central angle.

- •16. The Arc Addition Postulate.
- 17. SAS. (SA = SJ = SB and $\angle ASJ = \angle BSJ$.)
- 18. Equilateral.

Arc Illusion.

- •19. A semicircle.
- 20. Minor.
- 21. *(Student answer.)* (Many people assume that, because EF seems straightest, it comes from the largest circle.)
- 22. The three arcs seem to be part of congruent circles.

Chords and Arcs.

23. The Arc Addition Postulate.

•24. No. The sum of any two sides of a triangle is greater than the third side.

25.

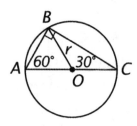

26. $m\widehat{AB} = 60°$, $m\widehat{BC} = 120°$, $m\widehat{ABC} = 180°$.

•27. $AB = r$, $BC = r\sqrt{3}$, $AC = 2r$. ($\triangle ABC$ is a 30°-60° right triangle.)

28. $m\widehat{ABC} = 3m\widehat{AB}$.

29. $BC = \sqrt{3}\,AB$.

Set II (pages 501–503)

J. L. Heilbron observes that ovals formed from pairs of equal interlaced or tangent circles were commonly used in designing plazas in sixteenth-century Italy. St. Peter's Square in Rome is the most famous example.

   Olympiodorus taught in Alexandria in the sixth century A.D. He wrote extensively on the rainbow, even attempting, as outlined in exercises 37 through 39, to determine the distance from the observer's eye to the center of the rainbow when the sun was on the horizon. The methods fails because, if SO were equal to OR, the rainbow would be well outside the atmosphere, which it is not.

The Square That Isn't.

30.

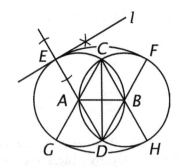

•31. $\triangle ABC$ and $\triangle ABD$ are equilateral and congruent.

32. A rhombus (and hence a parallelogram).

•33. Line *l* is tangent to circle A. If a line is perpendicular to a radius at its outer endpoint, it is tangent to the circle.

34. Line *l* is also perpendicular to radius DE of the circle D of which $\widehat{EF}$ is an arc.

35. $m\widehat{GE} = 120°$, $m\widehat{EF} = 60°$, $m\widehat{FH} = 120°$, $m\widehat{HG} = 60°$.

36. 360°.

Rainbow Distance.

37. $x^2 + y^2 = 1$.

38. $(x + 1)^2 + y^2 = z^2$.

39. From the first equation, $y^2 = 1 - x^2$ and, from the second equation, $y^2 = z^2 - (x + 1)^2$.
   So $1 - x^2 = z^2 - (x + 1)^2$,
   $1 - x^2 = z^2 - (x^2 + 2x + 1)$,
   $1 - x^2 = z^2 - x^2 - 2x - 1$,
   $2x = z^2 - 2$,
   $$x = \frac{z^2 - 2}{2}.$$

Night Sky.

40. The photograph is a time exposure. The stars seem to travel in circles about a point directly above the earth's axis because of the earth's rotation on its axis.

•41. All of the arcs have the same measure. (They all have equal central angles—namely, the angle through which the earth rotated during the time exposure.)

42. *(Student answer.)* (The earth rotates 360° in 24 hours, or 15° in one hour. The central angles of the arcs measure approximately 15°, so a guess of about one hour is reasonable.)

Original Proofs.

43. *Proof.*
   (1) Vertical angles ∠AOB and ∠COD are central angles of both circles. (Given.)
   (2) ∠AOB = ∠COD. (Vertical angles are equal.)
   (3) $m\widehat{AB}$ = ∠AOB and $m\widehat{CD}$ = ∠COD. (An arc is equal in measure to its central angle.)
   (4) $m\widehat{AB} = m\widehat{CD}$. (Substitution.)

44. *Proof.*
    (1) The vertices of trapezoid ABCD (AD ∥ BC) lie on a circle and $m\widehat{AB} = m\widehat{CD}$. (Given.)
    (2) AB = CD. (In a circle, equal arcs have equal chords.)
    (3) ABCD is isosceles. (A trapezoid is isosceles if its legs are equal.)
    (4) ∠B = ∠C. (The base angles of an isosceles trapezoid are equal.)

45. *Proof.*
    (1) AB = DE. (Given.)
    (2) $m\widehat{AB} = m\widehat{DE}$. (In a circle, equal chords have equal arcs.)
    (3) $m\widehat{AB} + m\widehat{BD} = m\widehat{BD} + m\widehat{DE}$. (Addition.)
    (4) $m\widehat{AB} + m\widehat{BD} = m\widehat{AD}$ and $m\widehat{BD} + m\widehat{DE} = m\widehat{BE}$. (Arc Addition Postulate.)
    (5) $m\widehat{AD} = m\widehat{BE}$. (Substitution.)
    (6) AD = BE. (In a circle, equal arcs have equal chords.)

46. *Proof.*
    (1) $m\widehat{AC} = m\widehat{AD}$ and $m\widehat{CB} = m\widehat{BD}$. (Given.)
    (2) AC = AD and CB = BD. (In a circle, equal arcs have equal chords.)
    (3) AB is the perpendicular bisector of CD. (In a plane, two points each equidistant from the endpoints of a line segment determine the perpendicular bisector of the line segment.)
    (4) AB contains the center of the circle. (The perpendicular bisector of a chord of a circle contains the center of the circle.)

47. *Proof.*
    (1) AD is a diameter of circle O, and AB ∥ OC. (Given.)
    (2) ∠A = ∠2. (Parallel lines form equal corresponding angles.)
    (3) ∠B = ∠1. (Parallel lines form equal alternate interior angles.)
    (4) OA = OB. (All radii of a circle are equal.)
    (5) ∠B = ∠A. (If two sides of a triangle are equal, the angles opposite them are equal.)
    (6) ∠1 = ∠2. (Substitution.)
    (7) $m\widehat{BC} = ∠1$ and $m\widehat{CD} = ∠2$. (An arc is equal in measure to its central angle.)
    (8) $m\widehat{BC} = m\widehat{CD}$. (Substitution.)

Set III (page 503)

Obtuse Ollie's curve is an example of a curve of constant width. Such curves can be constructed by starting with any number of mutually intersecting lines. The radius of the initial arc is arbitrary; each arc is centered on the point of intersection of the two lines that bound it. Interested students may enjoy proving that such a curve actually closes and that it has a constant width. (The "constant width" in regard to Ollie's curve is DG = EH = FI.) The proofs are based on the fact that all radii of a circle are equal and some simple algebra. Martin Gardner discussed curves of constant width in his "Mathematical Games" column in *Scientific American* (February 1963). This column is included in *The Unexpected Hanging and Other Mathematical Diversions* (Simon & Schuster, 1969).

Ollie is correct. The measures of the six arcs are equal to the measures of their central angles. Three of the central angles are the angles of ΔABC; so their sum is 180°. The other three central angles are vertical to the three angles of ΔABC; so their sum also is 180°.

Chapter 12, Lesson 4

Set I (pages 506–507)

Skeet shooting is a sport in which clay targets are thrown into the air from two locations, a "high house" on the left and a "low house" on the right. The shooters advance from station to station, starting at A and ending at H. Although shooting was included in the first modern Olympics in 1896, skeet shooting did not become an Olympic event until 1968.

Skeet Range.

1. 16.5°. (In ΔAOG, OA = OG; so ∠G = ∠A.)

•2. 147°. (In ΔAOG, ∠AOG = 180 − ∠A − ∠G = 180 − 16.5 − 16.5 = 147.)

•3. 24.5°. ($m\widehat{AG} = ∠AOG$; $\frac{147}{6} = 24.5$.)

•4. 33°. [$m\widehat{GX} = 180 − m\widehat{AG} = 180 − 147 = 33$, or $∠A = \frac{1}{2}m\widehat{GX}$, $m\widehat{GX} = 2∠A = 2(16.5) = 33$.]

5. 147°. ($m\widehat{YX} = ∠YOX = ∠AOG = 147$.)

Theorem 63.

6. Two points determine a line.

•7. All radii of a circle are equal.

8. If two sides of a triangle are equal, the angles opposite them are equal.

•9. An exterior angle of a triangle is equal to the sum of the remote interior angles.

10. Substitution.

11. Division (or multiplication).

•12. A central angle is equal in measure to its intercepted arc.

13. Substitution.

14. Two points determine a line.

•15. Betweenness of Rays Theorem.

16. Substitution.

•17. Arc Addition Postulate.

18. Substitution.

19. The center of the circle lies outside the angle. (The proof is similar to case 2 but uses subtraction.)

Corollary 1.

•20. An inscribed angle is equal in measure to half its intercepted arc.

21. Substitution.

Corollary 2.

•22. The measure of a semicircle is 180°.

23. An inscribed angle is equal in measure to half its intercepted arc.

24. Substitution.

•25. A 90° angle is a right angle.

Inscribed Triangle.

26.

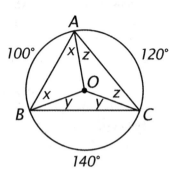

27. Central angles.

•28. Inscribed angles.

29. $\angle AOB = 100°$, $\angle BOC = 140°$, $\angle COA = 120°$.

•30. $x = 40°$, $y = 20°$, and $z = 30°$.

31. $\angle ABC = 60°$, $\angle BCA = 50°$, $\angle CAB = 70°$.
($\angle ABC = x + y = 40° + 20° = 60°$,
$\angle BCA = y + z = 20° + 30° = 50°$,
$\angle CAB = x + z = 40° + 30° = 70°$.)

32. $\angle ABC = \frac{1}{2}m\widehat{AC} = \frac{1}{2}120° = 60°$;

$\angle BCA = \frac{1}{2}m\widehat{BA} = \frac{1}{2}100° = 50°$;

$\angle CAB = \frac{1}{2}m\widehat{CB} = \frac{1}{2}140° = 70°$.

Set II (pages 508–509)

The figure for exercises 33 and 34 appears in a book on the number π because early methods for determining π were based on approximating the circumference of a circle with the perimeters of regular polygons. In the figure, 40 equal chords form a regular polygon having 40 sides that looks very much like a circle. More on this subject will be considered in Chapter 14.

Exercises 35 through 41 establish the theorem that an angle formed by a tangent and a chord is equal in measure to half its intercepted arc (Book III, Proposition 32, of Euclid's *Elements*).

Exercises 43 through 47 show that, if two similar triangles are inscribed in a circle, they must also be congruent.

The method for constructing the square root of a number by inscribing a right triangle in a semicircle appears in the *Elements* as Proposition 13 of Book VI, stated as follows: "To two given straight lines to find a mean proportional."

Exercises 55 through 58 are connected to the following problem: Find the longest line segment that can be drawn through one of the intersection points of two intersecting circles such that its endpoints lie on the circles. This problem is discussed by Mogens Esrom Larsen in an article in *The Lighter Side of Mathematics*, Richard K. Guy and Robert E. Woodrow, editors (Mathematical Association of America, 1994). Larsen appropriately refers to it as a good example illustrating the role of insight in problem solving, "the experience of seeing right through the problem as can happen in geometry."

Forty Angles.

•33. $9°$. ($\frac{360}{40}$.)

34. $63°$. [$\frac{1}{2}(14 \times 9)$.]

Tangents, Chords, and Intercepted Arcs.

35. $\angle$BAT is a right angle. If a line is tangent to a circle, it is perpendicular to the radius drawn to the point of contact.

•36. They are complementary.

•37. It is a right angle. An angle inscribed in a semicircle is a right angle.

38. They are complementary. The acute angles of a right triangle are complementary.

39. They are equal. Complements of the same angle are equal.

40. An inscribed angle is equal in measure to half its intercepted arc.

41. Substitution.

42. They prove that the measure of an angle formed by a tangent and a chord is equal to half the measure of its intercepted arc.

Similar Inscribed Triangles.

•43. $n°$. ($\triangle$ABC ~ $\triangle$DEF; so $\angle$F = $\angle$C.)

•44. $2n°$.

45. $2n°$.

46. AB = DE. In a circle, equal arcs have equal chords.

47. $\triangle$ABC $\cong$ $\triangle$DEF by ASA. ($\angle$A = $\angle$D and $\angle$B = $\angle$E because corresponding angles of similar triangles are equal and AB = DE.)

Square-Root Construction.

48.

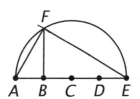

49. $\angle$AFE is a right angle because an angle inscribed in a semicircle is a right angle.

•50. It is an altitude (to the hypotenuse) of the triangle.

51. $\frac{AB}{BF} = \frac{BF}{BE}$.

•52. BF = $\sqrt{3}$ in. ($\frac{1}{BF} = \frac{BF}{3}$, $BF^2 = 3$, BF = $\sqrt{3}$.)

53. 1.73 in.

54. Yes. (BF $\approx 1\frac{3}{4}$ = 1.75 in.)

Intersecting Triangles and Circles.

55.

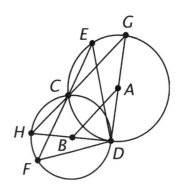

56. $\triangle$DEF ~ $\triangle$DGH by AA. ($\angle$E = $\angle$G and $\angle$F = $\angle$H because inscribed angles that intercept the same arc are equal.)

57. $\triangle$DGH is larger because diameter DG > DE and diameter DH > DF.

58. AB = $\frac{1}{2}$GH and AB $\parallel$ GH. AB is a midsegment of $\triangle$DGH, and a midsegment of a triangle is parallel to the third side and half as long.

Finding the Center.

Example figure:

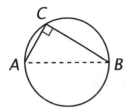

The carpenter should place the corner of the right angle of the set square on the edge of the wheel and mark the two points in which its sides intersect the wheel's edge. These two points must be the endpoints of a diameter of the wheel. (Because the right angle is inscribed in the circle, its measure is half its intercepted arc. $90° = \frac{1}{2}180°$; so the arc is a semicircle and its chord is a diameter.) Doing this a second time gives a second diameter and the diameters intersect in the center of the wheel.

(Alternatively, the carpenter could place the wheel *inside* the angle of the set square and mark successive quarter circles.)

Chapter 12, Lesson 5

Set I (pages 512–513)

A detailed discussion of danger angles is included by Elbert S. Maloney in *Dutton's Navigation and Piloting* (Naval Institute Press, 1985). A danger circle (not shown in the figure for exercises 17 through 21) is constructed with the middle point of the danger as its center and the distance from that center at which it is desired to pass as radius. A second circle is then constructed through points A and B tangent to the offshore side of the danger circle. Any inscribed angle such as ∠D that intercepts AB in this second circle provides the measure of the danger angle. The navigator of a passing ship keeps track of the measure of ∠S, knowing that the ship is safe if ∠S < ∠D and in danger if ∠S > ∠D.

Exercises 27 through 38 suggest that the sum of the corner angles of a symmetric star with an odd number of sides is 180°. The proof that this is true for any five-pointed star, no matter how irregular, was considered in the Set III exercise of Chapter 6, Lesson 5.

Theorem 64.

1. Two points determine a line.

•2. An exterior angle of a triangle is equal to the sum of the remote interior angles.

•3. An inscribed angle is equal in measure to half its intercepted arc.

4. Substitution.

•5. 36. ($x + 3x + 2x + 4x = 360$, $10x = 360$, $x = 36$.)

6. $m\overset{\frown}{AC} = 36°$, $m\overset{\frown}{CB} = 108°$, $m\overset{\frown}{BD} = 72°$, $m\overset{\frown}{DA} = 144°$.

•7. 54° and 126°. [$\frac{1}{2}(36 + 72) = \frac{1}{2}(108) = 54$; $180 - 54 = 126$.]

Theorem 65.

8. Two points determine a line.

9. An exterior angle of a triangle is equal to the sum of the remote interior angles.

•10. Subtraction.

11. An inscribed angle is equal in measure to half its intercepted arc.

12. Substitution.

13. 30°. ($360 - 100 - 110 - 120 = 30$.)

•14. 40°. [$\angle A = \frac{1}{2}(m\overset{\frown}{BC} - m\overset{\frown}{DE}) = \frac{1}{2}(110 - 30) = \frac{1}{2}(80) = 40$.]

•15. 75°. [$\angle B = \frac{1}{2}m\overset{\frown}{DEC} = \frac{1}{2}(30 + 120) = \frac{1}{2}(150) = 75$.]

16. 65°.

Danger Angle.

17. An inscribed angle.

•18. 55°.

•19. A secant angle.

•20. It is less than 55°. [$\angle S = \frac{1}{2}(m\overset{\frown}{AB} - $ a positive quantity); $\angle D = \frac{1}{2}m\overset{\frown}{AB}$.]

21. It would be more than 55°.

Perpendicular Chords.

22. ∠AED is a right angle (or, ∠AED = 90°).

•23. $m\widehat{AD} + m\widehat{CB} = 180°$.

$[AED = \frac{1}{2}(m\widehat{AD} + m\widehat{CB}).]$

24. $m\widehat{AC} + m\widehat{DB} = 180°$.

25. 70°. (180 − 110.)

26. 100°. (180 − 80.)

Star Problem 1.

•27. 72°. ($\frac{360}{5}$.)

28. 144°. (2 × 72.)

•29. 36°. [$\frac{1}{2}(144 − 72)$.]

30. 216°. (3 × 72.)

31. 108°. ($\frac{1}{2}216$.)

32. 180°. (5 × 36.)

Star Problem 2.

33. 40°. ($\frac{360}{9}$.)

•34. 20°. ($\frac{1}{2}40$.)

35. 40°. [$\frac{1}{2}(40 + 40)$.]

36. 60°. [$\frac{1}{2}(80 + 40)$.]

37. 100°. [$\frac{1}{2}(80 + 120)$.]

38. 180°. (9 × 20.)

Set II (pages 514–515)

The angle of view of a camera lens determines the width of the scene in front of the camera that will fit in the frame of the film. It is inversely related to the focal length of the lens. Wide-angle lenses consequently have shorter-than-normal focal lengths, whereas the focal lengths of telephoto lenses with very small angles of view are longer than normal.

The model of the motion of the sun created by the second-century Greek astronomer Hipparchus and used by Ptolemy is described in *The Cambridge Illustrated History of Astronomy*, Michael Hoskin, editor (Cambridge University Press, 1997): "Hipparchus's elegant and simple model of the motion of the sun worked well enough for Ptolemy to retain it unaltered, and so it has been handed down to us. In the model, the sun orbited the earth on a circle, moving about the center of the circle with a uniform angular speed that took it round in 365 1/4 days. But Hipparchus knew that the seasons are of different lengths: in particular, as seen from earth, the sun moved the 90° from spring equinox to summer solstice in 94 1/2 days, and the 90° from summer solstice to autumn equinox in 92 1/2 days. Since both these intervals were longer than one-quarter of 365 1/4 days, the sun appeared to move across the sky with a speed that was not uniform but varied.

The earth therefore could not be located at the center of the circle, but must be 'eccentric'. By how far, and in which direction? To generate these longer intervals, the earth had to be displaced from the center of the circle in the opposite direction, so that the corresponding 90° arcs seen from earth were each more than one-quarter of the circle, and it therefore took the sun intervals of more than one-quarter of 365 1/4 days to traverse them. Hipparchus's calculations showed that the eccentricity needed to be 1/24th of the radius of the circle, and that the line from earth to center had to make an angle of 65.5° with the spring equinox.

Granted these parameters, the model was completely defined. Luckily it reproduced the remaining two seasons well enough, and so this single eccentric circle was sufficient to reproduce the motion of the sun."

The figure for exercises 56 through 59 is related to "The Tantalus Problem" introduced in the Set III exercise of Chapter 3, Lesson 3. The original problem was, given isosceles ΔABC with ∠A = 20°, ∠DBC = 60°, and ∠HCB = 50°, to find the measures of ∠DHC and ∠HDB. Without the circle, the problem is notorious for being extremely difficult. With the circle, it is easily solved. For more on this, see pages 123–125 and pages 227–229 of *Tracking the Automatic Ant,* by David Gale (Springer, 1998), and pages 292–295 of *Geometry Civilized,* by J. L. Heilbron (Clarendon Press, 1998).

Angle of View.

39. 100°. ($\angle ACB = \frac{1}{2}m\widehat{AB}$.)

•40. If a line through the center of a circle bisects a chord that is not a diameter, it is also perpendicular to the chord.

41. SAS.

•42. 17 ft. ($\tan 25° = \frac{8}{CD}$, CD tan 25° = 8,

$CD = \frac{8}{\tan 25°} \approx 17$.)

43. To other points on the major arc AB.

44. The photograph could not include the people at both ends of the row, because ∠ACB would become larger than the camera's angle of view.

The Seasons.

•45. Each angle is 90°.
[$\angle AEB = \frac{1}{2}(93.14° + 86.86°) = \frac{1}{2}(180°) = 90°$,
etc.]

46. No. If E were the center of the circle, the angles at E would be central angles and be equal in measure to their intercepted arcs.

47. Spring.

48• About 92.5 days. ($\frac{91.17}{360} \times 365.25 \approx 92.5$.)

Place Kicker.

49.

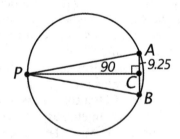

50. C is the midpoint of AB. If a line through the center of a circle is perpendicular to a chord, it also bisects the chord.

•51. 6°. ($\tan \angle APC = \frac{9.25}{90} \approx 0.103$, ∠APC ≈ 5.9°.)

52. 12°. (2 × 6.)

53. 24°. ($\angle APB = \frac{1}{2}m\widehat{AB}$.)

54. At other positions on the circle.

55. It would get smaller.

Angles Problem.

56. 20°. ($\frac{360}{18}$.)

57.

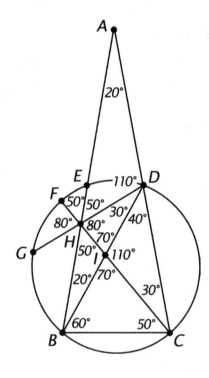

58. ΔABC, ΔABD, and ΔBCH.

59. ΔADH and ΔBIH, or ΔDHC and ΔDHI, or ΔBDH and ΔAHC.

Set III (page 515)

This problem was first posed by the fifteenth-century mathematician known as Regiomontanus in the following form: "At what point on the ground does a perpendicularly suspended rod appear largest (that is, subtend the greatest visual angle)?" Eli Maor, who includes the story of Regiomontanus as well as a discussion of the problem in his book titled *Trigonometric Delights* (Princeton University Press, 1998), notes that "it has been claimed that this was the first extreme problem in the history of mathematics since antiquity."

As the king walks toward the statue, the angle becomes larger until he gets to point B. Beyond this point, the angle becomes smaller again.

The angle is at its largest at B because it is an inscribed angle that intercepts just the arc of the circle corresponding to the statue. All other viewing angles are secant angles whose vertices are outside the circle; therefore, each is measured by half the difference of the statue arc and another arc of the circle.

Chapter 12, Lesson 6

Set I (pages 517–519)

Exercises 18 through 21 demonstrate the method usually taught for constructing the tangents to a circle from an external point. Euclid's method for doing so (Book III, Proposition 17) was considered in exercises 29 through 33 of Lesson 2.

The radius of a circle inscribed in a triangle is not ordinarily determined by the lengths of the triangle's sides. For a right triangle, however,

$r = \dfrac{1}{2}(a + b - c)$, in which a and b are the lengths

of the legs and c is the length of the hypotenuse. A special case is considered in exercises 22 through 25.

The fact that the sums of the lengths of the opposite sides of a quadrilateral circumscribed about a circle are equal is revealed in exercises 26 and 27.

Korean Proof.

1. That ∠PTO and ∠PT'O are (equal because they are) right angles.

2. All radii of a circle are equal.

•3. HL.

4. Corresponding parts of congruent triangles are equal.

5. The tangent segments to a circle from an external point are equal.

Theorem 67.

6. Two points determine a line.

•7. Inscribed angles that intercept the same arc are equal.

8. AA.

•9. Corresponding sides of similar triangles are proportional.

10. Multiplication.

Lunar Eclipse.

11. In a plane, two lines perpendicular to a third line are parallel to each other.

•12. The tangent segments to a circle from an external point are equal.

13. SSS (or SAS or HL).

14. ∠APS = ∠CPS because corresponding parts of congruent triangles are equal.

15. AA.

Intersecting Chords.

16. It gets larger.

17. It stays the same.

Tangent Construction.

18.

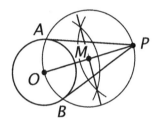

•19. Right angles. An angle inscribed in a semicircle is a right angle.

20. PA ⊥ OA and PB ⊥ OB.

21. If a line is perpendicular to a radius at its outer endpoint, it is tangent to the circle.

Tangent Triangle.

22.

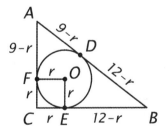

23. ΔABC is a right triangle because $9^2 + 12^2 = 15^2$. (If the sum of the squares of two sides of a triangle is equal to the square of the third side, it is a right triangle.)

•24. CFOE is a square.

25. $(9 - r) + (12 - r) = 15$, $21 - 2r = 15$, $2r = 6$, $r = 3$.

Tangent Quadrilateral.

26.

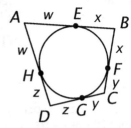

27. The sums of the lengths of the opposite
    sides of ABCD are equal.
    $AB + DC = (w + x) + (y + z) = w + x + y + z$
    and
    $BC + AD = (x + z) + (w + y) = w + x + y + z.$

Set II (pages 519–521)

In his *Mathematical Magic Show* (Knopf, 1977),
Martin Gardner wrote about the "ridiculous
question" of exercises 35 through 37: "This is one
of those curious problems that can be solved in a
different way on the assumption that they have
answers. Since P can be anywhere on the circle
from X to Y, we move P to a limit (either Y or X).
In both cases one side of triangle ABC shrinks to
zero as side AB expands to 10, producing a
degenerate straight-line 'triangle' with sides of
10, 10, and 0 and a perimeter of 20."

Grid Exercise 1.

28.

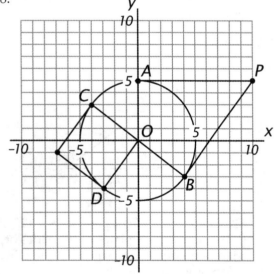

29. $d = \sqrt{(x-0)^2 + (y-0)^2} = 5$; so $x^2 + y^2 = 25.$

30. Point A (0, 5): $0^2 + 5^2 = 0 + 25 = 25$;
    point B (4, –3): $4^2 + (-3)^2 = 16 + 9 = 25.$

•31. PA is tangent to circle O because it is
    perpendicular to radius OA at its outer
    endpoint.

32. The slope of PB is $\frac{4}{3}$ and the slope of OB is
    $-\frac{3}{4}$. Because the product of these slopes is
    –1, PB ⊥ OB; so PB is tangent to the circle.

•33. PA = 10; PB = $\sqrt{(10-4)^2 + (5-(-3))^2}$ =
    $\sqrt{6^2 + 8^2}$ = $\sqrt{100}$ = 10.

34. (–7, –1).

Ridiculous Question.

•35. 20 units.

36. The perimeter of ΔABC = AB + BC + AC =
    (AP + PB) + BC + AC. Because tangent
    segments to a circle from an external point
    are equal, YA = AP and XB = PB. Substituting,
    the perimeter of ΔABC =
    YA + XB + BC + AC = YC + XC = 10 + 10 = 20.

37. Because, although it might at first seem
    impossible to figure out, the problem is
    actually very easy. (The solution is obvious
    if you think of folding AY onto AP and BX
    onto BP.)

Grid Exercise 2.

38.

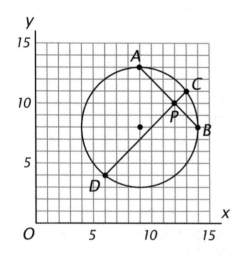

39. $d = \sqrt{(x-9)^2 + (y-8)^2}$ = 5;
    so $(x - 9)^2 + (y - 8)^2 = 25.$

40. Point A (9, 13): $(9 - 9)^2 + (13 - 8)^2 = 0 + 25 = 25$;
    point B (14, 8): $(14 - 9)^2 + (8 - 8)^2 = 25 + 0 = 25$;
    point C (13, 11): $(13 - 9)^2 + (11 - 8)^2 = 16 + 9 = 25$;
    point D (6, 4): $(6 - 9)^2 + (4 - 8)^2 = 9 + 16 = 25$.

•41. (12, 10).

42. PA = $\sqrt{18}$ or $3\sqrt{2}$; PB = $\sqrt{8}$ or $2\sqrt{2}$;
    PC = $\sqrt{2}$; PD = $\sqrt{72}$ or $6\sqrt{2}$.

43. PA · PB = 12; PC · PD = 12.
    (PA · PB = $3\sqrt{2} \cdot 2\sqrt{2}$ = 6 · 2 = 12;
    PC · PD = $\sqrt{2} \cdot 6\sqrt{2}$ = 6 · 2 = 12.)

Chord Division.

44.

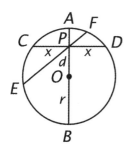

•45. If a line through the center of a circle is perpendicular to a chord, it also bisects the chord.

•46. BP.

47. AP.

48. AP · PB = CP · PD; so $(r - d)(r + d) = x \cdot x$,
    $r^2 - d^2 = x^2$, $x^2 = r^2 - d^2$.

49. EP · PF = CP · PD = $x^2 = r^2 - d^2$.

50. *r* represents the radius of the circle and *d* represents the distance of point P from its center.

51. Draw chord EF through the center of the circle and choose point P to be the center of the circle.

52. 25. ($r^2 - d^2 = 5^2 - 0^2 = 25$.)

Earth's Shadow.

53. Because $\triangle SAP \sim \triangle EBP$, $\dfrac{EP}{SP} = \dfrac{EB}{SA}$.

$$\frac{EP}{93,000,000 + EP} = \frac{3,960}{432,000},$$

432,000EP = 368,280,000,000 + 3,960EP,
428,040EP = 368,280,000,000,
EP ≈ 860,000 miles.

Set III (page 521)

The story "The Brick Moon" by Edward Everett Hale originally appeared in the *Atlantic Monthly* in 1869–1870. Its plot is summarized in considerable detail in *Space Travel—A History,* by Wernher von Braun and Frederick I. Ordway III (Harper & Row, 1985). According to this summary, "its 37 inhabitants signaled the Earth in Morse code by jumping up and down on the outside of the satellite. People on Earth threw them books and other objects, some of which missed the Moon and went into orbit around it." The geometry of this problem is the same as that for existing artificial satellites, except that BO is different.

The Brick Moon.

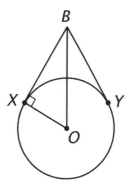

ΔBXO is a right triangle; so $BX^2 + XO^2 = BO^2$.
Because XO = 4,000 and BO = 4,000 + 4,000 = 8,000,
$BX^2 = 8,000^2 - 4,000^2$, $BX^2 = 48,000,000$,
BX ≈ 6,928 miles.

Chapter 12, Review

Set I (pages 523–524)

Friedrich von Martens, a German engraver-photographer living in Paris, built the camera with which he took the panoramic photograph of the city in 1844. As the overhead view suggests, the picture was taken on a curved plate.

Panoramic Camera.

1. Radii.

•2. A central angle.

3. A minor arc.

4. 150°.

Chords and Radius.

5. *Example figure:*

6. A diameter.

•7. 60°.

8. 180°.

Finding Lengths and Angles.

•9. 35°. ($\frac{1}{2}$70.)

•10. 16.5. ($8x = 11 \cdot 12$.)

11. 100°. [$\frac{1}{2}$(75 + 125).]

12. 8. (If a line through the center of a circle is perpendicular to a chord, it also bisects the chord.)

13. 60°. (Inscribed angles that intercept the same arc are equal.)

14. 50°. (90 − 40.)

•15. 30°. [$\frac{1}{2}$(105 − 45).]

16. 4. (10 − 6.)

Two Equilateral Triangles.

17. 4.

•18. 60°.

19. 6.

20. 60°.

21. BC < DE (or BC = $\frac{2}{3}$DE).

22. $m\widehat{BC} = m\widehat{DE}$.

Not Quite Right.

23. ∠B and ∠D are right angles. If a line is tangent to a circle, it is perpendicular to the radius drawn to the point of contact.

24. ∠A and ∠C are supplementary. Because the sum of the angles of a quadrilateral is 360° and ∠B = ∠D = 90°, ∠A + ∠C = 180°.

•25. AB = AD. The tangent segments to a circle from an external point are equal.

26. BC = CD. All radii of a circle are equal.

Three Ratios.

•27. They are similar to ΔABC.

28. $\frac{1}{1}$, or 1.

•29. $\frac{1}{2}$.

30. $\frac{1}{2}$.

31. $\frac{1}{4}$.

Set II (pages 524–526)

The dome of Santa Maria del Fiore was designed by Filippo Brunelleschi and built in 1420–1446. The arch used is called "a quinto acuto," referring to the fact that a radius 4 units long springs from a base 5 units long. The arch is so strong that the dome was built without a scaffold. Its construction is considered to be one of the most impressive architectural accomplishments of the Renaissance.

The most famous ring of standing stones built in prehistoric England is Stonehenge. Although the measures of the arcs in the egg-shaped ring considered in exercises 45 through 50 are determined by the angles of the 3-4-5 right triangles, the radii are not determined by their sides. As a student who makes some drawings of this ring can discover, the longer the radii are in comparison with the sides of the triangles, the rounder the ring.

Exercises 55 through 62 demonstrate an appearance of the golden ratio, $\frac{1+\sqrt{5}}{2}$, in an unexpected place. (As the figure suggests, circles A, B, and C are tangent to one another as well as being tangent to a diameter of circle O. Circles A and C are also tangent to circle O.) Students who have done the Set III exercise of Chapter 11, Lesson 1, may recall the interesting result repeated here: if r is the golden ratio, then $\frac{1}{r} = r - 1$.

A surprising aspect of exercise 63, the "pond problem," is that the dimensions of the garden cannot be determined from the given information, because they are *not* determined by it! This situation was previously encountered in Martin Gardner's "oil-well puzzle," included in exercises 47 through 50 of Chapter 11, Lesson 2. Both problems are connected to the fact that, given a point inside a rectangle, the sum of the squares of its distances from two opposite corners is equal to the sum of the squares of the distances from the other two corners.

Tangent Circles.

32.

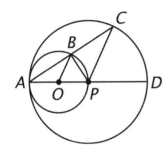

33. ∠ABP = 90°. ∠ABP is a right angle because it is inscribed in a semicircle.

•34. PB ⊥ AC.

35. AB = BC because, if a line through the center of a circle is perpendicular to a chord, it also bisects the chord.

•36. Isosceles. (In the figure, they are also obtuse.)

37. ΔAOB ~ ΔAPC by AA. (∠A = ∠A and, because OA = OB and PA = PC, ∠A = ∠ABO and ∠A = ∠C; so ∠ABO = ∠C.)

38. OB ∥ PC and OB = $\frac{1}{2}$PC because OB is a midsegment of ΔAPC.

Arch Problem.

39.

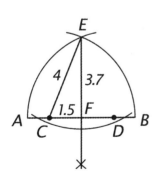

•40. 4 cm. (CE = CB = 5 – 1.)

41. 1.5 cm. (CF = $\frac{1}{2}$CD = $\frac{1}{2}$3.)

•42. Approximately 3.7 cm. (EF = $\sqrt{4^2 - 1.5^2}$ = $\sqrt{13.75}$ ≈ 3.7.)

43. Approximately 0.74. ($\frac{EF}{AB}$ ≈ $\frac{3.7}{5}$ ≈ 0.74.)

44. Approximately 68°. (In right ΔECF, cos ∠ECF = $\frac{CF}{CE}$ = $\frac{1.5}{4}$ = 0.375; so ∠ECF ≈ 68°. $m\widehat{AE}$ = $m\widehat{BE}$ = ∠ECF ≈ 68°.)

Prehistoric Ring.

45. 180°. ($\widehat{EFG}$ is a semicircle.)

•46. 53°. (In right ΔADC, from the fact that sin ∠ADC = $\frac{4}{5}$ or cos ∠ADC = $\frac{3}{5}$ or tan ∠ADC = $\frac{4}{3}$, we know that $m\widehat{GH}$ = ∠ADC ≈ 53°.)

47. 74°. [$m\widehat{HI}$ = ∠HCI = ∠BCD = 2∠ACD ≈ 2(90° – 53°) = 2(37°) = 74°.]

48. 53°. (From the symmetry of the figure, $m\widehat{IE}$ = $m\widehat{GH}$.)

49. They seem to be equal. ($m\widehat{EFG}$ = 180°, and $m\widehat{GH}$ + $m\widehat{HI}$ + $m\widehat{IE}$ ≈ 53° + 74° + 53° = 180°.)

50. In ΔBCD, ∠BDC + ∠BCD + ∠DBC = 180°. Because ∠HCI = ∠BCD, $m\widehat{GH}$ = ∠BDC, $m\widehat{HI}$ = ∠HCI, and $m\widehat{IE}$ = ∠DBC, it follows by substitution that $m\widehat{GH}$ + $m\widehat{HI}$ + $m\widehat{IE}$ = 180°.

Circular Saw.

•51. C is the midpoint of AB.

•52. Approximately 9.7 in. (In right ΔAOC, AO = 5 and OC = 5 – 3.75 = 1.25; so AC = $\sqrt{5^2 - 1.25^2}$ ≈ 4.84. AB = 2AC ≈ 9.68.)

•53. Approximately 75.5°. (In right ΔAOC, cos ∠AOD = $\frac{OC}{AO}$ = $\frac{1.25}{5}$ = 0.25; so ∠AOD ≈ 75.5°.)

54. Approximately 151°. ($m\widehat{ADB}$ = ∠AOB = 2∠AOD ≈ 151°.)

Golden Ratio.

55. 2.

•56. $\sqrt{5}$. ($\sqrt{2^2+1^2}$.)

57. $\sqrt{5}+1$.

•58. $\dfrac{\sqrt{5}+1}{2}$.

59. 1.618034.

60. $\dfrac{2}{\sqrt{5}+1}$.

61. 0.618034.

62. It appears to be exactly 1 less than the ratio for exercise 59.

Pond Problem.

63. Applying the Pythagorean Theorem to some of the right triangles formed, we get $(60 + r)^2 = w^2 + y^2$, $(52 + r)^2 = w^2 + z^2$, $(28 + r)^2 = x^2 + z^2$, $(40 + r)^2 = x^2 + y^2$. It follows that $(60 + r)^2 + (28 + r)^2 = (52 + r)^2 + (40 + r)^2$ because both sums are equal to $w^2 + x^2 + y^2 + z^2$. So $3{,}600 + 120r + r^2 + 784 + 56r + r^2 = 2{,}704 + 104r + r^2 + 1{,}600 + 80r + r^2$, $4{,}384 + 176r = 4{,}304 + 184r$, $80 = 8r$, $r = 10$. The radius of the pond is 10 yards.

Chapter 12, Algebra Review (page 528)

•1. $x = y$
$2x = y + 8$;
$2x = x + 8$,
$x = 8$,
$y = 8$.
(8, 8).

•2. $y = 2x$
$4x - y = 14$;
$4x - 2x = 14$,
$2x = 14$,
$x = 7$,
$y = 2(7) = 14$.
(7, 14).

•3. $y = x - 2$
$3(x + 1) = 4y$;
$3(x + 1) = 4(x - 2)$,
$3x + 3 = 4x - 8$,
$11 = x$,
$y = 11 - 2 = 9$.
(11, 9).

•4. $y = x - 7$
$y = 5x - 19$;
$5x - 19 = x - 7$,
$4x = 12$,
$x = 3$,
$y = 3 - 7 = -4$.
(3, –4).

•5. $x = y + 1$
$2x = 3y + 3$;
$2(y + 1) = 3y + 3$,
$2y + 2 = 3y + 3$,
$-1 = y$,
$x = -1 + 1 = 0$.
(0, –1).

•6. $y = 3x - 1$
$x + 2y = 33$;
$x + 2(3x - 1) = 33$,
$x + 6x - 2 = 33$,
$7x = 35$,
$x = 5$,
$y = 3(5) - 1 = 15 - 1 = 14$.
(5, 14).

•7. $x + 11 = 5y$
$x = 2(3y - 8)$;
$2(3y - 8) + 11 = 5y$,
$6y - 16 + 11 = 5y$,
$y = 5$,
$x = 2(3(5) - 8) = 2(15 - 8) = 2(7) = 14$.
(14, 5).

•8. $x + y = 14$
$y = 3x$;
$x + 3x = 14$,
$4x = 14$,
$x = 3.5$,
$y = 3(3.5) = 10.5$.
$(3.5, 10.5)$.

•9. $x = 5y$
$2x - 7y = 27$;
$2(5y) - 7y = 27$,
$10y - 7y = 27$,
$3y = 27$,
$y = 9$,
$x = 5(9) = 45$.
$(45, 9)$.

•10. $y = x + 3$
$x + 4y = 2$;
$x + 4(x + 3) = 2$,
$x + 4x + 12 = 2$,
$5x = -10$,
$x = -2$,
$y = -2 + 3 = 1$.
$(-2, 1)$.

•11. $2x + y = -40$
$8y - 3 = x$;
$2(8y - 3) + y = -40$,
$16y - 6 + y = -40$,
$17y = -34$,
$y = -2$,
$x = 8(-2) - 3 = -16 - 3 = -19$.
$(-19, -2)$.

•12. $9x = y + 5$
$4y - 3x = -20$;
$9x - 5 = y$,
$4(9x - 5) - 3x = -20$,
$36x - 20 - 3x = -20$,
$33x = 0$,
$x = 0$,
$9(0) - 5 = y$,
$-5 = y$.
$(0, -5)$.

•13. $x + y = 35$
$x - y = 67$;
$2x = 102$,
$x = 51$,
$51 + y = 35$,
$y = -16$.
$(51, -16)$.

•14. $5x - 7y = 92$
$5x + y = 4$;
$-8y = 88$,
$y = -11$,
$5x + (-11) = 4$,
$5x = 15$,
$x = 3$.
$(3, -11)$.

•15. $3x - 5y = 51$
$x + 5y = 23$;
$4x = 74$,
$x = 18.5$,
$18.5 + 5y = 23$,
$5y = 4.5$,
$y = 0.9$.
$(18.5, 0.9)$.

•16. $8x - 7y = 62$
$4x - 7y = 66$;
$4x = -4$,
$x = -1$,
$8(-1) - 7y = 62$,
$-8 - 7y = 62$,
$-7y = 70$,
$y = -10$,
$(-1, -10)$.

•17. $x + y = 7$
$3x + 2y = 25$;
$3x + 3y = 21$,
$y = -4$,
$x + (-4) = 7$,
$x = 11$.
$(11, -4)$.

•18. $4x + 3y = 31$
$2x - 9y = 5$;
$12x + 9y = 93$,
$14x = 98$,
$x = 7$,
$4(7) + 3y = 31$,
$28 + 3y = 31$,
$3y = 3$,
$y = 1$.
$(7, 1)$.

•19. $2x + 5y = 29$
$4x - y = 25$;
$4x + 10y = 58$,
$11y = 33$,
$y = 3$,
$2x + 5(3) = 29$,
$2x + 15 = 29$,
$2x = 14$,
$x = 7$.
$(7, 3)$.

•20. $5x + 4y = 53$
$x - 2y = 5;$
$2x - 4y = 10,$
$7x = 63,$
$x = 9,$
$9 - 2y = 5,$
$-2y = -4,$
$y = 2.$
$(9, 2).$

•21. $8x - 3y = 32$
$7x + 9y = 28;$
$24x - 9y = 96,$
$31x = 124,$
$x = 4,$
$8(4) - 3y = 32,$
$32 - 3y = 32,$
$-3y = 0,$
$y = 0.$
$(4, 0).$

•22. $6x + 6y = 24$
$10x - y = -15;$
$x + y = 4,$
$11x = -11,$
$x = -1,$
$10(-1) - y = -15,$
$-10 - y = -15,$
$-y = -5,$
$y = 5.$
$(-1, 5).$

•23. $5x - 7y = 54$
$2x - 3y = 22;$
$10x - 14y = 108,$
$10x - 15y = 110,$
$y = -2,$
$2x - 3(-2) = 22,$
$2x + 6 = 22,$
$2x = 16,$
$x = 8.$
$(8, -2).$

•24. $7x - 5y = 40$
$3x - 2y = 16;$
$14x - 10y = 80,$
$15x - 10y = 80,$
$x = 0,$
$7(0) - 5y = 40,$
$-5y = 40,$
$y = -8.$
$(0, -8).$

Set I (pages 532–534)

Exercises 15 through 21 illustrate a remarkable fact. Although it is easy to see why the lines through the pairs of intersections of three circles are concurrent if the circles have equal radii, the lines are concurrent even if the circles have different radii. If two circles intersect in two points, the line determined by those points is called the *radical axis* of the two circles. More on the theorem that the radical axes of three circles with noncollinear centers, taken in pairs, are concurrent can be found in two books by Howard Eves: *Fundamentals of Modern Elementary Geometry* (Jones and Bartlett, 1992) and *College Geometry* (Jones and Bartlett, 1995).

Exercises 22 through 25 lead to another surprise. If four lines intersect to form four triangles, the circumcircles of the triangles always intersect in a common point, named the Miquel point after an obscure nineteenth-century mathematician, Auguste Miquel. In *The Penguin Dictionary of Curious and Interesting Geometry* (Penguin, 1991), David Wells describes other remarkable properties of this point. The centers of the four circumcircles also lie on a circle that passes through it. It is also the focus of the unique parabola tangent to the four lines.

The regions into which a map is divided as in exercise 31 are called Dirichlet domains. According to Jay Kappraff, "a Dirichlet domain of a point from a set of points is defined to be the points of space nearer to that point than to any of the other points of the set." A detailed discussion of the geometry of Dirichlet domains and their connection to soap bubbles, spider webs, and phyllotaxis can be found in Kappraff's *Connections — The Geometric Bridge Between Art and Science* (McGraw-Hill, 1991).

Hot Tub.

1. Chords.

2. Inscribed angles.

3. Cyclic.

•4. Its circumcircle.

•5. ΔABC is inscribed in circle O.

•6. They are equidistant from it.

Circumcircles.

•7. Minor arcs.

8. A semicircle.

9. A major arc.

10. An acute triangle.

11. A right triangle.

12. An obtuse triangle.

•13. A diameter (or a chord).

14. The midpoint of its hypotenuse.

RGB Color.

15.

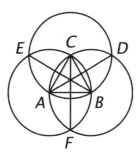

16. They are concurrent.

•17. The lines are the perpendicular bisectors of the sides of ΔABC, and the perpendicular bisectors of the sides of a triangle are concurrent.

18. *Example figure:*

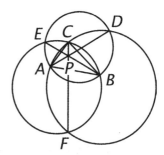

•19. No.

20. Yes.

21. By the Intersecting Chords Theorem. In the circle at the top, AP · PD = BP · PE and, in the circle on the left, BP · PE = CP · PF; so AP · PD = BP · PE = CP · PF.

Four Lines and Four Circles.

22. ΔABF, ΔACD, ΔBCE, ΔDEF.

23. They are their circumcircles.

•24. Three.

25. They intersect in a common point.

Equilateral Triangle.

26. Because AE, BF, and CD are altitudes, they are perpendicular to the sides of ΔABC. If a line through the center of a circle is perpendicular to a chord, it also bisects the chord.

27. 30°-60° right triangles.

•28. OA = 2OD because, in a 30°-60° right triangle, the hypotenuse is twice the shorter leg.

29. The radius is $\frac{2}{3}$ the length of one of the altitudes.

Nearest School.

30. The school at B.

31.

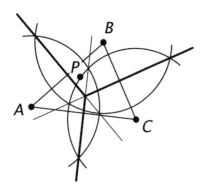

•32. The circumcenter of the triangle with A, B, and C as its vertices.

Set II (pages 534–535)

The triangle-folding exercises 33 through 40 are based on the fact that, if one point is reflected onto another, the line of reflection is the perpendicular bisector of the line segment connecting the two points.

Exercises 46 through 48 provide another example of a problem whose solution is based on Pythagorean triples and the Pythagorean Theorem yet was created more than a thousand years before Pythagoras was born. The tablet containing it was found in Susa, about 200 miles east of ancient Babylon.

Folding Experiment.

33.

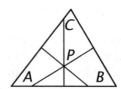

34. The folds are on the perpendicular bisectors of the sides of the triangle and they are concurrent.

35. It is equidistant from them.

•36. It is approximately 6.4 cm.

37.

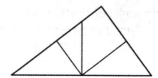

38. A right triangle because $9^2 + 12^2 = 15^2$; if the sum of the squares of two sides of a triangle is equal to the square of the third side, it is a right triangle.

•39. They are concurrent on the hypotenuse of the triangle.

•40. 7.5 cm.

Inscribed Triangles.

41.

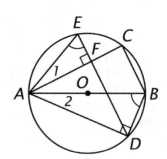

•42. Both ∠E and ∠ABD intercept $\overset{\frown}{AD}$. Inscribed angles that intercept the same arc are equal.

43. ∠ADB is a right angle because it is inscribed in a semicircle.

44. All right angles are equal.

45. If two angles of one triangle are equal to two angles of another triangle, the third pair of angles are equal.

Babylonian Problem.

46.

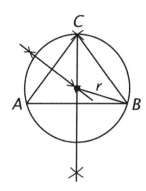

•47. Approximately 31 mm.

•48. 31.25 mm.

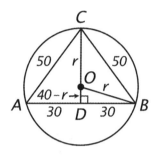

[Because ∆ACD is a right triangle with
AC = 50 and AD = 30, CD = 40. In right
∆BDO, $(40 - r)^2 + 30^2 = r^2$; so
$1{,}600 - 80r + r^2 + 900 = r^2$, $2{,}500 = 80r$,
$r = 31.25$.]

Diameters and Sines.

49. ∆BCD is a right triangle because ∠BCD is
inscribed in a semicircle.

50. Inscribed angles that intercept the same arc
are equal.

•51. $\sin D = \dfrac{a}{2r}$.

52. $\dfrac{a}{\sin A} = 2r$. (Because $\sin A = \sin D$,

$\sin A = \dfrac{a}{2r}$; so $2r \sin A = a$ and $\dfrac{a}{\sin A} = 2r$.)

53. It reveals that it is equal to the diameter of
the triangle's circumcircle.

The Slipping Ladder.

1.

2. Along an arc of a circle.

3. From the fact that the midpoint of the
hypotenuse of a right triangle is its
circumcenter, it follows that O is equidistant
from A, B, and C; that is, OA = OB = OC.
But OA and OB do not change in length as
the ladder slides down the wall, and so OC
does not change either. Because for every
position of the ladder Ollie's feet remain the
same distance from C, their path is along an
arc of a circle.

Chapter 13, Lesson 2

Set I (pages 538–539)

Theorem 69 of this lesson appears in the *Elements*
as Proposition 22 of Book III: "The opposite angles
of quadrilaterals in circles are equal to two right
angles." Strangely, although the converse is true
and very useful, Euclid does not state or prove it.

   A proof that the circumcircles of the
equilateral triangles constructed on the sides of a
triangle intersect in a common point (assumed in
exercises 24 through 28) can be found on pages
82 and 83 of *Geometry Revisited,* by H. S. M.
Coxeter and S. L. Greitzer (Random House, 1967).
The point in which the circles intersect is called
the Fermat point of the triangle. According to
David Wells in *The Penguin Dictionary of Curious
and Interesting Geometry* (Penguin, 1991), Fermat
challenged Torricelli (of barometer fame) to find
the point the sum of whose distances from the
vertices of a triangle is a minimum. (In other
words, to find the general solution to the Spotter's
Puzzle.) That point turns out to be the Fermat
point. According to Wells: "If all the angles of the
triangle are less than 120° the desired point . . . is
such that the lines joining it to the vertices meet
at 120°. If the angle at one vertex is at least 120°,
then the Fermat point coincides with that vertex."
Wells also remarks that "the Fermat point can be
found by experiment. Let three equal weights
hang on strings passing through holes at the

vertices of the triangle, the strings being knotted at one point. The knot will move to the Fermat point."

Quilt Quadrilaterals.

•1. By seeing if a pair of opposite angles are supplementary.

2. The diamond.

Cyclic and Noncyclic.

3. The lines appear to be their perpendicular bisectors.

•4. They must be concurrent.

5. It is equidistant from them.

Euclid's Proof.

6. Two points determine a line.

7. The sum of the angles of a triangle is 180°.

•8. Inscribed angles that intercept the same arc are equal.

9. Betweenness of Rays Theorem.

•10. Substitution.

11. Addition.

12. Substitution.

A Different Proof.

13.

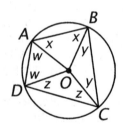

•14. They are isosceles because all radii of a circle are equal.

15. 360°. The sum of the angles of a quadrilateral is 360°.

16. 180°.

17. The sum of each pair of opposite angles of ABCD is equal to $w + x + y + z$.

Isosceles Trapezoid.

•18. AB ∥ DC because they are the bases of a trapezoid.

19. ∠A and ∠D are supplementary. Parallel lines form supplementary interior angles on the same side of a transversal.

20. ∠A = ∠B. The base angles of an isosceles trapezoid are equal.

•21. ∠B and ∠D are supplementary.

22. ABCD is cyclic. A quadrilateral is cyclic if a pair of its opposite angles are supplementary.

23. They prove that isosceles trapezoids are cyclic.

Equilateral Triangles on the Sides.

•24. They intersect in a common point, P.

25. They are cyclic.

26. They are each equal to 120°. They are supplementary to the angles opposite them in the quadrilaterals. The angles opposite them are 60° because they are also angles of the equilateral triangles.

27. *Example figure:*

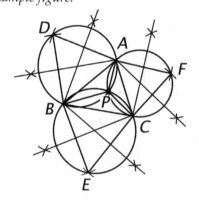

28. *(Student answer.)* (Yes.)

Set II (pages 539–541)

Brahmagupta's formula for the area of a cyclic quadrilateral, discovered in the seventh century, is an extension of the formula for the area of a triangle in terms of its sides discovered by Heron of Alexandria six centuries earlier. (Heron's Theorem was introduced in the Set III exercises of Chapter 9, Lesson 3.) H. S. M. Coxeter and S. L. Greitzer include a derivation of Brahmagupta's formula on page 58 of *Geometry Revisited*. Although they refer to it as "one of the simplest methods for obtaining Brahmagupta's formula," the method is based on several trigonometric identities including the Law of Cosines and is not easy.

J. L. Heilbron in *Geometry Civilized—History, Culture and Technique* (Clarendon Press, 1998), refers to Claudius Ptolemy as the astronomer "who was to astronomy what Euclid was to geometry" and refers to the proof of his theorem as "one of the most elegant in all plane geometry." From a special case of Ptolemy's Theorem comes the trigonometric identity

$$\sin(x - y) = \sin x \cos y - \cos x \sin y.$$

Cyclic or Not.

29. ABCD is cyclic, because its vertices are equidistant from point E. A circle is the set of all points in a plane equidistant from a given point; so a circle can be drawn with its center at E that contains all of the vertices.

•30. ABCD is not cyclic, because its opposite angles are not supplementary. (The sum of a right angle and an obtuse angle is more than 180°, and the sum of a right angle and an acute angle is less than 180°.)

31. ABCD is cyclic. Because ∠BCE is an exterior angle of ABCD, it forms a linear pair with ∠DCB; so ∠BCE and ∠DCB are supplementary. Because ∠BCE = ∠A, it follows that ∠A and ∠DCB are also supplementary; so ABCD is cyclic.

Brahmagupta's Theorem.

32. *Example figure:*

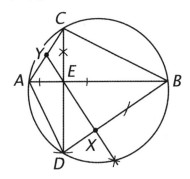

•33. Vertical angles are equal.

•34. Supplements of the same angle are equal. (∠7 and ∠5 are supplements of ∠6[∠8] and ∠6 and ∠8 are supplements of ∠7[∠5].)

35. Inscribed angles that intercept the same arc are equal.

36. Substitution.

37. Point Y is the midpoint of AC. Because ∠2 = ∠1, AY = YE and, because ∠3 = ∠4, YC = YE; so AY = YC.

38. If a cyclic quadrilateral has perpendicular diagonals, then any line through their point of intersection that is perpendicular to a side of the quadrilateral *bisects the opposite side.*

Area Formula.

39. The perimeter of a rectangle with sides *a* and *b* is 2*a* + 2*b*; so, for a rectangle, *s* = *a* + *b*.
$$A = \sqrt{(a+b-a)(a+b-b)(a+b-a)(a+b-b)} = \sqrt{baba} = \sqrt{a^2 b^2} = ab.$$

•40. The perimeter of this quadrilateral is 176, so for it, *s* = 88.
$$A = \sqrt{(88-25)(88-39)(88-52)(88-60)} = \sqrt{63 \cdot 49 \cdot 36 \cdot 28} = 1,764.$$

41. No. It is not cyclic and its area is evidently much smaller.

Ptolemy's Theorem.

•42. The Protractor Postulate.

43. Inscribed angles that intercept the same arc are equal.

44. AA.

45. Corresponding sides of similar triangles are proportional.

•46. Multiplication.

47. Addition.

48. Betweenness of Rays Theorem and substitution.

49. Inscribed angles that intercept the same arc are equal.

50. AA.

51. Corresponding sides of similar triangles are proportional.

52. Multiplication.

53. Addition.

•54. Substitution.

55. Multiplication.

Two Applications.

56. Because a rectangle is a cyclic quadrilateral and its opposite sides are equal, Ptolemy's Theorem becomes $a \cdot a + b \cdot b = c \cdot c$, or $a^2 + b^2 = c^2$, the Pythagorean Theorem!

57. Applying Ptolemy's Theorem to cyclic quadrilateral APBC,
PA · BC + PB · AC = PC · AB. Because ΔABC is equilateral, AB = BC = AC. By substitution, PA · AB + PB · AB = PC · AB; dividing by AB gives PA + PB = PC.

Set III (page 541)

The Set III puzzle is adapted from one by Stephen Barr, who has created many clever puzzles. Barr's puzzle was originally included by Martin Gardner in one of his "Mathematical Games" columns for *Scientific American* and is reprinted on page 110 of *Mathematical Carnival* (Knopf, 1975). Readers of Gardner's column discovered an extra set of points through which a circle could be drawn that neither Barr nor Gardner had noticed!

Overlapping Cards Puzzle.

1. These points are the vertices of the two rectangles. Because the opposite angles of a rectangle are supplementary, all rectangles are cyclic.

2.

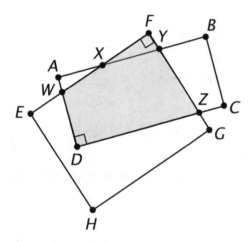

One set of points consists of W, F, Z, and D. Drawing the hidden parts of the edges WX and YZ produces quadrilateral WFZD. Because ∠F and ∠D are right angles, they are supplementary; so WFZD is cyclic.

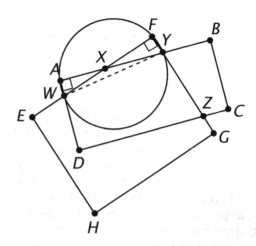

Another set of points consists of W, Y, F, and A. If WY is drawn and a circle is drawn with WY as its diameter, then it can be proved that points A and F lie on this circle owing to the fact that ∠A and ∠F are right angles.

Chapter 13, Lesson 3

Set I (pages 544–545)

Exercises 27 through 30 are related to a result known as Poncelet's porism. Jean Victor Poncelet, a nineteenth-century French mathematician, is credited for establishing the development of projective geometry as an independent subject. As Howard Eves explains in *An Introduction to the History of Mathematics* (Saunders, 1990), a porism is "a proposition stating a condition that renders a certain problem solvable, and then the problem has infinitely many solutions. For example, if *r* and *R* are the radii of two circles and *d* is the distance between their centers, the problem of inscribing a triangle in the circle of radius *R*, which will be circumscribed about the circle of radius *r*, is solvable if and only if $R^2 - d^2 = 2Rr$, and then there are infinitely many triangles of the desired sort." The general version of Poncelet's porism states that, if given two conics, an *n*-gon can be inscribed in one conic and circumscribed about the other, there are an infinite number of such *n*-gons.

Postage Stamp.

•1. The angle bisectors of a triangle are concurrent.

2. The incenter.

Circumcircle and Incircle.

•3. Two.

•4. They are perpendicular bisectors of the sides.

•5. The vertices of the triangle.

6. Three.

•7. Two angle bisectors.

8. A perpendicular from the triangle's incenter to one of its sides.

9. The sides of the triangle.

Circumscribed Quadrilateral.

•10. Its incircle.

•11. Its incenter.

12.

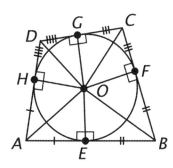

•13. They are perpendicular to the sides. If a line is tangent to a circle, it is perpendicular to the radius drawn to the point of contact.

14. They bisect the angles. They form pairs of triangles that are congruent by HL.

•15. They must be concurrent.

16. They must be equal.

•17. Only if the rectangle is a square.

18. Only if the parallelogram is a rhombus.

19. Only if the sum of the lengths of the bases is equal to the sum of the lengths of the legs.

Equilateral Triangle.

20. That circles have the same center.

•21. They are the perpendicular bisectors of the sides.

22. They are the bisectors of the angles.

23. Isosceles (obtuse) triangles and 30°-60° right triangles.

24. 2.

•25. 2π and 4π units.

26. π and 4π square units.

Construction Problem.

27. *Example figure:*

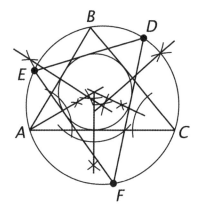

•28. EF seems to be tangent to the incircle.

29. *(Student answer.)* (Probably not.)

30. Yes.

Set II (pages 545–547)

In *Dutton's Navigation and Piloting* (Naval Institute Press, 1985), Elbert S. Maloney remarks that the LOP (line of position) is probably the most important concept in navigation. When three LOPs are taken, they are usually not concurrent, but intersect to form a triangle. The navigator ordinarily takes the point that is equidistant from the sides of this triangle (its incenter) as the "fix" or position of the ship. If there is a "constant error" in the LOPs, the method of "LOP bisectors" outlined in exercises 31 through 39 is used. As exercises 31 through 33 suggest and exercises 34 through 39 prove, the bisectors of an angle and the two remote exterior angles of a triangle are concurrent. In the method of LOP bisectors, the fix of the ship is taken at the point of concurrency.

In exercises 40 through 58, two special cases of a remarkable theorem are considered and, in exercises 59 through 62, the general case is developed. The theorem states that, for *any* polygon and its inscribed circle, the ratio of the areas is equal to the ratio of the perimeters. A very nice treatment of this theorem is included in *Geometrical Investigations—Illustrating the Art of Discovery in the Mathematical Field,* by John Pottage (Addison-Wesley, 1983). Pottage presents his material in the form of a dialogue between Galileo's characters Salviata, Sagredo, and Simplicio. His book, like the work of George Polya, is a wonderful example of teaching mathematics by the Socratic method.

Ship Location.

31. l_5.

•32. Exterior angles.

33. They appear to be concurrent.

34.

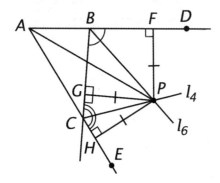

35. $\triangle PFB \cong \triangle PGB$ (AAS) and $\triangle PGC \cong \triangle PHC$ (AAS).

36. They are equal because PF = PG and PG = PH.

37. $\triangle FAP \cong \triangle HAP$ (HL). (These triangles are right triangles whose hypotenuse is AP and whose legs PF and PH are equal.)

•38. It proves that AP bisects $\angle DAE$ (because $\angle DAP = \angle PAE$).

39. They prove that the three bisector lines are concurrent.

Incircle Problem 1.

•40. FDEO is a square (because it is equilateral and equiangular).

41. AB = 2r.

•42. $2\pi r$ units.

43. $8r$ units.

44. πr^2 square units.

45. $4r^2$ square units.

•46. $\dfrac{\pi}{4}$. ($\dfrac{2\pi r}{8r} = \dfrac{\pi}{4}$.)

47. $\dfrac{\pi}{4}$. ($\dfrac{\pi r^2}{4r^2} = \dfrac{\pi}{4}$.)

Incircle Problem 2.

48. ECDO is a square (because it is equilateral and equiangular).

•49. AE = 3 − r.

50. DB = 4 − r.

•51. AB = 7 − 2r. (3 − r + 4 − r.)

52. 7 − 2r = 5, 2r = 2, r = 1.

53. 2π units.

54. 12 units.

•55. π square units.

56. 6 square units. ($\dfrac{1}{2}3 \cdot 4$.)

57. $\dfrac{\pi}{6}$. ($\dfrac{2\pi}{12} = \dfrac{\pi}{6}$.)

58. $\dfrac{\pi}{6}$.

Incircle Problem 3.

•59. The area of a triangle is half the product of its base and altitude.

60. $\alpha ABCD =$
$$\frac{1}{2}AB \cdot r + \frac{1}{2}BC \cdot r + \frac{1}{2}CD \cdot r + \frac{1}{2}DA \cdot r =$$
$$\frac{1}{2}r(AB + BC + CD + DA) = \frac{1}{2}rp.$$

•61. $\dfrac{2\pi r}{p}$.

62. $\dfrac{2\pi r}{p}$. ($\dfrac{\pi r^2}{\frac{1}{2}rp} = \dfrac{2\pi r}{p}$.)

Trisector Challenge.

63. $\angle 1 = \angle 2$. Because BE and CE bisect two angles of $\triangle BDC$, E is its incenter. Because the angle bisectors of a triangle are concurrent, DE must bisect $\angle BDC$.

Set III (page 547)

Every triangle has not only an incircle, but also three excircles. Each excircle is tangent to one side of the triangle and the lines of the other two sides. The center of each excircle is the point of concurrence of the bisectors of an angle and the two remote exterior angles of the triangle. The centers of the excircles are the vertices of a larger triangle whose altitudes lie on the lines that bisect the angles of the original triangle.

Excircles.

1. They appear to be its altitudes.

2. AY bisects the exterior angles of △ABC at A; so ∠1 = ∠2. Because O is the incenter of △ABC, AO bisects ∠CAB and so ∠3 = ∠4. ∠1 + ∠2 + ∠3 + ∠4 = 180°; so ∠2 + ∠2 + ∠3 + ∠3 = 180°, 2(∠2 + ∠3) = 180°, ∠2 + ∠3 = 90°, and so AO ⊥ YA. A line segment from a vertex of a triangle that is perpendicular to the line of the opposite side is an altitude of the triangle.

Chapter 13, Lesson 4

Set I (pages 550–551)

Martin Gardner points out a rather surprising fact (*Mathematical Circus*, Knopf, 1979): that the altitudes of a triangle are concurrent does not appear in Euclid's *Elements*. Gardner observes that, "although Archimedes implies it, Proclus, a fifth-century philosopher and geometer, seems to have been the first to state it explicitly." The fact that the medians of a triangle are concurrent does not appear in the *Elements* either, but Archimedes knew that the point in which they are concurrent is the triangle's center of gravity. More specifically, it is the center of gravity of a triangular plate of uniform thickness and density; it is usually *not* the center of gravity of a wire in the shape of the triangle.

"Ortho" Words.

1. Orthodontist.

2. Orthodox.

3. Orthopedic.

•4. The fact that the altitudes of a triangle form right angles with the lines of its sides.

Theorem 71.

•5. Two points determine a line.

•6. The Ruler Postulate.

7. Two points determine a line.

•8. A midsegment of a triangle is parallel to the third side.

9. A quadrilateral is a parallelogram if its opposite sides are parallel.

10. The diagonals of a parallelogram bisect each other.

•11. A line segment that connects a vertex of a triangle to the midpoint of the opposite side is a median.

12. Lines that contain the same point are concurrent.

Theorem 72.

13. An altitude of a triangle is perpendicular to the line of the opposite side.

•14. Through a point not on a line, there is exactly one line parallel to the line.

15. A quadrilateral is a parallelogram if its opposite sides are parallel.

16. The opposite sides of a parallelogram are equal.

17. Substitution.

•18. In a plane, a line perpendicular to one of two parallel lines is also perpendicular to the other.

19. The perpendicular bisectors of the sides of a triangle are concurrent.

Median Construction.

20. *(The construction lines and arcs have been omitted from the figure below.)*

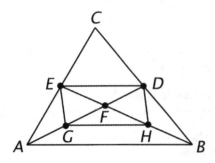

21. ED ∥ AB and ED = $\frac{1}{2}$AB. GH ∥ AB and

    GH = $\frac{1}{2}$AB. A midsegment of a triangle is parallel to the third side and half as long.

22. EDHG is a parallelogram. A quadrilateral is a parallelogram if two opposite sides are both parallel and equal.

•23. GD and EH bisect each other. The diagonals of a parallelogram bisect each other.

24. AG = GF and BH = HF because G and H are the midpoints of AF and FB, respectively. GF = FD and HF = FE because GD and EH bisect each other.

25. $\frac{AF}{FD} = \frac{2}{1} = 2$ and $\frac{BF}{FE} = \frac{2}{1} = 2$.

26. Points of trisection.

Altitude Construction.

27.

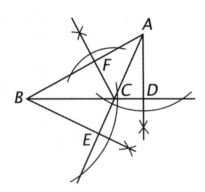

•28. ΔABC is obtuse.

•29. No. The altitude *segments* do not intersect.

30. Yes. We have proved that the lines that contain the altitudes of a triangle are concurrent.

31. Its orthocenter and its circumcenter.

Set II (pages 551–553)

Exercises 32 through 37 reveal Carnot's observation concerning the special relation between four points that are the three vertices of a triangle and its orthocenter: each point is the orthocenter of the triangle whose vertices are the other three. According to Howard Eves in *An Introduction to the History of Mathematics* (Saunders, 1990), Lazare Carnot was a general in the French Army and supported the French Revolution. In 1796, however, he opposed Napoleon's becoming emperor and had to flee to Switzerland. While living in exile, he wrote two important books on geometry. In one of them, he introduced several notations still used today, including ΔABC to represent the triangle having the vertices A, B, and C and $\overparen{AB}$ to represent arc AB. One of Carnot's sons became a celebrated physicist for whom the Carnot cycle of thermodynamics is named, and one of his grandsons became President of the French Republic.

Other Triangles, Other Orthocenters.

•32. GE, AF, and CD.

•33. At point B.

34. GF, AE, and BD.

35. At point C.

36. GD, BF, and CE.

37. At point A.

Medians as Bisectors.

38. The green line segments appear to be parallel to BC.

39. It appears to bisect them.

40. They are similar by AA. (Each pair of triangles has a common angle, and parallel lines form equal corresponding angles.)

•41. Corresponding sides of similar triangles are proportional.

•42. Substitution.

43. BM = MC.

44. Substitution.

45. Multiplication.

46. They suggest that the triangle would balance on the edge of the ruler.

Doing Without a Compass.

47. *Example figure:*

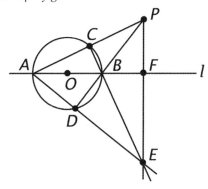

48. ∠ACB and ∠ADB are right angles. An angle inscribed in a semicircle is a right angle.

•49. PD and EC are altitudes of ∆APE.

50. Its orthocenter.

51. AF is the third altitude of ∆APE. The lines that contain the altitudes of a triangle are concurrent.

52. PE ⊥ *l*.

53. *Example figure:*

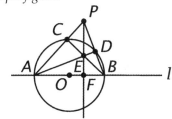

54. All of them.

Engineering Challenge.

55.

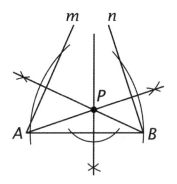

56. Constructing perpendiculars from A to line *n* and from B to line *m* produces two altitudes of the triangle formed by AB, *m*, and *n*. The point in which these altitudes intersect is the orthocenter of the triangle. A line drawn through this point perpendicular to AB therefore contains the third altitude of the triangle and passes through the point in which lines *m* and *n* intersect.

Set III (page 553)

The centroid of a uniform sheet in the shape of a quadrilateral was discovered by a German mathematician, Ferdinand Wittenbauer (1857–1922). The parallelogram whose sides lie in the lines through the points that trisect the sides of a quadrilateral is known as its "Wittenbauer parallelogram."

Balancing Point.

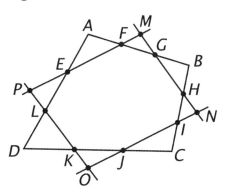

1. It seems to be a parallelogram.

2. The point in which the diagonals of the parallelogram intersect.

Chapter 13, Lesson 5

Set I (pages 557–558)

Although Ceva's Theorem is usually stated in terms of ratios, it appears in Nathan Altshiller Court's *College Geometry* (Barnes & Noble, 1952), in the following form: "The lines joining the vertices of a triangle to a given point determine on the sides of the triangle six segments such that the product of three nonconsecutive segments is equal to the product of the remaining three segments." This version of the theorem makes it immediately obvious that Obtuse Ollie's variations of it in exercises 25 and 26 are correct.

Which is Which?

•1. CF.

2. GF.

3. CD.

4. CE.

•5. No. GF is not a cevian, because neither of its endpoints is a vertex of the triangle.

Using Ceva's Theorem.

•6. 3.2. $(\frac{2}{3} \cdot \frac{4}{5} \cdot \frac{6}{x} = 1, \frac{48}{15x} = 1, 15x = 48, x = 3.2.)$

7. 2.4. $(\frac{3}{x} \cdot \frac{4}{x} \cdot \frac{x}{5} = 1, \frac{12}{5x} = 1, 5x = 12, x = 2.4.)$

•8. Not concurrent.
$\frac{7}{13} \cdot \frac{11}{13} \cdot \frac{11}{5} = \frac{847}{845} \approx 1.002 \neq 1.$

9. Not concurrent.
$\frac{14}{5} \cdot \frac{7}{17} \cdot \frac{13}{15} = \frac{1,274}{1,275} \approx 0.999 \neq 1.$

Equilateral Triangle.

10.

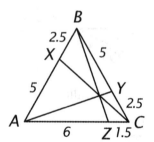

•11. 2.

12. 2.

13. Because the cevians are concurrent,
$2 \cdot 2 \cdot \frac{CZ}{ZA} = 1$; so $\frac{CZ}{ZA} = \frac{1}{4}$.

14. Because $\frac{CZ}{ZA} = \frac{1}{4}$, ZA = 4CZ. Also,
CZ + ZA = 7.5; so CZ + 4CZ = 7.5,
5CZ = 7.5, CZ = 1.5, and AZ = 6.

Ratio Relations.

•15. $\frac{AP}{PY} \cdot \frac{BP}{PZ} \cdot \frac{CP}{PX} = 2 \cdot 2 \cdot 2 = 8.$

16. $\frac{PX}{CX} + \frac{PY}{AY} + \frac{PZ}{BZ} = \frac{1}{3} + \frac{1}{3} + \frac{1}{3} = 1.$

17. $\frac{CX}{PX} + \frac{AY}{PY} + \frac{BZ}{PZ} = 3 + 3 + 3 = 9.$

•18. $\frac{AP}{AY} + \frac{BP}{BZ} + \frac{CP}{CX} = \frac{2}{3} + \frac{2}{3} + \frac{2}{3} = 2.$

19. $\frac{AX}{XB} \cdot \frac{BY}{YC} \cdot \frac{CZ}{ZA} = 1$ (by Ceva's Theorem.)

20. $\frac{AB}{AX} + \frac{BC}{BY} + \frac{CA}{CZ} = 2 + 2 + 2 = 6.$

Right Triangle.

•21. $\frac{BY}{YC} = \frac{6}{2} = 3.$

22. $\frac{CZ}{ZA} = \frac{2}{4} = \frac{1}{2}.$

23. $\frac{AX}{XB} \cdot \frac{BY}{YC} \cdot \frac{CZ}{ZA} = 1, \frac{AX}{XB} \cdot 3 \cdot \frac{1}{2} = 1,$
$\frac{AX}{XB} = \frac{2}{3}.$

•24. Because △ABC is a right triangle with legs 6 and 8, its hypotenuse is 10. Because $\frac{AX}{XB} = \frac{2}{3}$ and AX + XB = 10, AX = 4 and XB = 6.

Ollie's Equations.

25. Alice is wrong. Ollie's equation is correct. One way to show this is to start with Alice's equation, $\frac{a}{b} \cdot \frac{c}{d} \cdot \frac{e}{f} = 1$. Clearing it of fractions gives $ace = bdf$. Dividing both sides by ace gives $1 = \frac{bdf}{ace} = \frac{b}{c} \cdot \frac{d}{e} \cdot \frac{f}{a}.$

26. Ollie's second equation also is correct. One way to show this is to start with the expression $\frac{a}{f} \cdot \frac{e}{d} \cdot \frac{c}{b}$, getting $\frac{aec}{fdb} = \frac{ace}{bdf} = \frac{a}{b} \cdot \frac{c}{d} \cdot \frac{e}{f}.$

27. They suggest that you can begin with any one of the six segments and go in either direction around the triangle.

Set II (pages 559–560)

In his *A Course of Geometry for Colleges and Universities* (Cambridge University Press, 1970),

Dan Pedoe wrote: "The theorems of Ceva and Menelaus naturally go together, since the one gives the condition for lines through vertices of a triangle to be concurrent, and the other gives the condition for points on the sides of a triangle to be collinear." Menelaus is known for his book *Sphaerica*, in which the concept of a spherical triangle appears for the first time. The spherical version of the theorem of Menelaus considered in exercises 28 through 33 is, in fact, the basis for the development of much of spherical trigonometry.

Exercises 52 through 59 demonstrate, for the case of an acute triangle, how Ceva's Theorem can be used to prove that the altitudes of a triangle are concurrent.

The Theorem of Menelaus.

•28. $\triangle AXZ \sim \triangle CPZ$ and $\triangle BXY \sim \triangle CPY$.

29. Corresponding sides of similar triangles are proportional.

30. Multiplication.

31. Multiplication and division.

•32. Division (and substitution).

33. The equation in Ceva's Theorem.

Concurrent or Not?

•34. If a line parallel to one side of a triangle intersects the other two sides in different points, it divides the sides in the same ratio.

35. $\dfrac{AX}{XB} \cdot \dfrac{BY}{YC} \cdot \dfrac{CZ}{ZA} = \dfrac{ZA}{CZ} \cdot \dfrac{BY}{BY} \cdot \dfrac{CZ}{ZA} = 1.$ (By substituting $\dfrac{ZA}{CZ}$ for $\dfrac{AX}{XB}$ and BY for YC.)

36. It proves that AY, BZ, and CX are concurrent.

What Kind of Triangle?

37.

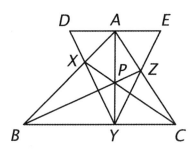

38. Isosceles.

39. AA. (The parallel lines form equal alternate interior angles in the triangles. Also, the vertical angles are equal.)

40. Corresponding sides of similar triangles are proportional.

•41. $\dfrac{AX}{BX} \cdot \dfrac{BY}{CY} \cdot \dfrac{CZ}{AZ} = 1$ (by Ceva's Theorem).

42. $\dfrac{AD}{BY} \cdot \dfrac{BY}{CY} \cdot \dfrac{CY}{AE} = 1$ by substitution.
    $(\dfrac{AX}{BX} = \dfrac{AD}{BY}$ and $\dfrac{CZ}{AZ} = \dfrac{CY}{AE}.)$

43. $\dfrac{AD}{AE} = 1$ because $\dfrac{AD}{BY} \cdot \dfrac{BY}{CY} \cdot \dfrac{CY}{AE} = 1.$

44. Because $\dfrac{AD}{AE} = 1$, AD = AE by multiplication.

•45. In a plane, a line perpendicular to one of two parallel lines is also perpendicular to the other.

46. $\triangle DYA \cong \triangle EYA$ by SAS. (AD = AE, $\angle DAY = \angle EAY$, and AY = AY.)

47. It proves that $\triangle DYE$ is isosceles. DY = EY because corresponding sides of congruent triangles are equal.

The Gergonne Point.

48.

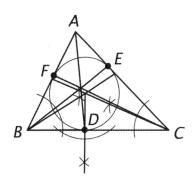

49. AD, BE, and CF appear to be concurrent.

50. According to Ceva's Theorem, AD, BE, and CF are concurrent if $\dfrac{AF}{FB} \cdot \dfrac{BD}{DC} \cdot \dfrac{CE}{EA} = 1.$

The tangent segments to a circle from an external point are equal; so AF = AE, BF = BD, and CD = CE. Substituting in the product $\dfrac{AF}{FB} \cdot \dfrac{BD}{DC} \cdot \dfrac{CE}{EA}$ gives

$\dfrac{AE}{BD} \cdot \dfrac{BD}{DC} \cdot \dfrac{CD}{EA} = 1$; so the three cevians are concurrent.

51. The point of concurrence of AD, BE, and CF (or, the point in which the cevians connecting the vertices of a triangle to the points of tangency of the incircle to the opposite sides are concurrent).

Another Look at Altitudes.

•52. ΔAXC ~ ΔAZB.

53. ΔBYA ~ ΔBXC.

54. ΔCZB ~ ΔCYA.

55. $\dfrac{AX}{AZ} = \dfrac{XC}{ZB}.$

56. $\dfrac{BY}{BX} = \dfrac{YA}{XC}.$

57. $\dfrac{CZ}{CY} = \dfrac{ZB}{YA}.$

58. $\dfrac{AX}{XB} \cdot \dfrac{BY}{YC} \cdot \dfrac{CZ}{ZA} = \dfrac{AX \cdot BY \cdot CZ}{XB \cdot YC \cdot ZA} =$

$\dfrac{AX \cdot BY \cdot CZ}{AZ \cdot BX \cdot CY} = \dfrac{AX}{AZ} \cdot \dfrac{BY}{BX} \cdot \dfrac{CZ}{CY} =$

$\dfrac{XC}{ZB} \cdot \dfrac{YA}{XC} \cdot \dfrac{ZB}{YA} = 1$; so AY, BZ, and CX are concurrent.

59. The lines containing the altitudes of a triangle are concurrent.

Set III (page 560)

If we go either clockwise or counterclockwise around a triangle and three cevians divide each side in the ratio $\dfrac{1}{2}$, then the smaller triangle formed by them has an area that is one-seventh the area of the original triangle. A generalization of this result is that, if the three cevians divide each side in the ratio $\dfrac{1}{x}$, then the smaller triangle has an area that is $\dfrac{(x-1)^2}{x^2 + x + 1}$ times the area of the original triangle. In fact, in the chapter on affine geometry in his *Introduction to Geometry* (Wiley, 1969), H. S. M. Coxeter considers an even broader generalization. If the three cevians divide the sides in the ratios $\dfrac{1}{x}, \dfrac{1}{y}, \dfrac{1}{z}$, then the smaller triangle has an area that is

$\dfrac{(xyz-1)^2}{(xy+x+1)(yz+y+1)(xz+z+1)}$ times the area

of the original triangle! The "seven congruent triangles" approach to the version considered in the text comes from Hugo Steinhaus's *Mathematical Snapshots* (Oxford University Press, 1969).

Area Puzzle.

1. αΔABD = 7x.
   (αΔABD = αΔAFG + αFGHB + αΔBDH = x + 5x + x = 7x.)

2. αΔADC = 14x. (Because ΔABD and ΔADC have equal altitudes and the base of ΔADC is twice as long as the base of ΔABD, αΔADC = 2αΔABD = 14x.)

3. αΔGHI = 3x.
   (αΔGHI = αΔADC – αAGIE – αDHIC – αΔEIC = 14x – 5x – 5x – x = 3x.)

4. Because αΔGHI = 3x and αΔABC = αΔABD + αΔADC = 7x + 14x = 21x, αΔGHI = $\dfrac{1}{7}$αΔABC.

5. The pairs of regions with the same number appear to be congruent and hence have the same area. From this, it appears that ΔABC is equal in area to the sum of the areas of the seven colored triangles. If these triangles are congruent, then αΔGHI (the area of the yellow triangle) is one-seventh of the area of ΔABC.

Set I (pages 562–563)

Exercises 6 through 34 lead to the discovery of the Euler line. Euler, the most prolific writer on mathematics in history, noted the remarkable relation of the circumcenter, orthocenter, and centroid of every nonequilateral triangle in 1765. Not only are the three points always collinear, but the distance from the centroid to the orthocenter is always twice the distance from the centroid to the circumcenter. It is interesting to note that Euler's proof of this relation was analytic. It was not until 1803 that Lazare Carnot published the first proof by Euclidean methods. The exercises are developed around a triangle chosen on a coordinate grid so that the coordinates of every point in the figure (except W) are integers. This was done for two reasons: (1) to help the student draw the figure accurately enough that the Euler line theorem can be discovered and (2) to review basic ideas such as distance and slope. Unfortunately, this approach obscures the fact that the results obtained are not limited to this particular triangle but are generally true.

Tilted Square.

•1. AB and DC.

•2. $-\dfrac{3}{2}$.

3. Because the sides are perpendicular.

•4. $\sqrt{13}$ units. ($\sqrt{3^2+2^2}$.)

5. B(10, 6), C(8, 9), D(5, 7).

Euler's Discovery.
The Centroid.

6.

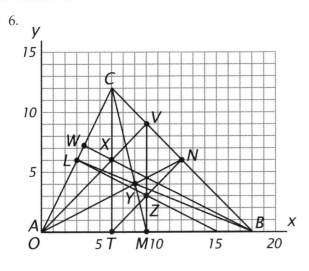

•7. M(9, 0), L(3, 6).

•8. (8, 4).

9. (12, 6).

10. The coordinates of N are the averages of (or midway between) the corresponding coordinates of B and C.

•11. CN = $\sqrt{(6-12)^2+(12-6)^2}$ = $\sqrt{36+36}$ = $\sqrt{72}$ or $6\sqrt{2}$;
NB = $\sqrt{(12-18)^2+(6-0)^2}$ = $\sqrt{36+36}$ = $\sqrt{72}$ or $6\sqrt{2}$.

12. Its centroid.

13. Medians.

The Circumcenter.

14. (9, 9).

15. (6, 0).

•16. mNT = 1 and mCB = –1.

17. Yes. NT ⊥ CB; so their slopes are the opposites of the reciprocals of each other (or the product of their slopes is –1).

•18. (9, 3).

19. Its circumcenter.

20. mAC = 2 and mZL = $-\dfrac{1}{2}$.

21. Their slopes indicate that AC ⊥ ZL.

22. The perpendicular bisectors of its sides.

The Orthocenter.

23. CT ⊥ AB.

•24. At V.

25. mAV = 1.

26. Yes, because mCB = –1 and AV ⊥ CB. (Also, AV ∥ NT because AV ⊥ CB and NT ⊥ CB; so mAV = mNT.)

•27. (6, 6).

28. Its orthocenter.

29. Its altitudes.

30. They appear to be collinear.

31. $mXY = -1$ and $mYZ = -1$ (so, X, Y, and Z *are* collinear).

32. XY = 2YZ because XY = $\sqrt{8}$ = $2\sqrt{2}$ and YZ = $\sqrt{2}$.

33. Z.

34. ZA = ZB = ZC = $\sqrt{90}$ = $3\sqrt{10}$.

Set II (pages 564–565)

Students who did the Set III exercise of Chapter 6, Lesson 1 (page 218), may recall that the figure for exercises 35 through 42 is remarkable in that it shows an arrangement of eight points in the plane such that the perpendicular bisector of the line segment connecting any two of the points passes through exactly two other points of the figure. For most pairs of points in the figure, such as A and B or G and C, this is obvious from the symmetries of its parts.

Exercises 43 through 51 explore another property possessed by all triangles. If squares are constructed on the sides of a triangle as in the familiar figure for the Pythagorean Theorem, the line segments connecting the centers of the squares to the opposite vertices of the triangle are always concurrent. Furthermore, these line segments have a special relation to the triangle whose vertices are the centers of the squares. Each one not only is perpendicular to one of its sides but also has the same length. Again, the use of the coordinate grid makes it easier to draw the figure accurately (especially the squares and their centers) but obscures the fact that the results are true for all triangles.

Exercises 52 through 60 concern a similar result for parallelograms. If squares are constructed outward on the sides of any parallelogram, the two line segments connecting the centers of the opposite squares are both equal and perpendicular to each other. This is connected to the fact that the quadrilateral whose vertices are the centers of the squares also is a square.

A partial proof of Napoleon's Theorem is considered in exercises 61 through 63. On the assumption that the three line segments connecting the vertices of the original triangle to the opposite vertices of the equilateral triangles are equal, are concurrent, and form equal angles with each other, it is fairly easy to prove that the triangle determined by the centers of the three equilateral triangles also is equilateral. In *Geometry Civilized* (Clarendon Press, 1998), J. L. Heilbron presents a complete trigonometric proof of the theorem based on the Law of Cosines. In *Hidden Connections, Double Meanings* (Cambridge University Press, 1988), David Wells wrote concerning Napoleon's Theorem: "To discover pattern and symmetry where there appears to be neither is always delightful and intriguing. It also suggests that we are not seeing all there is to be seen in the original diagram. There must certainly be another way of looking at it which will be more symmetrical from the start, and therefore make the conclusion more natural." On pages 49–52 of his book, Wells presents another way, using tessellations, to understand this remarkable theorem.

Triangles on Four Sides.

35.

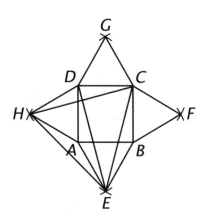

• 36. 48 units. (8 × 6.)

37. 150°. (∠CBE = 90° + 60° = 150°.)

• 38. 150°. (∠HAE = 360° − 60° − 90° − 60° = 150°.)

39. They appear to be perpendicular. (Also, DE appears to be the perpendicular bisector of HC.)

40. They are congruent by SAS. (They are also isosceles.)

41. It proves that ΔHCE is equilateral.

42. They prove that HC and DE are perpendicular because, in a plane, two points each equidistant from the endpoints of a line segment determine the perpendicular bisector of the line segment.

Squares on Three Sides.

43.

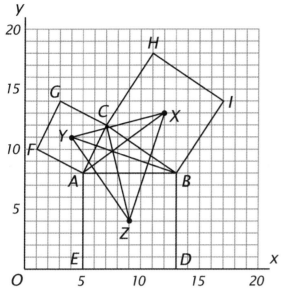

44. mAB = 0, mAC = 2, mCB = $-\dfrac{2}{3}$.

•45. No. AC and CB are not perpendicular, because the product of their slopes is not –1.

•46. G(3, 14), I(17, 14).

47. X(12, 13), Y(4, 11), Z(9, 4).

48. They appear to be concurrent.

49. They appear to be perpendicular to the sides of ΔXYZ.

50. It is true because $(m$AX$)(m$YZ$) = (\dfrac{5}{7})(-\dfrac{7}{5}) =$
    -1, $(m$BY$)(m$XC$) = (-\dfrac{1}{3})(3) = -1$, and
    $(m$CZ$)(m$YX$) = (-4)(\dfrac{1}{4}) = -1$.

51. They are equal to the sides of ΔXYZ to which they are perpendicular.
    AX = YZ = $\sqrt{7^2 + 5^2}$ = $\sqrt{74}$;
    BY = XZ = $\sqrt{9^2 + 3^2}$ = $\sqrt{90}$;
    CZ = YX = $\sqrt{2^2 + 8^2}$ = $\sqrt{68}$.

Squares on Four Sides.

52.

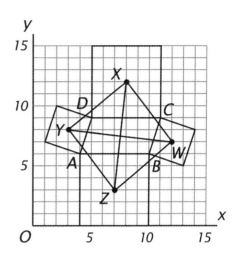

•53. AB = DC = 6; AD = BC = $\sqrt{10}$.

54. It is a parallelogram.

55. Both pairs of opposite sides are equal (or, two opposite sides are equal and parallel).

56. W(12, 7), X(8, 12), Y(3, 8), Z(7, 3).

•57. mXZ = 9, mWY = $-\dfrac{1}{9}$.

58. That XZ and WY are perpendicular.

59. XZ = WY = $\sqrt{1^2 + 9^2}$ = $\sqrt{82}$.

60. WXYZ is a square.

Napoleon Triangles.

61. They are cyclic because their opposite angles are supplementary. (For example, in APCE, ∠E = 60° and ∠APC = 120°.)

62. They are perpendicular to the triangle's sides because, in a plane, two points each equidistant from the endpoints of a line segment determine the perpendicular bisector of the line segment. (For example, XA = XP and YA = YP because all radii of a circle are equal; so XY is the perpendicular bisector of AP.)

63. ΔXYZ is equilateral because it is equiangular. It is equiangular because each of its angles is 60°. (For example, ∠X = 60° because, in quadrilateral PGXI, ∠XGP = ∠XIP = 90° and ∠GPI = 120°.)

Here, a surprising variation of Napoleon's Theorem is considered.

Alternating Triangles.

1. If equilateral triangles are drawn alternately outward and inward on the sides of a quadrilateral, their vertices determine a parallelogram.

2. Yes. The theorem still seems to apply. *Example figure:*

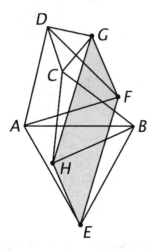

Chapter 13, Review

Set I (pages 566–568)

Exercises 1 through 4 illustrate the fact that the feet of the three perpendiculars to the sides of a triangle from a point on its circumcircle are collinear. The line that they determine is called the Simson line, named after Robert Simson, a professor of mathematics at the University of Glasgow. Simson produced an important edition of Euclid's *Elements* in 1756. Proofs of the existence of the Simson line based on cyclic quadrilaterals can be found in J. L. Heilbron's *Geometry Civilized* and Nathan Altshiller Court's *College Geometry*. The Simson line leads to so many other ideas that it is the subject of an entire chapter of Court's book.

   The configuration in exercises 28 through 31 has additional interesting properties. The perpendicular bisectors of the sides of the rectangle whose vertices are the incenters of the four overlapping triangles also bisect the four arcs of the circle corresponding to the sides of ABCD. Furthermore, each vertex of the rectangle is

collinear with a midpoint of one of these arcs and a vertex of ABCD. A proof of the theorem that the incenters of the four triangles determined by the vertices of a cyclic quadrilateral determine a rectangle can be found on page 133 of Nathan Altshiller Court's *College Geometry*.

Cyclic Triangle.

1. That there exists a circle that contains all of its vertices.

•2. ΔABC is inscribed in the circle.

3. They appear to be perpendicular to them.

4. They are collinear.

Circumcircle and Incircle.

5. *Example figure:*

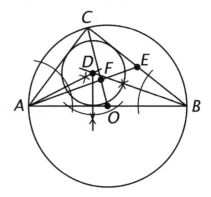

•6. ΔABC is a right triangle because ∠ACB is inscribed in a semicircle.

7. AE (or CO).

8. AC (or BC).

9. A and D (or B and D or C and D).

10. D.

11. O.

•12. F.

13. C.

Three Trapezoids.

•14. IJKL.

15. EFGH.

16. A quadrilateral is cyclic iff a pair of its opposite angles are supplementary. (For ABCD, we don't know whether a pair of opposite angles is supplementary or not.)

Double Identity.

17. Its circumcenter.

•18. Midsegments.

19. The sides of ΔDEF are parallel to the sides of ΔABC and half as long.

20. Its altitudes.

21. Its orthocenter.

22. A, B, and C.

Ceva's Theorem.

23.

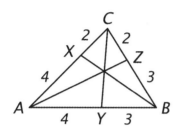

•24. AX = 4, YB = 3, and ZC = 2.

25. $\dfrac{CX}{XA} = \dfrac{1}{2}$, $\dfrac{AY}{YB} = \dfrac{4}{3}$, and $\dfrac{BZ}{ZC} = \dfrac{3}{2}$.

26. $\dfrac{CX}{XA} \cdot \dfrac{AY}{YB} \cdot \dfrac{BZ}{ZC} = 1.$

27. It indicates that AZ, BX, and CY are concurrent.

Five Circles.

•28. Cyclic.

29. It is circumscribed about it.

30. They are the incircles of the (overlapping) triangles whose sides are the sides and diagonals of the quadrilateral.

31. They seem to be the vertices of a rectangle. (This is surprising because ABCD, the quadrilateral from which it comes, has no special property other than being cyclic.)

Centroid.

32.

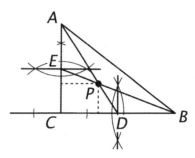

33. 1 in.

•34. $\dfrac{3}{4}$ in.

35. They are one-third of the lengths.

Set II (pages 568–570)

The nineteenth-century Swiss mathematician Jacob Steiner has been called by Howard Eves "one of the greatest synthetic geometers the world has ever known." The "Steiner problem" named after him concerns the problem of connecting a set of coplanar points with a set of line segments having the least possible total length. (The Spotter's problem is the special case of this problem for a set of three points situated at the vertices of an equilateral triangle.) In *The Parsimonious Universe* (Copernicus, 1996), Stefan Hildebrandt and Anthony Tromba describe solving the Steiner problem by means of soap films: "Suppose we make a frame consisting of two parallel glass or clear plastic plates; these are connected by *n* parallel pins of the same size that meet both plates perpendicularly. If this framework is immersed in a soap solution and withdrawn, a system of planar soap films is formed. These films are attached to the pins, and they have two kinds of liquid edges, both of which are straight lines. One type adheres to the glass (or plastic), which it meets at 90° angles. . . . The other type is where three films meet, forming three angles of 120°. . . . Since the soap-film system minimizes area, the subsystem of edges on one plate must minimize length among all connections between the *n* given points." The exercises demonstrate that, for four suitably located points, equilateral triangles drawn on opposite sides of the quadrilateral determined by them can be used to construct a network of lines that meet at equal (120°) angles. Repeating the method with the other pair of opposite sides produces a different network with a possibly different total length. One of the two, however, is the absolute minimum, the one obtained by the soap-film method.

Exercises 47 through 49 imply that two angle bisectors of a triangle meet at an angle determined entirely by the measure of the third angle of the triangle. In terms of the figure and with the use of a similar but more general argument, it can be shown that $\angle D = \frac{1}{2}\angle A + 90°$.

The two quadrilaterals in exercises 51 through 54 have another interesting relation. If perpendiculars are drawn from the point of intersection of the diagonals of a cyclic quadrilateral to its sides, their feet are the vertices of the inscribed quadrilateral of minimum perimeter. EFGH is therefore the quadrilateral of minimum perimeter that can be inscribed in ABCD.

Exercises 60 through 63 explore a more general case of the relations introduced in exercises 52 through 60 of Lesson 6. Martin Gardner included it in one of his "Mathematical Games" columns for *Scientific American*, reprinted in *Mathematical Circus* (Knopf, 1979). The column was on the subject of "simplicity" in science and mathematics, and Gardner wrote: "Mathematicians usually search for theorems in a manner not much different from the way physicists search for laws. They make empirical tests. In pencil doodling with convex quadrilaterals—a way of experimenting with physical models—a geometer may find that when he draws squares outwardly on a quadrilateral's sides and joins the centers of opposite squares, the two lines are equal and intersect at 90 degrees. He tries it with quadrilaterals of different shapes, always getting the same results. Now he sniffs a theorem. Like a physicist, he picks the simplest hypothesis. He does not, for example, test first the conjecture that the two lines have a ratio of one to 1.00007 and intersect at angles of 89 and 91 degrees, even though this conjecture may equally well fit his crude measurements. He tests first the simpler guess that the lines are always perpendicular and equal. His 'test,' unlike the physicist's, is a search for a deductive proof that will establish the hypothesis with certainty." Gardner goes on to remark that this result is known as Von Aubel's theorem and mentions several remarkable generalizations of it.

Exercises 64 through 67 establish the nice result that the area of a triangle is equal to the product of the lengths of its three sides divided by four times the radius of its circumcircle. (It is easier to begin with this conclusion and show that it is equivalent to the familiar formula for the area of a triangle than it is to do it the other way around.)

Soap-Film Geometry.

• 36. $\angle AED + \angle AGD = 180°$. $\angle AED$ and $\angle AGD$ are supplementary because they are opposite angles of a cyclic quadrilateral.

37. $\angle AGD = 120°$ because $\angle AED = 60°$.

• 38. $\angle EGD = \angle EAD$ because they are inscribed angles that intercept the same arc.

• 39. $\angle EGD = 60°$ because $\angle EGD = \angle EAD$.

40. $\angle DGH = 120°$ because $\angle DGH$ forms a linear pair with $\angle EGD$.

41. $\angle AGH = 120°$ because $\angle AGH = 360° - \angle AGD - \angle DGH$.

• 42. $\angle BHC = 120°$ because it is supplementary to $\angle BFC$.

43. $\angle FHC = 60°$ because $\angle FHC = \angle FBC$.

44. $\angle GHC = 120°$ because $\angle GHC$ forms a linear pair with $\angle FHC$.

45. $\angle BHG = 120°$ because $\angle BHG = 360° - \angle GHC - \angle BHC$.

46. It suggests that they form equal angles.

Irrelevant Information.

47. 125°. (In $\triangle ABC$, $\angle A = 70°$ and $\angle ABC = 60°$; so $\angle ACB = 50°$. Because D is the incenter of $\triangle ABC$, BD bisects $\angle ABC$ and so $\angle DBC = 30°$, and CD bisects $\angle ACB$ and so $\angle DCB = 25°$. In $\triangle DBC$, $\angle DBC = 30°$ and $\angle DCB = 25°$; so $\angle D = 125°$.)

• 48. 125°. (In $\triangle ABC$, $\angle A = 70°$ and $\angle ABC = 40°$; so $\angle ACB = 70°$. Reasoning as in exercise 47, we find that $\angle DBC = 20°$ and $\angle DCB = 35°$; so $\angle D = 125°$.)

49. 125°. (In $\triangle ABC$, $\angle ACB = 180° - 70° - \angle ABC = 110° - \angle ABC$. $\angle DBC = \frac{1}{2}\angle ABC$ and $\angle DCB = \frac{1}{2}(110° - \angle ABC) = 55° - \frac{1}{2}\angle ABC$; so $\angle D = 180° - \frac{1}{2}\angle ABC - (55° - \frac{1}{2}\angle ABC) = 125°$.)

Medians Theorem.

50. Because the medians bisect the sides of the triangle, AX = XB, BY = YC, and CZ = ZA;

    so $\dfrac{AX}{XB} = \dfrac{BY}{YC} = \dfrac{CZ}{ZA} = 1$. Therefore,

    $\dfrac{AX}{XB} \cdot \dfrac{BY}{YC} \cdot \dfrac{CZ}{ZA} = 1 \cdot 1 \cdot 1 = 1$; so the

    medians are concurrent.

Irregular Billiard Table.

51. Cyclic.

52. They appear to be perpendicular to its sides.

•53. They appear to bisect its angles.

54. It would travel around the sides of quadrilateral EFGH (because its angles of incidence and reflection at the points in which it hits the sides of ABCD are equal).

Perimeter Problem.

•55. DB and FB.

56. AE and AF.

57. AB.

58. AB.

59. The perimeter of ΔABC is
    AB + BC + CA = 2R + a + b.
    AB = (a − r) + (b − r) = 2R; so a + b = 2R + 2r.
    Therefore, the perimeter of ΔABC is
    2R + (2R + 2r) = 2r + 4R.

Squares on the Sides.

60.

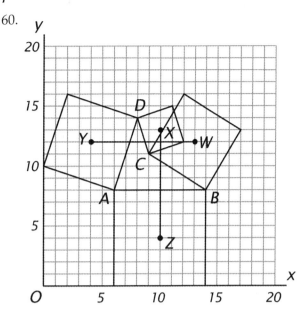

•61. ABCD is concave.

62. W(13, 12), X(10, 13), Y(4, 12), Z(10, 4).

63. XZ ⊥ WY and XZ = WY. (XZ = WY = 9.)

Area Problem.

64. ΔABD ~ ΔAEC.

65. ∠B = ∠E because they are inscribed angles that intercept the same arc. ∠ADB is a right angle because AD ⊥ BC, and ∠ACE is a right angle because it is inscribed in a semicircle; so ∠ADB = ∠ACE.

66. Because ΔABD ~ ΔAEC, $\dfrac{AB}{AE} = \dfrac{AD}{AC}$

    (corresponding sides of similar triangles are proportional). So AB · AC = AD · AE by multiplication.

67. $\dfrac{AB \cdot AC \cdot BC}{4r} = \dfrac{AD \cdot AE \cdot BC}{4r}$ by substitution.

    AE = 2r; so $\dfrac{AD \cdot 2r \cdot BC}{4r} = \dfrac{AD \cdot BC}{2}$.

    $\dfrac{AD \cdot BC}{2}$ is the area of ΔABC.

Set I (pages 574–576)

The two lithographs of Max Bill's *Fifteen Variations on a Single Theme* (1935–1938) shown in the text are Variations 14 and 8. The first picture in the series, the "Theme," consists of a continuous black spiral of 22 equal segments which form two sides of an equilateral triangle, three sides of a square, four sides of a regular pentagon, and so on, to seven sides of a regular octagon. In each of the 15 variations, the artist plays with various geometric possibilities inherent in this figure. Many of Bill's works were influenced by the figures of plane and solid geometry as well as topology.

In *Mathematical Models,* by H. Martyn Cundy and A. P. Rollett (Oxford University Press, 1961), the authors say concerning polygonal knots: "If a strip of paper is knotted once and carefully pressed flat the folds will form a regular pentagon. All polygons with an odd number of sides [starting with the pentagon] may be produced in this way. . . . The even-sided polygons require two strips of equal width; a reef-knot leads to the hexagon. . . ." Pictures of the first four polygonal knots can also be found in *The Penguin Dictionary of Curious and Interesting Geometry,* by David Wells (Penguin Books, 1991).

Geometry in Art.

1. Equilateral triangle, square, regular pentagon, regular hexagon, regular heptagon, regular octagon.

2. Each successive polygon shares a side with the preceding polygon.

•3. The radii of the polygons.

4. The circles that can be circumscribed about the polygons.

Geometry in Nature.

•5. Yes. If a triangle is equilateral, it is also equiangular.

6. Yes.

7. No. Every triangle is cyclic, but not every triangle is regular.

•8. No.

9. Yes. (If it is equiangular, it is a rectangle; so its opposite angles are supplementary.)

•10. Yes. (*All* regular polygons are cyclic.)

11. No.
 Example figure:

Cell Pattern.

12. Regular hexagons.

•13. 60°. ($\frac{360°}{6}$.)

•14. Equilateral.

15. The sides are equal to the radius.

16. *Example figure:*

17. The perimeter of the hexagon is three times the diameter of the circle.

Regular Dodecagon.

•18. Three squares.

19. Four equilateral triangles.

•20. They are factors of it.

•21. Equilateral triangles and regular pentagons.

22. Squares and regular octagons.

23. No regular polygons.

Polygonal Knot.

24. A regular heptagon.

25. No.

26. Yes.

•27. Regular polygons having an even number of sides.

28. Regular polygons having at least six sides.

The approximate construction of the regular pentagon explored in exercises 29 through 32 is discussed in detail by J. L. Heilbron on pages 226 and 227 of *Geometry Civilized* (Clarendon Press, 1998). For a pentagon to be regular, each of its angles must have a measure of 108°. Unfortunately, in this particular construction, ∠A = ∠B ≈ 108.37°, ∠C = ∠E ≈ 107.04°, and ∠D ≈ 109.17°. The method, popular with medieval masons, was also included by Renaissance artist Albrecht Dürer in his book titled *Course in the Art of Measurement with Compasses and Rulers* (1525).

Hidden in the figure for exercises 51 through 61 is the golden ratio. The appearance of the 36°-72°-72° isosceles triangles in the figure is the reason. Bisecting ∠ABF produces a triangle similar to ΔABF. From this, it can be shown that the ratio of a leg of ΔABF to its base is the golden ratio. It follows immediately from the figure that the diagonals of a regular pentagon cut each other in this same ratio.

Mason's Pentagon.

29.

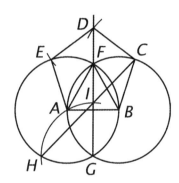

30. Yes. All five sides have been constructed equal to AB, the radius of the circles.

•31. If it isn't regular, it must not be equiangular.

32. Yes. ΔABF is regular because it is equilateral and therefore equiangular.

Folding an Octagon.

•33. Isosceles right triangles.

34. SAS.

35. Corresponding sides of congruent triangles are equal.

36. Isosceles (and obtuse).

37. ASA. (All of their acute angles equal 22.5°.)

38. They are equal.

•39. 22.5°. (∠AEI = $\frac{1}{2}$45° = 22.5°.)

40. 135°. [∠EIH = 180° – 2(22.5°) = 135°.]

•41. 135°. [∠IEJ = 90° + 2(22.5°) = 135°.]

42. They are equal.

43. They prove that it is regular.

Hexagonal Fastener.

44. Radii.

•45. Apothems.

46. PA = PB (all radii of a circle are equal) and ∠APC = ∠BPD (vertical angles are equal); so ΔAPC ≅ ΔBPD by HA.

•47. 30°-60° right triangles.

48. AC = $\frac{0.25 \text{ in}}{2}$ = 0.125 in and
PC = 0.125√3 in ≈ 0.2165 in.

49. AB = 2PA = 0.50 in and
CD = 2PC = 0.25√3 in ≈ 0.433 in.

50. CD, because the wrench is used to grip the sides of the head, not its corners.

Inscribed Pentagon.

51.

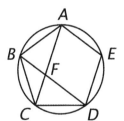

•52. They are equal because the sides of the pentagon are chords of the circle, and equal arcs in a circle have equal chords.

53. 72°. ($\frac{360°}{5}$.)

•54. 72°. [∠ABD = $\frac{1}{2}m\widehat{AD}$ = $\frac{1}{2}$(2 · 72°) = 72°.]

55. 72°. [∠AFB = $\frac{1}{2}(m\widehat{AB} + m\widehat{CD})$ =
$\frac{1}{2}$(72° + 72°) = 72°.]

56. 36°. ($\angle BAC = \frac{1}{2}m\widehat{BC} = \frac{1}{2}72° = 36°$.)

57. 36°.

58. Five.

59. They are all isosceles.

60. AEDF is a rhombus: AF = AB and FD = CD; so all four sides of AEDF are equal to the sides of the regular pentagon. Because AEDF is a rhombus, it follows that it is also a parallelogram.

61. AEDB and AEDC are isosceles trapezoids.

Set III (page 578)

A chapter of Robert Dixon's book titled *Mathographics* (Dover, 1987) deals with "Euclidean approximations" to some of the impossible constructions. As Dixon explains concerning the regular heptagon: "To divide a circle into exactly seven equal parts would require us to be able to construct the angle of $\frac{360°}{7} \approx 51.428571°$, which we cannot do." He then presents and analyzes three comparatively simple constructions that are amazingly close approximations of the regular heptagon based on constructing angles of approximately 51.317813°, 51.340192°, and 51.470701°.

A Close Construction.

1.

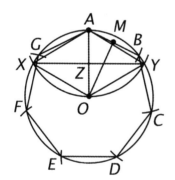

2. Equilateral.

3. AO is the perpendicular bisector of XY (because A and O are equidistant from X and Y.)

4. A 30°-60° right triangle.

5. $OZ = \frac{1}{2}r$. ($OZ = \frac{1}{2}OY$.)

6. $ZY = \frac{1}{2}r\sqrt{3}$. ($ZY = \sqrt{3}\,OZ$.)

7. $AB = \frac{1}{2}r\sqrt{3}$. (AB = ZY by construction.)

8. $AM = \frac{1}{4}r\sqrt{3}$. ($AM = \frac{1}{2}AB$.)

9. 25.658906...°. ($\sin \angle AOM = \frac{AM}{AO} = \frac{\frac{1}{4}r\sqrt{3}}{r} = \frac{\sqrt{3}}{4} = 0.433012702...$)

10. 51.317812...°. ($\angle AOB = 2\angle AOM$.)

11. 359.22468...°.

12. That ABCDEFG is not regular. If ABCDEFG were regular, $7\angle AOB$ would equal 360°.

Chapter 14, Lesson 2

Set I (pages 581–582)

The figure showing the design of Palma Nuova suggests that the city was surrounded by nine bastions (the spade-shaped regions) for its defense, as well as possibly by a moat.

Students taking chemistry may wonder how each hydrogen atom in the hydrogen fluoride rings can have two bonds. In his book titled *The Architecture of Molecules* (W. H. Freeman and Company, 1964), Linus Pauling explains: "Many properties of substances can be easily explained by the assumption that the hydrogen atom, which is normally univalent, can sometimes assume ligancy two, and form a bridge between two atoms. This bridge is called the hydrogen bond. . . . Hydrogen fluoride gas has been found to contain not only molecules HF, but also polymers, especially $(HF)_5$ and $(HF)_6$ . . . . Each hydrogen atom is strongly bonded to one fluorine atom (bond length 1.00 Å) and less strongly to another (bond length 1.50 Å)." "Å" is the symbol for Angstrom, a unit of length equal to 10^{-8} cm (10^{-10} m).

The Susan B. Anthony dollar coin never became popular, owing in part to the fact that it was too easily confused with a quarter in both look and feel.

Perimeter Equations.

1. The radius.

•2. The number of sides.

3. 3.

•4. 3.0902. (10 sin 18°.)

5. 3.1411. (100 sin 1.8°.)

6. 3.1416. (1,000 sin 0.18°.)

7. 3.1416. (10,000 sin 0.018°.)

8. It increases.

9. It becomes more and more circular.

Angles and Radii.

•10. Radii.

11. △ABP ≅ △CBP by SSS.

12. ∠1 = ∠2 because corresponding parts of congruent triangles are equal.

13. No.

•14. PB bisects ∠ABC.

Italian City.

15. Nine.

•16. 9 sin 20°.

•17. 3.078.

18. 6,772 ft. [2(3.078)(1,100).]

Hydrogen Fluoride.

19. A regular pentagon and a regular hexagon.

20. They seem to be equal.

•21. The radius of the hexagon appears to be longer.

•22. 108°. ($\frac{3 \cdot 180°}{5}$.)

23. 120°. ($\frac{4 \cdot 180°}{6}$, or 2·60°.)

24. 12.5 angstroms. (5 × 2.5.)

25. 15 angstroms. (6 × 2.5.)

•26. 5 sin 36° ≈ 2.94.

27. 6 sin 30° = 3.

•28. Approximately 2.13 angstroms. [12.5 = 2(2.94)r, r ≈ 2.13.]

29. 2.5 angstroms. [15 = 2(3)r, r = 2.5.]

Dollar Coin.

30. 75 mm. [c = 2πr = 2π(12) = 24π ≈ 75.]

•31. 74 mm. [p = 2(11 sin $\frac{180}{11}$)12 ≈ 74.]

32. 7 mm. ($\frac{74}{11}$ ≈ 7.)

Set II (pages 583–584)

Most people who use them probably have no idea that the name "cell phone" refers to the cells into which the areas in which they are used are divided or that the cells have the same shape that bees use in building honeycombs: the regular hexagon.

Although regular pentagons cannot be used to fill a plane as hexagons can be, the figure for exercises 41 through 48 shows that 10 of them can fit perfectly around a regular decagon. In his book titled *Geometry in Architecture* (Wiley, 1984), William Blackwell observes that this is a nice arrangement for surrounding a large 10-sided pavilion with 10 smaller galleries.

Exercises 49 through 58 lead to Proposition 10 of Book XIII of the *Elements*: "If an equilateral pentagon is inscribed in a circle, the side of the pentagon is equal in square to that of the hexagon and that of the decagon inscribed in the same circle." A good discussion of how this proposition is proved is included by Benno Artmann in *Euclid—The Creation of Mathematics* (Springer, 1999). Artmann remarks concerning this theorem: "In other words, the side of the pentagon is the hypotenuse of a right triangle that has the sides of the hexagon and of the decagon as its legs— a really unexpected and beautiful insight!" After explaining that Euclid may have discovered this remarkable theorem in his investigation of the icosahedron, Artmann also observes: "It is common in mathematics that certain parts of complicated proofs become of independent interest, and afterwards one wonders how anybody could have thought of them."

Phone Cells.

33. p = 2Nr = 2(6 sin 30°)r = 2(3)r = 6r.

34.

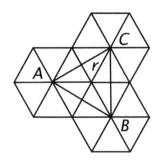

35. Equilateral triangles. The sides of each cell are equal to its radius; so the perimeter of each cell is $6r$.

36. Equilateral.

•37. $r\sqrt{3}$. [r is the hypotenuse of each 30°-60° right triangle included by $\triangle ABC$; so the length of the shorter leg is $\dfrac{r}{2}$ and the length of the longer leg is $\dfrac{r}{2}\sqrt{3}$.
$AB = 2(\dfrac{r}{2}\sqrt{3}) = r\sqrt{3}$.]

38. $3r\sqrt{3}$.

•39. $p = 2Nr = 2(3\sin 60°)r \approx 5.196r$.

40. Yes. $3r\sqrt{3} \approx 5.196r$.

Ten Pentagons.

•41. $\dfrac{10}{5} = 2$.

42. 36°. $(\angle X = \dfrac{1}{2}\cdot\dfrac{360°}{5} = 36°.)$

43. 18°. $(\angle XYZ = \dfrac{1}{2}\cdot\dfrac{360°}{10} = 18°.)$

•44. By the Law of Sines.

•45. $\dfrac{R}{r} = \dfrac{\sin 36°}{\sin 18°} \approx 1.9$.

(Because $\dfrac{\sin \angle X}{R} = \dfrac{\sin \angle XYZ}{r}$,

$\dfrac{R}{r} = \dfrac{\sin \angle X}{\sin \angle XYZ} = \dfrac{\sin 36°}{\sin 18°}.)$

46. X and Y are equidistant from V and Z. (Two points each equidistant from the endpoints of a line segment determine the perpendicular bisector of the line segment.)

47. The tangent of an acute angle of a right triangle is the ratio of the length of the opposite leg to the length of the adjacent leg.

•48. $\dfrac{A}{a} = \dfrac{\tan 36°}{\tan 18°} \approx 2.2$. (Because $\tan \angle X = \dfrac{WZ}{a}$

and $\tan \angle XYZ = \dfrac{WZ}{A}$,

$WZ = a \tan \angle X = A \tan \angle XYZ$;

so $\dfrac{\tan \angle X}{\tan \angle XYZ} = \dfrac{A}{a}.)$

Euclid's Discovery.

•49. 18°. $(\dfrac{1}{2}\cdot\dfrac{360°}{10} = 18°.)$

50. 30°. $(\dfrac{1}{2}\cdot\dfrac{360°}{6} = 30°.)$

51. 36°. $(\dfrac{1}{2}\cdot\dfrac{360°}{5} = 36°.)$

•52. $AX = 3.0901699\ldots$. $(\sin 18° = \dfrac{AX}{10}$;
so $AX = 10 \sin 18°.)$

53. $BY = 5$. $(\sin 30° = \dfrac{BY}{10}$; so $BY = 10 \sin 30°.)$

54. $CZ = 5.8778525\ldots$. $(\sin 36° = \dfrac{CZ}{10}$;
so $CZ = 10 \sin 36°.)$

55. $a^2 = (2\,AX)^2 = 38.19660\ldots$.

56. $b^2 = (2\,BY)^2 = 100$.

57. $c^2 = (2\,CZ)^2 = 138.19660\ldots$.

58. Euclid discovered that $a^2 + b^2 = c^2$. If a regular decagon, hexagon, and pentagon are inscribed in circles of a given radius (or the same circle), the sum of the squares of a side of the decagon and a side of the hexagon is equal to the square of a side of the pentagon.

Set III (page 584)

This exercise suggests that the ratio of the length of a diagonal to the length of a side of a regular pentagon is the golden ratio. We are more accustomed to seeing this ratio in its algebraic form, $\frac{1+\sqrt{5}}{2}$, than in its trigonometric form, $2 \sin 54°$.

Golden-Heather.

1. $\angle AOB = \frac{360°}{5} = 72°$; so $\angle AOX = 36°$.

   $\sin \angle AOX = \frac{AX}{OA}$; so $\sin 36° = \frac{AX}{1}$.

   $AB = 2AX = 2 \sin 36°$.

2. $\angle COE = 2\angle AOB = 144°$; so $\angle COY = 72°$.

   $\sin \angle COY = \frac{CY}{OC}$; so $\sin 72° = \frac{CY}{1}$.

   $CE = 2CY = 2 \sin 72°$.

3. $\frac{2 \sin 72°}{2 \sin 36°} = 1.6180339\ldots$ and

   $2 \sin 54° = 1.6180339\ldots.$

4. The golden ratio.

Chapter 14, Lesson 3

Set I (pages 587–588)

The expression $M = n \sin \frac{180}{n} \cos \frac{180}{n}$ has meaning only when n is an integer larger than 2, because n represents the number of sides of a polygon. Although a calculator readily finds sines and cosines for any angle, our right triangle definitions are valid only for acute angles. Nevertheless, with the use of a calculator when $n = 1$ or 2 and there is no proper polygon, M turns out to equal 0, suggesting that no area is enclosed.

Exercises 20 through 28 provide another preview of the fact that our work with regular polygons will be useful in measuring the circle.

Area Equations.

1. The radius.

2. The number of sides.

•3. 0. $(1 \sin 180° \cos 180°.)$

4. 0. $(2 \sin 90° \cos 90°.)$

•5. 1.30. $(3 \sin 60° \cos 60°.)$

6. 3.14. $(180 \sin 1° \cos 1°.)$

7. 3.

Inscribed Polygon.

8. The area of the circle.

•9. The perimeter of the nonagon.

10. The length of a side of the nonagon.

11. The circumference of the circle.

The Area of a Square.

12. A line segment that connects the center of the square to a vertex, or the distance from its center to one of its vertices.

13. s^2.

•14. $4a^2$. $[s = 2a;$ so $s^2 = (2a)^2 = 4a^2.]$

15. An isosceles right triangle.

•16. $r = a\sqrt{2}$.

17. $r^2 = 2a^2$.

•18. $2r^2$. $(\alpha ABCD = 4a^2$ and $r^2 = 2a^2$; so $\alpha ABCD = 2r^2.)$

19. $(4 \sin 45° \cos 45°)r^2 = 2r^2$.

Close But Not Quite.

20. Its circumference.

21. 628 units. $[c = 2\pi(100) \approx 628.]$

•22. 628 units. $[p = 2(60 \sin 3°)100 \approx 628.]$

23. The circle.

24. 3.14. $[\frac{628}{2(100)}.]$

25. 3.14.

26. The circle.

27. 31,416 square units. $(a = \pi 100^2 \approx 31,416.)$

•28. 31,359 square units. $[A = 60(\sin 3° \cos 3°)100^2 \approx 31,359.]$

József Kürschák, the man associated with the geometric proof that the area of a regular dodecagon having radius r is $3r^2$, was a professor of mathematics at the Polytechnic University in Budapest for many years. He is also remembered as the author of the *Hungarian Problem Books I and II*, still available in the New Mathematics Library of the Mathematical Association of America.

Honeycomb Geometry.

29. Equilateral.

30. An apothem.

•31. 30°-60° right triangles.

•32. $\dfrac{r}{2}$.

33. $\dfrac{r}{2}\sqrt{3}$.

•34. $\dfrac{\sqrt{3}}{4}r^2$.

35. $\dfrac{3\sqrt{3}}{2}r^2$.

   [$\alpha ABCDEF = 6\alpha\Delta AOB = 6(\dfrac{\sqrt{3}}{4}r^2) = \dfrac{3\sqrt{3}}{2}r^2.$]

36. $(6 \sin 30° \cos 30°)r^2$.

37. In ΔBOP, $\sin \angle BOP = \dfrac{PB}{OB} = \dfrac{\frac{r}{2}}{r} = \dfrac{1}{2}$.

38. In ΔBOP, $\cos \angle BOP = \dfrac{OP}{OB} = \dfrac{\frac{r}{2}\sqrt{3}}{r} = \dfrac{\sqrt{3}}{2}$.

39. $(6 \sin 30° \cos 30°)r^2 = (6(\dfrac{1}{2})(\dfrac{\sqrt{3}}{2}))r^2 = \dfrac{3\sqrt{3}}{2}r^2$.

40. Yes. It agrees with the answer to exercise 35.

Kürschák Triangles.

41. ΔFHI and ΔBHI. Because they are equilateral and share a side, they are congruent by SSS. (Or, because they share a side and have 60° angles, they are congruent by ASA or SAS.)

•42. They are isosceles.

•43. That they are 15°. ($\dfrac{90° - 60°}{2}$.)

44. That they are 150°. [$180° - 2(15°)$.]

45. $\angle AHI = \angle HIC = 150°$. [$2(60°) + 2(15°)$.]

46. r. (DH = DI = AB = BC = CD = DA = r.)

•47. AH = HI = IC = $(2 \sin 15°)r \approx 0.518r$. [AH, HI, and IC are sides of a regular dodecagon with radius r. Its perimeter is $2(12 \sin 15°)r$; so the length of each side is $(2 \sin 15°)r$.]

48. DE = DF = DG = $(2 \sin 15°)r \approx 0.518r$. (The yellow triangles are isosceles.)

•49. $\dfrac{1}{4}r^2$. (BAHIC contains one-fourth of the examples of each type of triangle that make up the square.)

50. $\dfrac{3}{4}r^2$. ($r^2 - \dfrac{1}{4}r^2$.)

51. $\dfrac{1}{4}r^2$. ($\alpha\Delta DAH = \dfrac{1}{3}\alpha DAHIC$.)

Rats!

52. (*Student answer.*) (The square has the larger radius.)

53. (*Student answer.*) (The square has the larger perimeter.)

•54. 7.1 cm. [$100 = 4(\sin 45° \cos 45°)r^2$, $100 = 2r^2$, $r^2 = 50$, $r = \sqrt{50} \approx 7.1$.]

55. 6.5 cm. [$100 = 5(\sin 36° \cos 36°)r^2$, $100 \approx 2.38r^2$, $r^2 \approx 42.0$, $r \approx 6.5$.]

56. 40 cm. (Because $s^2 = 100$, $s = 10$, and $p = 4s = 40$.)

•57. 38 cm. [$p \approx 2(5 \sin 36°)(6.5) \approx 38$.]

58. (*Student answer.*) (Both the radius and the perimeter of the pentagon are smaller than those of the square.)

Hebrew Exercise.

59. The problem is to prove that the area of a regular octagon having a radius of R is $2\sqrt{2}\, R^2$.

60. $A = (8 \sin 22.5° \cos 22.5°)R^2 \approx 2.828R^2$, and $2\sqrt{2}\, R^2 \approx 2.828R^2$.

61.

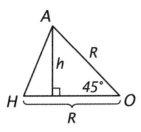

In $\triangle AOH$, $h = \dfrac{R}{\sqrt{2}} = \dfrac{\sqrt{2}R}{2}$; so

$$\alpha\triangle AOH = \frac{1}{2}HO \cdot h = \frac{1}{2}R(\frac{\sqrt{2}R}{2}) = \frac{\sqrt{2}}{4}R^2.$$

The area of the octagon is

$$8\alpha\triangle AOH = 8(\frac{\sqrt{2}}{4}R^2) = 2\sqrt{2}\,R^2.$$

Set III (page 590)

More on the work of Marjorie Rice, including several beautiful Escher-like tessellations reproduced in full color, can be found in the chapter titled "In Praise of Amateurs" by Doris Schattschneider in *The Mathematical Gardner*, edited by David A. Klarner (Prindle, Weber & Schmidt, 1981) and by googling "Marjorie Rice Tessellation Database" on the internet. The central tessellation with its sixfold rotation symmetry that is the subject of the exercises was discovered by Michael Hirschhorn, a teacher in New South Wales.

Congruent Pentagons.

1. They are equilateral. (The colors along the sides of the pentagons can be used to establish this.)

2. The boundary of the figure is a regular 18-gon because it is both equilateral and equiangular. (From the measures of the angles of the pentagonal pieces, we can determine that each of the angles of the 18-gon has a measure of 160°.)

3. Approximately 2.88 units. Because the perimeter of the 18-gon is 18 and

   $$p = 2Nr = 2n \sin\frac{180}{n}r, \quad 18 = 2(18 \sin 10°)r.$$

   So $r = \dfrac{1}{2\sin 10°} \approx 2.88$.)

4. Approximately 25.53 square units.
   $[A \approx (18 \sin 10° \cos 10°)(2.88)^2 \approx 25.53.]$

5. Approximately 1.42 square units.

   $(\dfrac{25.53}{18} \approx 1.42.)$

Chapter 14, Lesson 4

Set I (pages 593–595)

Hats are the only instance of a clothing size in which circumference is measured to determine diameter. A tape measure is wrapped around the widest part of the head (about an inch above the eyebrows) to get the circumference of the inner band of the appropriate hat. Dividing this number by π gives the hat size. This method works quite well because the circle is a reasonable approximation of the cross section of a person's head.

The time that it takes the moon to travel once around Earth, about 27.3 days, is called the "sidereal" month. The word "sidereal" means "determined by the stars," and that is how the period of revolution was found. A "lunar" or "synodic" month is the time between two successive new moons, about 29.5 days, and a "solar" month is one-twelfth of a solar year (about 30.4 days).

In *The Innovators* (Wiley, 1996), David Billington wrote: "It has been argued that the steamboat was the first great American contribution to modern technology." Billington, a professor of civil engineering at Princeton, includes a wealth of information on Robert Fulton's calculations, including his patent formulas and the mathematics of his drag, power, and paddlewheel calculations.

The Babylonian-tablet exercises provide another example of Pythagorean triples appearing many centuries before Pythagoras was born. (If π is taken as 3.14, the convenient relation of the sides of the triangle, surely intended by the problem's creator, is lost.) It seems strange, however, that the writer of a problem of this level would think that π was equal to 3.

Francois Viète has been called the greatest French mathematician of the sixteenth century. In *A History of π* (Golem Press, 1971), Petr Beckmann reports that Viète was the first to discover a way to represent π as an expression of an infinite sequence of mathematical operations:

$$\pi = \cfrac{2}{\sqrt{\frac{1}{2}} \cdot \sqrt{\frac{1}{2} + \frac{1}{2}\sqrt{\frac{1}{2}}} \cdot \sqrt{\frac{1}{2} + \frac{1}{2}\sqrt{\frac{1}{2} + \frac{1}{2}\sqrt{\frac{1}{2}}}} \cdots}$$

Unfortunately, the expression is too cumbersome and time consuming to be of practical use in calculating π, and so Viète was also one of the last mathematicians to resort to using polygons to do so. The number of sides used, 393,216, results from 16 successive doublings of Archimedes' original hexagon.

Circumference.

1. Because $d = 2r$ and $c = 2\pi r = \pi(2r)$, $c = \pi d$ by substitution.

•2. π.

3. 2π.

Hat Sizes.

4. $\frac{22}{7} \approx 3.143$ and $\frac{22.75}{7.25} \approx 3.138$. (Both ratios are approximately the value of π.)

•5. The circumference of the hat.

6. The diameter of the hat.

The Moon's Orbit.

•7. Approximately 1,500,780 mi. [$c = 2\pi(238,857) \approx 1,500,783$.]

8. Approximately 656 hours. ($\frac{1,500,783}{2,287} \approx 656$.)

9. Approximately 27.3 days. ($\frac{656}{24} \approx 27.3$.)

Semicircles.

•10. πa.

11. πb.

12. $\pi(a + b)$.

13. Because $\pi(a + b) = \pi a + \pi b$, the length of semicircle C is equal to the sum of the lengths of semicircles A and B.

14. No.

Steamboat Geometry.

15. 42,240 ft/hour. (8 × 5,280.)

•16. 704 ft/minute. ($\frac{42,240}{60}$.)

•17. 14π ft.

18. 16. ($\frac{704 \text{ ft}/\text{minute}}{14\pi \text{ ft}/\text{revolution}} \approx$ 16 revolutions/minute.)

Babylonian Problem.

19.

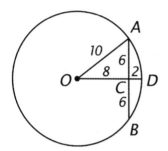

•20. 10 units. [$c = 2\pi r$, $60 = 2(3)r$, $r = 10$.]

•21. 8 units. (OC = OD – CD = 10 – 2 = 8.)

22. 6 units. (AC = $\sqrt{10^2 - 8^2} = \sqrt{36} = 6$.)

23. With the assumption that $\pi = 3$, the sides of ΔOAC form a Pythagorean triple: 6-8-10. (Or, the sides of ΔOAC are all integers.)

•24. AC = $\frac{1}{2}$AB. If a line through the center of a circle is perpendicular to a chord, it also bisects the chord.

25. 12 units.

Viète's Calculation.

•26. 3.1415927.
($N = 393,216 \sin \frac{180}{393,216} \approx 3.1415927$.)

27. No. The value must be less than π because the perimeter of a regular polygon is always less than the circumference of a circle having the same radius.

Perimeters and Diameters.

•28. $2\sqrt{2}$. ($p = 4s$ and $d = s\sqrt{2}$; so $\frac{p}{d} = \frac{4s}{s\sqrt{2}} = \frac{4}{\sqrt{2}} = 2\sqrt{2}$.)

•29. 2.83.

30. 3. ($p = 6s$ and $d = 2s$; so $\frac{p}{d} = \frac{6s}{2s} = 3$.)

31. π.

32. 3.14.

33. Close to 3.14 (or close to, but less than, π.)

34. $p = 2Nr$ and $d = 2r$; so $\dfrac{p}{d} = \dfrac{2Nr}{2r} = N$.

   $N = 100 \sin \dfrac{180}{n} \approx 3.1410759$.

Set II (pages 595–597)

Owing to Earth's rotation, the speed at which someone living on the equator is traveling, 1,040 miles per hour, is much faster than the official world land-speed record! Our speed due to Earth's motion around the sun, 67,000 miles per hour, is more than three times as great as that of the Space Shuttle.

   Adam Kochansky published his construction as "An Approximate Geometrical Construction for Pi," and so he knew that it was not exact. The exercises exploit the calculator in finding the various distances. It isn't difficult to show (without using a calculator) that the exact length of BE is

$\sqrt{\dfrac{40-6\sqrt{3}}{3}}$. In *Geometry Civilized* (Clarendon

Press, 1998), J. L. Heilbron, noting that Kochansky liked to calculate, quotes him as saying: "The periphery thus found differs from the closest approximation to the true Archimedean value [of pi] by less than the ratio of one to ten times the current year 1685. . . ." Doing the calculation,

$\sqrt{\dfrac{40-6\sqrt{3}}{3}} + \dfrac{1}{16,850} = 3.14159268 \ldots ,$

gives the value of π, 3.14159265 . . . correctly to seven decimal places.

Going in Circles.

•35. Approximately 24,880 mi. [Student rounding may differ. $c = 2\pi(3,960) \approx 24,881$.]

•36. Approximately 1,040 mi.
   $(\dfrac{24,881}{24} \approx 1,037.)$

•37. 1,040 mi/hour.

38. Approximately 584,000,000 mi.
   $[c = 2\pi(93,000,000) \approx 584,336,234.]$

39. Approximately 1,600,000 mi.
   $(\dfrac{584,336,234}{365.25} \approx 1,599,825.)$

40. Approximately 67,000 mi.
   $(\dfrac{1,599,825}{24} \approx 66,659.)$

41. Approximately 67,000 mi/hour.

•42. Approximately 19 mi/second.
   $(\dfrac{66,659}{3,600} \approx 18.5.)$

Two Squares.

43. HL. $\triangle$ABO and $\triangle$DCO are right triangles with AB = DC because ABCD is a square; OB = OC because all radii of a circle are equal.

44. OD = $\dfrac{1}{2}x$. (OD = $\dfrac{1}{2}$AD = $\dfrac{1}{2}x$.)

•45. $r^2 = \dfrac{5}{4}x^2$. [In right $\triangle$DCO,
   $r^2 = x^2 + (\dfrac{1}{2}x)^2 = \dfrac{5}{4}x^2$.]

•46. $r^2 = \dfrac{1}{2}y^2$. (In right $\triangle$GPH, $y^2 = r^2 + r^2 = 2r^2$;
   so $r^2 = \dfrac{1}{2}y^2$.)

47. $x^2 = \dfrac{2}{5}y^2$. (Because $r^2 = \dfrac{5}{4}x^2$ and $r^2 = \dfrac{1}{2}y^2$,
   $\dfrac{5}{4}x^2 = \dfrac{1}{2}y^2$; so $x^2 = \dfrac{2}{5}y^2$.)

48. The area of ABCD is two-fifths of the area of EFGH.

SAT Problem.

•49. $2a + 2b + 2c$.

50. $2\pi a + 2\pi b + 2\pi c$.

51. $\dfrac{1}{\pi}$. $[\dfrac{2a+2b+2c}{\pi(2a+2b+2c)} = \dfrac{1}{\pi}.]$

Videotape.

•52. 825 ft. $(\dfrac{1.375 \cdot 60 \cdot 60 \cdot 2}{12} = 825.)$

•53. $\pi \approx 3.14$ in. $[c = 2\pi(0.5) = \pi.]$

54. $3\pi \approx 9.42$ in. $[c = 2\pi(1.5) = 3\pi.]$

55. $2\pi \approx 6.28$ in. $(\dfrac{\pi+3\pi}{2} = 2\pi.)$

56. About 1,576 times. ($\frac{825 \cdot 12}{6.28} \approx 1{,}576.$)

Kochansky's Construction.

57.

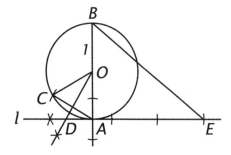

58. Equilateral.

59. A 30°-60° right triangle.

60. DA ≈ 0.5773503.

$(\text{DA} = \frac{\text{OA}}{\sqrt{3}} = \frac{1}{\sqrt{3}} \approx 0.5773503.)$

61. AE ≈ 2.4226497.
(AE = DE – DA ≈ 3 – 0.5773503 = 2.4226497.)

62. BE ≈ 3.1415333. (BE² = AB² + AE² so
BE = $\sqrt{2^2 + \text{AE}^2} \approx 3.1415333.$)

63. It is a good approximation of π, correct to
five digits (or four decimal places).

Set III (page 597)

Tire Change.

The ratio of the circumference of the larger tire to
the circumference of the smaller tire is

$\frac{33\pi}{28.9\pi} \approx 1.14$; so, for each revolution of the tires,

the truck goes 1.14 times as far as before.
70 miles × 1.14 ≈ 79.8 miles. The truck would be
going about 80 miles per hour.

Chapter 14, Lesson 5

Set I (pages 600–602)

The area calculated in exercise 3 for a typical
hurricane is more meaningful when compared
with the areas of states such as Florida (59,000
mi²) and Louisiana (48,000 mi²), both of which
were hit hard by Hurricane Andrew in 1992.

Another unit of circular area measure is the
"circular mil," defined as the area of a circle

having a diameter of one "mil" (0.001 in).
Circular units of area are obviously of little value
in working with anything other than circles.
Circular mils are used chiefly in the measurement
of wire.

Exercises 13 through 16 provide an example of
how a good foundation in geometry is important
in calculus. Related rates and maximum/minimum
problems are particularly obvious examples in
that nearly every problem requires the use of
plane or solid geometry.

Exercise 17 is actually the first theorem of
Archimedes' *Measurement of a Circle*: "The area
of a circle is equal to that of a right triangle in
which one leg is equal to the radius and the other
to the circumference of the circle."

Although the result of exercises 18 through
21, that in central pivot irrigation the number of
subdivisions of a square region makes no difference
in the area watered, may seem at first surprising,
the reason is obvious when considered from the
viewpoint of a dilation. No matter how small or
large the cells might be, the ratio of the areas of

each circle and its corresponding square, $\frac{\pi r^2}{4r^2}$,

remains constant.

The use of the circle on a square grid to
estimate π was first investigated by Gauss.
According to David Hilbert and Stefan Cohn-
Vossen in *Geometry and the Imagination*
(Chelsea, 1952), Gauss "tried to determine the
number $f(r)$ of lattice points in the interior and
on the boundary of a circle of radius r, where the
center of the circle is a lattice point and r is an
integer. Gauss found the value of this number
empirically for many values of n. . . . His interest
was prompted by the fact that an investigation of
this function yields a method for approximating
the value of π." The details and an explanation of
the procedure can be found on pages 33–35 of
Hilbert's book.

Hurricanes.

•1. 300 mi.

•2. 940 mi. [$c = \pi(300) \approx 942.$]

•3. 71,000 mi². [$A = \pi(150)^2 \approx 70{,}686.$]

•4. 15 mi. ($700 = \pi r^2$, $r = \sqrt{\frac{700}{\pi}} \approx 15.$)

•5. 90 mi. [$c = 2\pi(15) \approx 94.$]

Square and Circular Inches.

6. The area of a square with sides of length 1 inch.

7. The area of a circle with a diameter of 1 inch.

8. A square inch.

•9. x^2. ($\dfrac{x^2}{1^2} = x^2$.)

10. 100 square inches.

11. x^2. [$\dfrac{\pi(\frac{x}{2})^2}{\pi(\frac{1}{2})^2} = x^2$.]

•12. 100 circular inches.

Ripples.

13. 60 cm.

•14. $3{,}600\pi$ cm². [$A = \pi(60)^2 = 3{,}600\pi$.]

•15. $60t$ cm.

16. $3{,}600\pi t^2$ cm². [$A = \pi(60t)^2 = 3{,}600\pi t^2$.]

Equal Areas.

17. $2\pi r$ (or, the circumference of the circle). If the areas are equal, $\pi r^2 = \dfrac{1}{2}$ABr; so AB = $2\pi r$.

Central-Pivot Irrigation.

18. *(Student answer.)*

•19. 36π square units. [The diameter of each circle is 6; so the radius is 3. The total area of the four circles is $4(\pi 3^2) = 36\pi$.]

20. 36π square units. [The diameter of each circle is 4; so the radius is 2. The total area of the nine circles is $9(\pi 2^2) = 36\pi$.]

21. They are equal.

Pi Square.

22. $\sqrt{\pi}$. ($\pi = s^2$; so $s = \sqrt{\pi}$.)

•23. π^2. [$A = \pi r^2 = \pi(\sqrt{\pi})^2 = \pi^2$.]

24. π. ($\dfrac{\pi^2}{\pi} = \pi$.)

Circle on a Grid.

•25. 6 units.

26. 3.13888 . . . . ($\dfrac{113}{6^2} = 3.13888 \ldots$.)

•27. 3.1425. ($\dfrac{1{,}257}{20^2} = 3.1425$.)

28. About 3.1411. ($\dfrac{282{,}697}{300^2} \approx 3.14107 \ldots$.)

29. They suggest that the limit is π.

30. About 31,416. ($\dfrac{p}{100^2} \approx \pi$,

$p \approx 10{,}000\pi \approx 31{,}416$. The number of points is actually 31,417.)

Set II (pages 602–604)

Exercises 39 through 46 look at an early precursor of integration. Sawaguchi Kazuyuki published his method for finding the area of a circle in 1670, at about the time that Newton and Leibniz were discovering the rules of the calculus. As Petr Beckmann observes in *A History of π* (Golem Press, 1971), in theory the method can be used to find π to any desired degree of accuracy by choosing a sufficiently large number of strips. Beckmann also points out that, in practice, the series converges very slowly in addition to requiring the inconvenient extraction of square roots.

Area Problems.

•31. $3\pi x^2$. [The radius of circle A is $2x$; so its area is $\pi(2x)^2 = 4\pi x^2$. The radius of circle B is x; so its area is πx^2. The shaded area is $4\pi x^2 - \pi x^2 = 3\pi x^2$.]

32. $5\pi x^2$. [The radius of the larger circle is $3x$; so its area is $\pi(3x)^2 = 9\pi x^2$. The radius of the smaller circle is $2x$; so its area is $\pi(2x)^2 = 4\pi x^2$. The shaded area is $9\pi x^2 - 4\pi x^2 = 5\pi x^2$.]

•33. $(4 - \pi)x^2$. [The side of square ABCD is $2x$; so its area is $(2x)^2 = 4x^2$. The radius of circle O is x; so its area is πx^2. The shaded area is $4x^2 - \pi x^2 = (4 - \pi)x^2$.]

34. $(\pi - \frac{\sqrt{3}}{2})x^2$. [Because $\triangle ABC$ is a 30°-60° right triangle, $AB = 2x$ and $AC = x\sqrt{3}$. Because $AB = 2x$, $OB = x$; so the area of the circle is πx^2. Because the legs of $\triangle ABC$ are x and $x\sqrt{3}$, its area is $\frac{1}{2}x(x\sqrt{3}) = \frac{\sqrt{3}}{2}x^2$. The shaded area is $\pi x^2 - \frac{\sqrt{3}}{2}x^2 = (\pi - \frac{\sqrt{3}}{2})x^2$.]

•35. $(5\pi - 8)x^2$. [The legs of each small right triangle are $2x$ and x, and so $OB = \sqrt{(2x)^2 + x^2} = \sqrt{5x^2} = x\sqrt{5}$. The radius of circle O is $x\sqrt{5}$; so its area is $\pi(x\sqrt{5})^2 = 5\pi x^2$. The area of the rectangle is $(4x)(2x) = 8x^2$. The shaded area is $5\pi x^2 - 8x^2 = (5\pi - 8)x^2$.]

•36. $(3\sqrt{3} - \pi)x^2$. [The small triangle is a 30°-60° right triangle with shorter leg x, and so its longer leg is $x\sqrt{3}$ and $AB = 2x\sqrt{3}$. The area of equilateral $\triangle ABC$ is $\frac{\sqrt{3}}{4}AB^2 = \frac{\sqrt{3}}{4}(2x\sqrt{3})^2 = 3\sqrt{3}\,x^2$. The radius of the circle is x and so its area is πx^2. The shaded area is $3\sqrt{3}\,x^2 - \pi x^2 = (3\sqrt{3} - \pi)x^2$.]

Cable Disaster.

37. *Scale drawing:*

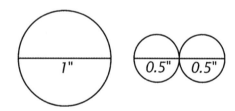

38. The strength of the cable depends, not on its diameter, but on the area of its cross section. The area of a cross section of a 1-inch cable is $\pi(\frac{1}{2})^2 = \frac{1}{4}\pi$ in². The total area of the cross sections of two $\frac{1}{2}$-inch cables is $2\pi(\frac{1}{4})^2 = \frac{1}{8}\pi$ in². The 1-inch cable should have been replaced by four $\frac{1}{2}$-inch cables.

Slicing a Circle.

39.

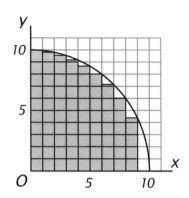

40. The height of the fifth strip is a leg of a right triangle with hypotenuse 10 and base 5. $h^2 + 5^2 = 10^2$; so $h = \sqrt{10^2 - 5^2} = \sqrt{100 - 25} = \sqrt{75}$. The width of the strip is 1; so its area is $1 \cdot \sqrt{75} = \sqrt{75}$.

41. $\sqrt{99}$, $\sqrt{96}$, $\sqrt{91}$, $\sqrt{84}$, $\sqrt{64} = 8$, $\sqrt{51}$, $\sqrt{36} = 6$, and $\sqrt{19}$. ($\sqrt{10^2 - 1^2} = \sqrt{99}$, $\sqrt{10^2 - 2^2} = \sqrt{96}$, $\sqrt{10^2 - 3^2} = \sqrt{91}$, etc.)

•42. 72.6.

•43. 290. [4(72.6) ≈ 290.]

44. It would be less than the actual area of the circle because the rectangular strips do not fill the quarter circle.

45. Narrower strips should fill more of the circle.

46. Keep dividing the circle into narrower and narrower strips.

Tangent Circles.

•47. $a + b$. [Its diameter is $2a + 2b = 2(a + b)$; so its radius is $a + b$.]

•48. $\pi a^2 + \pi ab$ or $\pi a(a + b)$. [The blue area is half the area of the largest circle plus half the area of the left circle minus half the area of the right circle. $\frac{1}{2}\pi(a + b)^2 + \frac{1}{2}\pi a^2 - \frac{1}{2}\pi b^2 = \frac{1}{2}\pi(a^2 + 2ab + b^2) + \frac{1}{2}\pi a^2 - \frac{1}{2}\pi b^2 = \pi a^2 + \pi ab = \pi a(a + b)$.]

49. $\pi ab + \pi b^2$ or $\pi b(a + b)$. [The yellow area is found in a similar way.

$$\frac{1}{2}\pi(a + b)^2 + \frac{1}{2}\pi b^2 - \frac{1}{2}\pi a^2 =$$

$$\frac{1}{2}\pi(a^2 + 2ab + b^2) + \frac{1}{2}\pi b^2 - \frac{1}{2}\pi a^2 =$$

$\pi ab + \pi b^2 = \pi b(a + b)$.]

50. $\dfrac{a}{b}$. $\left[\dfrac{\pi a(a+b)}{\pi b(a+b)} = \dfrac{a}{b}.\right]$

51. Yes. If $a = b$, the two areas should be equal. We have shown that $\dfrac{\text{blue area}}{\text{yellow area}} = \dfrac{a}{b}$; so,

if $a = b$, $\dfrac{a}{b} = 1$, $\dfrac{\text{blue area}}{\text{yellow area}} = 1$, and so

blue area = yellow area.

52. 1. (Each border consists of three semicircles, one from each of the three circles.)

Set III (page 604)

In *The Age of Faith* (Simon & Schuster, 1950), Will Durant called Ramon Lull "one of the strangest figures of the many-sided thirteenth century." Lull wrote 250 books in Catalan, Latin, and Arabic on such subjects as love poetry, theology, warfare, education, philosophy, and science. According to Durant, "Amid all these interests he was fascinated by an idea that has captured brilliant minds in our own time—that all the formulas and processes of logic could be reduced to mathematical or symbolical form." More on Lull can be found in chapter 3 of *Science—Good, Bad, and Bogus,* by Martin Gardner (Prometheus Books, 1981).

Lull's Claim.

1. $2\pi r = 4s$.

2. $\pi r^2 = s^2$.

3. Solving the first equation for s, $s = \dfrac{\pi r}{2}$.

Substituting for s in the second equation,

$\pi r^2 = \left(\dfrac{\pi r}{2}\right)^2$, $\pi r^2 = \dfrac{\pi^2 r^2}{4}$, $4\pi r^2 = \pi^2 r^2$, $\pi = 4!$

Set I (pages 606–608)

In *Animal Navigation* (Scientific American Library, 1989), Talbot H. Waterman remarks that dolphins, like bats, "use echolocation to find food, avoid obstacles, and maybe even to navigate over longer distances. Repeated sound bursts . . . emanate from the head in a 20 to 30° beam. These directional high intensity clicks are reflected back by objects not completely absorbing them. Such echoes, which give a detailed sound picture of the environment at ranges exceeding 300 meters, are markedly better than the best underwater visibility."

Franz Reuleaux was a French engineer and mathematician who wrote an important book in 1876 on mechanisms in machinery. One of Martin Gardner's "Mathematical Games" columns in *Scientific American* deals with the subject of curves of constant width, of which, other than the circle, the Reuleaux triangle is the simplest example. This column is included in Gardner's book titled *The Unexpected Hanging and Other Mathematical Diversions* (Simon & Schuster, 1969). The curve was known to earlier mathematicians, but Reuleaux was the first to demonstrate its constant-width properties, including the fact that it can be rotated inside a square without any space to spare. Exercises 18 through 22 prove that the perimeter of the Reuleaux triangle is the same as the circumference of a circle having the same width. It is also true that, of all curves of constant width having a given width, the Reuleaux triangle is the curve with the smallest area.

In *Human Information Processing—An Introduction to Psychology* (Harcourt Brace Jovanovich, 1977), Peter H. Lindsay and Donald A. Norman write concerning auditory space perception: "The cues used to localize a sound source are the exact time and intensity at which the tones arrive at the two ears. Sounds arrive first at the ear closer to the source and with greater intensity. The head tends to cast an acoustic shadow between the source and the ear on the far side. With some simple calculations, it is possible to determine the approximate maximum possible time delay between signals arriving at the two ears." These calculations are done in exercises 23 through 25. Lindsay and Norman observe that the human nervous system is able to distinguish the time at which a sound reaches the ear within an accuracy of 0.00003 second! They also explain that sound localization varies with frequency: low frequency sounds bend easily around a person's head, whereas high frequency sounds, whose wavelength is short compared with the size of the head, do not.

Although the degree seems to have been used to measure angles since angles were first measured, the radian was invented in 1871 by James Thomson, the brother of the physicist Lord Kelvin (William Thomson). Unlike the degree, based on the arbitrary division of a circle into 360 parts, the radian is a natural unit of angle measure. In *Trigonometric Delights* (Princeton University Press, 1998), Eli Maor remarks that the "reason for using radians is that it simplifies many formulas. For example, a circular arc of angular width θ (where θ is in radians) subtends an arc length given by $s = r\theta$; but if θ is in degrees, the corresponding formula is $s = \dfrac{\theta}{360} 2\pi r$. Similarly, the area of a circular sector of angular width is $A = \dfrac{1}{2}\theta r^2$ for θ in radians and $A = \dfrac{\theta}{360}\pi r^2$ for θ in degrees. The use of radians rids these formulas of the 'unwanted' factor $\dfrac{\pi}{180}$." Maor adds: "Even more important, the fact that a small angle and its sine are nearly equal numerically—the smaller the angle, the better the approximation—holds true only if the angle is measured in radians. . . .

It is this fact, expressed as $\lim\limits_{\theta \to 0} \dfrac{\sin \theta}{\theta} = 1$, that makes the radian measure so important in calculus."

Orange Slices.

• 1. A sector.

• 2. 45°. ($\dfrac{360°}{8}$.)

• 3. $\dfrac{1}{4}\pi r$. ($\dfrac{2\pi r}{8}$.)

• 4. $\dfrac{1}{8}\pi r^2$.

Sonar Beams.

• 5. The length of $\overset{\frown}{AB}$ in meters.

6. The area of sector DAB in square meters.

7. 105 m. (From exercise 5.)

• 8. 15,700 m². (From exercise 6.)

Latitude.

• 9. About 6,220 mi. [$\dfrac{1}{4}2\pi(3{,}960) \approx 6{,}220$.]

10. About 69 mi. ($\dfrac{6{,}220}{90} \approx 69$.)

11. About 1,380 mi. ($20 \cdot 69 = 1{,}380$.)

Driveway Design.

• 12. 2,009 ft². ($41 \cdot 49 = 2{,}009$.)

• 13. 887 ft². [$2(\dfrac{1}{4}\pi(18)^2) + 8 \cdot 18 + 13 \cdot 18 \approx 887$.]

14. 1,122 ft². ($2{,}009 - 887 = 1{,}122$.)

Two Sectors.

15. $\dfrac{1}{2}\pi r^2$.

• 16. πr^2. [$\dfrac{1}{4}\pi(2r)^2 = \pi r^2$.]

17. The yellow and blue areas are equal.

Reuleaux Triangle.

• 18. 60°. ($m\overset{\frown}{AB} = \angle C = 60°$.)

• 19. $\dfrac{1}{3}\pi x$. ($\dfrac{60}{360}2\pi x = \dfrac{1}{3}\pi x$.)

20. $3x$.

21. πx. ($3 \cdot \dfrac{1}{3}\pi x = \pi x$.)

• 22. x. ($c = \pi d = \pi x$; so $d = x$.)

Sound Delay.

• 23. About 9 in. [About $\dfrac{1}{2}(7) + \dfrac{1}{4}(7\pi) \approx 9$.]

24. 13,200 in. ($1{,}100 \cdot 12 = 13{,}200$.)

• 25. About 0.0007 second.

($\dfrac{9 \text{ in}}{13{,}200 \text{ in / second}} \approx 0.0007$ second.)

The Radian.

26. $2\pi r$.

• 27. 360°.

•28. $\frac{1}{2\pi}$. ($\frac{r}{2\pi r} = \frac{1}{2\pi}$.)

29. 57.3°. ($\frac{1}{2\pi} \cdot 360° \approx 57.3°$.)

•30. 57.3°. ($\angle AOB = m\widehat{AB}$.)

Set II (pages 608–611)

In a section titled "Lunatics" in *Geometry Civilized—History, Culture, and Technique* (Clarendon Press, 1998), J. L. Heilbron wrote: "Despite repeated failures, adventuresome geometers over the ages have tried their hands at squaring the circle. One of the earliest and most promising attempts was the work of Hippocrates of Chios, not the great physician but a man distinguished enough in his own line to be named in the old texts as the first to compile an *Elements of Geometry.* Hippocrates' attempt at circle-squaring involved 'lunes,' moon-shaped areas bounded by arcs of circles." Exercises 47 through 56 consider one version of Hippocrates' lunes: four of them are equal in area to the very square on which they are constructed.

Archimedes' "salt cellar" problem is theorem 14 from his *Liber Assumptorum,* or *Book of Lemmas*, a collection of theorems that has survived only in an Arabic translation.

Windshield Wipers.

•31. 46 in. $[\frac{120}{360}2\pi(22) \approx 46.]$

•32. 469 in². $[\frac{120}{360}\pi(22)^2 - \frac{120}{360}\pi(6)^2 \approx 469.]$

33. No, because the areas wiped usually overlap.

Land Area.

•34. $\angle AOC \approx 53°$. ($\sin \angle AOC = \frac{8}{10}$, $\angle AOC \approx 53.13°$.)

35. $\angle AOB \approx 106°$. ($2 \cdot 53° = 106°$.)

36. $m\widehat{AB} \approx 106°$.

•37. About 93 square units. $[\frac{106}{360}\pi(10)^2 \approx 93.]$

38. 48 square units. (Because OD = 10 – 4 = 6, $a\triangle ABO = \frac{1}{2}AB \cdot OD = \frac{1}{2}16 \cdot 6 = 48.$)

•39. About 45 square units. (93 – 48 = 45.)

40. Yes.

Running Track.

•41. 63.66 m. ($100 = \frac{1}{2}\pi d, d = \frac{200}{\pi} \approx 63.66.$)

42. 66.06 m. $[63.66 + 2(1.2) = 66.06.]$

•43. 103.77 m. $[\frac{1}{2}\pi(66.06) \approx 103.77.]$

44. 25. ($\frac{10,000}{400} = 25.$)

•45. About 188 m. (103.77 – 100 = 3.77; 25 laps is 50 semicircles; 50 · 3.77 ≈ 188.)

46. About 24 m! $[\frac{1}{2}2\pi(R + 0.15) - \frac{1}{2}2\pi R = 0.15\pi \approx 0.47;$ $50 \cdot 0.47 \approx 24.]$

Four Crescents.

•47. $2\sqrt{2}$. ($\triangle AOB$ is an isosceles right triangle with legs equal to 2.)

48. $\sqrt{2}$.

•49. 4π. $[\pi(2)^2 = 4\pi.]$

•50. 8. $[(2\sqrt{2})^2 = 8.]$

•51. $4\pi - 8$.

52. $\pi - 2$. ($\frac{4\pi - 8}{4} = \pi - 2.$)

•53. π. $[\frac{1}{2}\pi(\sqrt{2})^2 = \pi.]$

54. 2. $[\pi - (\pi - 2) = 2.]$

55. 8.

56. They prove that the four yellow crescents are together equal in area to the green square.

Drop-Leaf Table.

57. 20 in.

•58. 1,256.64 in². $[\pi(20)^2 = 400\pi \approx 1,256.64.]$

59. 60°. (Because CO = 2OH, △COH is a 30°-60° right triangle; so ∠COH = 60°.)

60. 120°.

61. 418.88 in². $[\frac{120}{360}\pi(20)^2 \approx 418.88.]$

•62. 34.64 in.
[CD = 2CH = 2($10\sqrt{3}$) = $20\sqrt{3}$ ≈ 34.64.]

63. 173.2 in².
[α△COD ≈ $\frac{1}{2}$(34.64)(10) = 173.2.]

•64. 246 in². (418.88 – 173.2 = 245.68.)

65. 765 in². [1,256.64 – 2(245.68) = 765.28.]

Salt Cellar.

•66. $2a + b$.

•67. $2a + 2b$.
[CD = (2a + b) + b = 2a + 2b.]

68. $a + b$.

69. $\pi(a + b)^2$.

70. The area of the "salt cellar" is equal to
$$\frac{1}{2}\pi(2a + b)^2 - 2(\frac{1}{2}\pi a^2) + \frac{1}{2}\pi b^2 =$$
$$\frac{1}{2}\pi(4a^2 + 4ab + b^2) - \pi a^2 + \frac{1}{2}\pi b^2 =$$
$$2\pi a^2 + 2\pi ab + \frac{1}{2}\pi b^2 - \pi a^2 + \frac{1}{2}\pi b^2 =$$
$$\pi a^2 + 2\pi ab + \pi b^2 = \pi(a^2 + 2ab + b^2) = \pi(a + b)^2.$$

Set III (page 611)

This exercise is based on the "pizza problem" discussed in detail by Joseph D. E. Konhauser, Dan Yelleman, and Stan Wagon in *Which Way Did the Bicycle Go?* (Mathematical Association of America, 1996): "Suppose four cuts are made through a point P in a disk so that the eight angles at P are all equal to 45°. If the resulting eight pizza slices are colored alternately black and white, must the white area equal the black area?" That the answer is "yes" was first discovered and proved by L. J. Upton in 1968. As Konhauser, Yelleman, and Wagon explain: "It turns out that the equal-area property holds if and only if the number of chords is even and greater than or equal to 4 (equivalently, the number of pizza slices is 8, 12, 16, . . .). They include a proof by

dissection for the 4-chords case and a proof using calculus and polar coordinates of the general case. There is a nice photograph of Stan Wagon on the last page of the book with a granite sculpture illustrating the dissection solution to the problem.

To prove, then, that Acute Alice's division of the pizza is a fair one is clearly not something of which a beginning geometry student is capable. To find a fairly obvious counterexample to Obtuse Ollie's claim, however, is fairly easy to do, as the figure below illustrates.

Pizza Puzzle.

Example figure for two cuts:

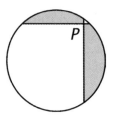

This figure shows that Ollie's method may easily result in one pair of opposite pieces having a greater area than the other pair. Strange as it may seem, regardless of where point P is chosen, Alice's method results in two sets of alternating pieces that are equal in area!

Chapter 14, Review

Set I (pages 612–614)

Of cholesterol, P. W. Atkins wrote in *Molecules* (Scientific American Library, 1987): "Although this molecule has an elaborate, rigid, hydrocarbon framework, its business end (more formally, its *functional group*) is primarily the –OH group. In other words, cholesterol is chemically an elaborate alcohol (hence the –*ol* in its name.)" In a line formula for an organic molecule, the line segments represent the bonds between the carbon atoms, which are understood to be located at their endpoints unless otherwise marked. The hydrogen atoms are usually ignored. The cholesterol molecule contains 27 carbon atoms, 1 oxygen atom, and 46 hydrogen atoms.

In Book IV of the *Elements*, Euclid discusses inscribed and circumscribed polygons—the triangle (Propositions 4 and 5), the square (Propositions 6 through 9), the regular pentagon

(Propositions 11 through 14), the regular hexagon (Proposition 15)—concluding with the regular 15-gon (Proposition 16). In his commentary on the *Elements* (Cambridge University Press, 1926), Sir Thomas Heath wrote: "Proclus refers to this proposition in illustration of his statement that Euclid gave proofs of a number of propositions with an eye to their use in astronomy. 'With regard to the last proposition in the fourth Book in which he inscribes the side of the fifteen-angled figure in a circle, for what object does anyone assert that he propounds it except for the reference of this problem to astronomy? For, when we have inscribed the fifteen-angled figure in the circle through the poles, we have the distance from the poles both of the equator and the zodiac, since they are distant from one another by the side of the fifteen-angled figure.' This agrees with what we know from other sources, namely that up to the time of Eratosthenes (c.284–204 B.C.) 24° was generally accepted as the correct measurement of the obliquity of the ecliptic."

Although the penny farthing bicycle reached its peak of popularity in about 1885, it is still popular in some parts of Europe. At the time of this writing, Hammacher Schlemmer was offering a version of the bicycle handmade in the Czech Republic for $4,999.95!

Concerning the Gothic arch, J. L. Heilbron wrote in *Geometry Civilized—History, Culture, and Technique* (Clarendon Press, 1998): "The construction of the equilateral triangle, which figures so prominently in Euclid's proofs, leads to a figure of great beauty and importance. . . . The curvilinear shape ACB is that of the basic Gothic arch. You can see it everywhere in churches built during the later middle ages. . . . At the threshold of the study of geometry, in connection with the very first proposition of Euclid, you have encountered one of the most important and versatile elements of architecture."

Molecule.

1. Regular hexagons and a regular pentagon.

•2. It must be convex, equilateral, and equiangular.

3. It becomes more circular.

A Regular 15-gon.

•4. $m\widehat{AB} = 120°$. ($\frac{360°}{3} = 120°$.)

5. $m\widehat{AD} = 72°$. ($\frac{360°}{5}$.)

•6. $m\widehat{DB} = 48°$.
($m\widehat{DB} = m\widehat{AB} - m\widehat{AD} = 120° - 72° = 48°$.)

7. $m\widehat{BE} = 24°$.
($m\widehat{BE} = m\widehat{DE} - m\widehat{DB} = 72° - 48° = 24°$.)

8. B and E, C and F.
($m\widehat{BE} = m\widehat{CF} = 24° = \frac{1}{15}360°$. Because $m\widehat{EF} = 72° = 3(24°)$, BE and CF are sides of the *same* regular 15-gon.)

•9. 3.12. ($N = 15 \sin \frac{180}{15} = 15 \sin 12° \approx 3.12$.)

•10. 3.05. ($M = 15 \sin 12° \cos 12° \approx 3.05$.)

11. π.

•12. 6.24r. [$p = 2Nr \approx 2(3.12)r = 6.24r$.]

13. 3.05r^2. ($A = Mr^2 \approx 3.05r^2$.)

14. 0.42r. ($s \approx \frac{6.24r}{15} \approx 0.42r$.)

Penny Farthing Bicycle.

•15. About 420 turns. [$\frac{5,280}{2\pi(2)} \approx 420$.]

16. About 3.77 ft. [$c = 2\pi(0.6) = 1.2\pi \approx 3.77$.]

17. About 1,401 turns. ($\frac{5,280}{3.77} \approx 1,401$.)

SAT Questions.

•18. $2\pi^2$. [$c = 2\pi(\pi) = 2\pi^2$.]

19. $\frac{1}{\pi}$. ($1 = \pi d$, $d = \frac{1}{\pi}$.)

•20. 4π. [$4\pi = 2\pi r$, $r = 2$, $A = \pi(2)^2 = 4\pi$.]

Semicircles on the Sides.

•21. $\frac{\pi}{8}a^2$, $\frac{\pi}{8}b^2$, and $\frac{\pi}{8}c^2$. [$A = \frac{1}{2}\pi(\frac{a}{2})^2 = \frac{\pi}{8}a^2$, etc.]

22. Yes. Because $\triangle ABC$ is a right triangle, $a^2 + b^2 = c^2$. Multiplying each side of this equation by $\frac{\pi}{8}$ gives $\frac{\pi}{8}a^2 + \frac{\pi}{8}b^2 = \frac{\pi}{8}c^2$.

Area Comparisons.

23. $\frac{1}{2}\pi r^2$.

•24. $\frac{1}{2}\pi r^2$. $[\frac{1}{8}\pi(2r)^2 = \frac{1}{2}\pi r^2.]$

25. They are equal. Because α1 + α2 = α2 + α3, α1 = α3 by subtraction.

26. The yellow and red areas are equal. (This follows from the result of exercise 25 and the symmetry of the figure.)

Gothic Arch.

27.

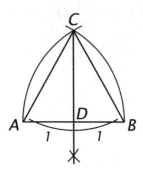

•28. $\sqrt{3} \approx 1.732$. $[\frac{\sqrt{3}}{4}s^2 = \frac{\sqrt{3}}{4}(2)^2.]$

•29. $\frac{2}{3}\pi \approx 2.094$. $[\frac{60}{360}\pi(2)^2 = \frac{2}{3}\pi.]$

30. $\frac{2}{3}\pi - \sqrt{3} \approx 0.362$.

•31. $\frac{4}{3}\pi - \sqrt{3} \approx 2.457$. $[\sqrt{3} + 2(\frac{2}{3}\pi - \sqrt{3}) = \sqrt{3} + \frac{4}{3}\pi - 2\sqrt{3} = \frac{4}{3}\pi - \sqrt{3}.]$

Set II (pages 614–616)

The answer to exercise 43, the width of the United States, is, of course, *very* approximate. The air distance between San Francisco and New York City, both with latitudes close to 40°, is 2,572 miles; the coast of Maine is roughly 400 miles east of New York City.

Exercises 50 through 54 are based on the surprising fact that, given two concentric circles in which a chord of the larger circle is tangent to the smaller circle, the area of the ring between the circles is determined solely by the length of the chord, regardless of the sizes of the circles! As the smaller circle shrinks to a point, the chord becomes the diameter of the larger circle and the expression for the area of the ring, $\pi R^2 - \pi r^2$, becomes πR^2, the area of the larger circle.

In exercises 55 through 62, the students discover a surprising property of the heart-shaped curve: *any* line through point O divides its border into two equal parts. We would expect this result of a line through the center of a figure that has point symmetry, but not of a figure that does not.

Cup Problem.

32. (*Student answer.*) (From the symmetry of the figure, $\overset{\frown}{AG}$ and $\overset{\frown}{BH}$ evidently have equal lengths, as do $\overset{\frown}{GD}$ and $\overset{\frown}{HC}$. The two lower arcs appear to be longer than the two upper ones.)

•33. 45°. ($m\overset{\frown}{AG} = \angle ABG = 45°.$)

•34. 0.79. $[\frac{45}{360}2\pi(1) = \frac{\pi}{4} \approx 0.79.]$

•35. $\sqrt{2} - 1$. (Because ABFE is a square, BE = $\sqrt{2}$. BG = BA = 1; so EG = BE − BG = $\sqrt{2} - 1$.)

36. 135°. ($m\overset{\frown}{GD} = \angle GED = 45° + 90° = 135°.$)

•37. 0.98. $[\frac{135}{360}2\pi(\sqrt{2} - 1) = \frac{3}{4}\pi(\sqrt{2} - 1) \approx 0.98.]$

38. (*Student answer.*)

Time Zones.

•39. About 1,037 mi. $[\frac{2\pi(3,960)}{24} \approx 1,037.]$

•40. About 3,034 mi. [In △ABO, $\cos 40°$ (or $\sin 50°$) = $\frac{AB}{3,960}$; so AB = 3,960 cos 40° ≈ 3,034.]

41. About 19,060 mi. $[2\pi(3,034) \approx 19,060.]$

•42. About 794 mi. $(\frac{19,060}{24} \approx 794.)$

43. Approximately 3,000 mi. (4 · 794 = 3,176.)

From Dodecagon to Square.

44. It is equilateral. The vertices of the dodecagon divide the circle into 12 equal arcs; so each arc has a measure of $\frac{360°}{12} = 30°$.

Two of the angles of the triangle are inscribed angles of the circle and each intercepts an arc equal to four of these arcs; so each of these angles has a measure of $\frac{1}{2}(4 \cdot 30°) = 60°$.

•45. $\sqrt{2}$ (or about 1.414). ($A = Mr^2$; so
$6 = 12 \sin\frac{180}{12}\cos\frac{180}{12}r^2$,
$6 = 12 \sin 15° \cos 15°r^2$, $6 = 3r^2$, $r^2 = 2$,
$r = \sqrt{2}$.)

46. $\sqrt{3}$ (or about 1.732). ($A = Mr^2$; so
$6 = 4 \sin\frac{180}{4}\cos\frac{180}{4}r^2 = 4 \sin 45° \cos 45°r^2$,
$6 = 2r^2$, $r^2 = 3$, $r = \sqrt{3}$.)

•47. 8.78. [$p = 2Nr = 2(12 \sin 15°)\sqrt{2} \approx 8.78$.]

48. 9.80. [$p = 2Nr = 2(4 \sin 45°)\sqrt{3} \approx 9.80$.]

49. It increases.

Chord and Ring.

50.

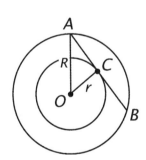

51. OC $\perp$ AB and OC bisects AB. (If a line is tangent to a circle, it is perpendicular to the radius drawn to the point of contact and, if a line through the center of a circle is perpendicular to a chord, it also bisects it.)

•52. $\pi R^2 - \pi r^2$ or $\pi(R^2 - r^2)$.

53. π. (Because AB = 2, AC = 1; so $r^2 + 1^2 = R^2$ and $R^2 - r^2 = 1$.)

54. The radii of the two circles are not needed to find the area of the ring. The area depends only on the length of the chord. (In general, if AB = 2x, then AC = x, $r^2 + x^2 = R^2$, $R^2 - r^2 = x^2$; so the area of the ring is πx^2.)

Half a Heart.

•55. 4π. $\{2[\frac{1}{2}2\pi(1)] + \frac{1}{2}2\pi(2) = 2\pi + 2\pi = 4\pi.\}$

56. $\angle$AOC = $(180 - x)°$.

57. $\angle$OCE = $(180 - x)°$.
(In $\triangle$CEO, EO = EC; so $\angle$OCE = $\angle$AOC.)

•58. $\angle$AEC = $(360 - 2x)°$.
($\angle$AEC is an exterior angle of $\triangle$CEO; so $\angle$AEC = $\angle$AOC + $\angle$OCE.)

•59. $\frac{180-x}{90}\pi$. $[\frac{360-2x}{360}2\pi(1) = \frac{180-x}{90}\pi.]$

60. $\frac{x}{90}\pi$. $[\frac{x}{360}2\pi(2) = \frac{x}{90}\pi.]$

61. 2π. $(\frac{180-x}{90}\pi + \frac{x}{90}\pi = \frac{180}{90}\pi = 2\pi.)$

62. Line CD bisects the border of the heart because the length of $\overset{\frown}{\text{CAD}}$, 2π, is half the length of the border of the heart, 4π.

Biting Region.

•63. About 83 ft.
$[\frac{1}{4}2\pi(15 - 10) + \frac{3}{4}2\pi(15) + \frac{1}{4}2\pi(15 - 12) = \frac{106}{4}\pi \approx 83.3.]$

64. About 557 ft^2.
$[\frac{1}{4}\pi(15 - 10)^2 + \frac{3}{4}\pi(15)^2 + \frac{1}{4}\pi(15 - 12)^2 = \frac{709}{4}\pi \approx 556.8.]$

Set I (pages 620–622)

The figures of exercises 1 through 6 are of special interest in the field of computational vision, a subfield of artificial intelligence. Martin A. Fischler and Oscar Firschein observe in their book titled *Intelligence—The Eye, the Brain, and the Computer* (Addison-Wesley, 1987): "We know that the shading and texture present in a single image can produce a vivid impression of three-dimensional structure. What is the computational basis of this effect? A crucial source of information about three-dimensional structure is provided by the spatial distribution of surface markings in an image. Since projection distorts texture geometry in a manner that depends systematically on surface shape and orientation, isolating and measuring this projective distortion in the image allows recovery of the three-dimensional structure of the textured surface."

In *Visual Explanations* (Graphics Press, 1997), Edward R. Tufte points out that, "For centuries, the profound, central issue in depicting information has been how to represent three or more dimensions of data on the two-dimensional display surfaces of walls, stone, canvas, paper, and recently, computer screens." About the figure of exercises 7 through 10, he remarks: "As early as 1642, pole-people were active in scaling recession for landscapes." Tufte also includes a nice "before and after" illustration for the Brighton pavilion (1808) that uses people holding poles to give a sense of scaling to a picture in which the perspective is ambiguous.

Ted Adelson's "corrugated Mondrian," the figure of exercises 13 through 20 is discussed in detail by Donald D. Hoffman in *Visual Intelligence* (Norton, 1998). According to Hoffman, the two grays in the numbered regions "measure precisely the same luminance on a photometer." He continues: "You also see this figure in three dimensions. It's possible, suggests psychologist Alan Gilchrist, that you group each surface with those that lie in the same 3D plane with it. Thus each vertical column would be a group. If so, then rectangle 2 would be the darkest in its group, and 1 the lightest in its group, explaining why A looks lighter than B." This hypothesis is supported by a flattened version of the figure in which the apparent difference of the two grays is not as pronounced.

Visual Planes.

•1. The second.

2. The right side.

3. Rectangles.

4. A horizontal plane.

5. The bottom part.

6. Circles.

Pole People.

•7. It appears to be oblique to the ground.

•8. A vertical plane.

9. That they are the same length.

•10. That they lie in a horizontal plane.

Posts and Walls.

•11. No. AB is perpendicular to the floor if it is perpendicular to *every* line in the floor that passes through point B.

12. Yes. The floor is perpendicular to DEFG because the floor contains a line, EH, that is perpendicular to DEFG.

3-D Shading.

•13. Parallelograms.

14. Rectangles.

•15. EFGH.

16. CDIJ.

17. BCJK and DEHI.

18. It can be viewed as being either behind or in front of the rest of the figure.

19. Rectangle 2.

20. (*Student answer.*) (The grays in the two rectangles are actually identical.)

Polarized Light.

•21. They intersect in line *l* (or they both contain it).

22. The planes appear to be perpendicular.

•23. One plane must contain a line that is perpendicular to the other plane.

Kilroy in Space.

24. CD ∥ FE. (CD and FE lie in the same plane, CDEF, and can't intersect, because they lie in parallel planes.)

•25. CF is perpendicular to plane EFGH.

26. Planes ABCD and EFGH are parallel.

•27. If a plane intersects two parallel planes, it intersects them in *parallel lines.*

28. If a line is perpendicular to one of two parallel planes, it *is perpendicular to the other.*

29. Two planes perpendicular to the same line are *parallel to each other.*

30. *(Student answer.)* (It depends on whether the figure is interpreted as seen from above or below.)

Set II (pages 622–624)

Desargues's Theorem, a basic theorem of projective geometry, is considered in exercises 31 through 37. Gérard Desargues was a French architect, engineer and army officer who wrote an important book on the conic sections as well as a book on how to teach children to sing! Howard Eves calls him "the most original contributor to synthetic geometry in the 17th century." His theorem says that, "If two triangles, in the same plane or not, are so situated that lines joining pairs of corresponding vertices are concurrent, then the points of intersection of pairs of corresponding sides are collinear, and conversely." As the exercises show, this theorem is proved quite easily by treating the figure for the theorem as if it were a three-dimensional "pyramid" of lines.

Triangles in Perspective.

31. They appear to be collinear.

•32. Three noncollinear points determine a plane.

•33. If two points lie in a plane, the line that contains them lies in the plane.

•34. They lie in it.

35. They lie in it.

36. A line.

37. That they are collinear.

Impossible Slice.

38. If two planes intersect, their intersection is a line.

39. That three noncollinear points determine a plane. (If points A, B, and C lie on a circle, then they are noncollinear, but they don't determine a plane, because they lie in two different planes.)

Uneven Bars.

•40. No. Two horizontal lines can be skew. They can also intersect.

41. Yes. Part of the definition of parallel lines is that they lie in the same plane.

•42. About 91 cm. (In right △ABE, $AB^2 = 43^2 + 80^2$; so $AB = \sqrt{8,249} \approx 91$.)

•43. About 62°. (In right △ABE, $\tan \angle BAE = \dfrac{BE}{AE} = \dfrac{80}{43}$; so $\angle BAE \approx 62°$.)

Pyramid.

44. If a line is perpendicular to a plane, it is perpendicular to every line in the plane that passes through the point of intersection.

•45. SAS.

46. Corresponding parts of congruent triangles are equal.

•47. A quadrilateral is a parallelogram if its diagonals bisect each other.

48. The opposite sides of a parallelogram are equal.

49. SSS.

50. SSS (or SAS).

Lopsided Box.

51. ABEH ⊥ BCDE and ABCF ⊥ BCDE (because AB ⊥ BCDE), and CDGF ⊥ ABCF (because CD ⊥ ABCF).

52. ∠ABE and ∠ABC (because AB ⊥ BCDE), and ∠DCF, ∠DCA, and ∠DCB (because DC ⊥ ABCF).

53. Yes; AD > AC. Because ∠ACD is a right angle in △ACD, ∠ACD > ∠ADC; so AD > AC.

Cube and Hypercube.

54. Six.

•55. Nine.

56. Eight.

57. 13.

Set III (page 625)

In a talk given in Amsterdam in 1963, Escher said: "If you want to focus the attention on something nonexistent, then you have to try to fool first yourself and then your audience, by presenting your story in such a way that the element of impossibility is veiled, so that a superficial listener doesn't even notice it. There has to be a certain enigma in it, which does not immediately catch the eye." (*The Magic of M. C. Escher*, Abrams, 2000).

The impossibility is in the way in which the columns of the first floor are connected to the second floor. As a result, someone climbing the ladder starts inside the building but ends up outside on reaching the top. The top floor is also at an angle with the second floor, which is impossible. The seated figure holds a distorted box that mirrors the impossible structure of the building.

Chapter 15, Lesson 2

Set I (pages 630–631)

The beetle on the box is another example of the Necker cube illusion, named after the Swiss scientist, Louis Necker, who first wrote about it in 1832. Martin Gardner included it in a *Scientific American* "Mathematical Games" column on optical illusions, republished in his *Mathematical Circus* (Knopf, 1979). The idea of adding a beetle to the figure was that of the mathematical physicist Roger Penrose and his father, L. S. Penrose. Of the Penrose version, Gardner wrote: "The beetle appears to be on the outside. Stare at the back corner of the box and imagine it to be the corner nearest you. The box will suddenly flip-flop, transporting the beetle to the floor *inside*."

It seems strange to think of the edge of a card (or a sheet of paper) as a rectangle having an area, but, as exercise 10 makes obvious, these objects have a thickness that can be easily calculated.

Concerning exercises 14 through 17, Martin Gardner wrote in his *Wheels, Life and Other Mathematical Amusements* (W. H. Freeman and Company, 1983): "Many geometric problems are solved by finding integral solutions for Diophantine equations. . . . Among the many geometrical Diophantine problems that are still unsolved, one of the most difficult and notorious is known as the problem of the 'integral brick' or 'rational cuboid.' The 'brick' is a rectangular parallelepiped. There are seven unknowns: The brick's three edges, its three face diagonals, and the space diagonal. . . . Can a brick exist for which all seven variables have integer values?. . . The problem has not been shown to be impossible, nor has it been solved. . . . The smallest brick with integral edges and face diagonals (only the space diagonal is nonintegral) has edges of 44, 117 and 240. This was known by Leonhard Euler to be the minimum solution. If all values are integral except a face diagonal, the smallest brick has edges of 104, 153 and 672 [the brick in exercises 14 through 17], a result also known to Euler."

Five different cross sections of a cube are considered in exercises 20 through 28. A good project for students who find this group of exercises interesting would be to try to discover (and possibly build models of) other possible cross sections. In *Curiosities of the Cube* (Crowell, 1977), Ernest R. Ranucci and Wilma E. Rollins observe: "The intersection of a plane with a cube can be an isosceles trapezoid. . . . Other cross sections include: a point, a line segment, scalene triangle, isosceles triangle, equilateral triangle, trapezoid, rectangle, square, parallelogram (are these always rectangles?), rhombus, pentagon, and hexagon."

Inside or Outside?

•1. Six.

•2. Eight.

•3. 12.

•4. F.

5. CD, EF, and GH.

6. AD, AE, BC, and BF.

•7. CG, DH, EH, and FG.

8. (*Student answer.*) (The box can be seen in two ways; so both answers are possible.)

Deck of Cards.

9. Changing the rounded corners of the cards to square ones.

•10. 0.025 cm. ($\frac{1.3}{52}$.)

11. 52.2 cm². (5.8 × 9.0.)

•12. 0.225 cm². ($\frac{1.3 \times 9.0}{52}$.)

13. 0.145 cm². ($\frac{1.3 \times 5.8}{52}$.)

Integer Lengths.

•14. 185 units.
($\sqrt{104^2 + 153^2} = \sqrt{34,225} = 185$.)

•15. 697 units.
($\sqrt{104^2 + 153^2 + 672^2} = \sqrt{485,809} = 697$.)

16. 680 units.
($\sqrt{104^2 + 672^2} = \sqrt{462,400} = 680$.)

17. About 689 units.
($\sqrt{153^2 + 672^2} = \sqrt{474,993} \approx 689$.)

SAT Problem.

•18. $8x + 10 = 106$.

19. 12. ($8x = 96$, $x = 12$.)

Slicing a Cube.

•20. A regular hexagon. (This slice is the perpendicular bisector of a diagonal of the cube. With the use of this fact and symmetry, it can be proved that the hexagon is regular.)

•21. ΔHEM and ΔHEN are congruent right triangles (SAS).

22. Isosceles.

23. ΔHDA and ΔFBA are congruent isosceles right triangles.

24. Equilateral.

•25. ∠EAC is a right angle. (EA is perpendicular to the plane of ABCD.)

26. A rectangle. No. (AE < EG.)

27. OP ∥ EG and OP = $\frac{1}{2}$EG. (OP is a midsegment of ΔHEG.)

•28. An isosceles trapezoid.

Set II (pages 631–633)

The surprising result of cutting a hole in a cube discovered in exercises 38 through 42 is from a puzzle created by nineteenth-century English puzzlist Henry Ernest Dudeney. Included in his *536 Puzzles & Curious Problems* (Scribners, 1967), Dudeney called it "A Cube Paradox" and posed it in the following way: "I had two solid cubes of lead, one very slightly larger than the other. Through one of them I cut a hole (without destroying the continuity of its four sides) so that the other cube could be passed right through it. On weighing them afterwards it was found that the larger cube was still the heavier of the two! How was this possible?" Dudeney's answer: "It is a curious fact that a cube can be passed through another cube of smaller dimensions. Suppose a cube to be raised so that its diagonal is perpendicular to the plane on which it rests. Then the resulting projection will be a regular hexagon. [Dudeney shows a square hole cut through this hexagon to allow the passage of a cube of the same dimensions.] But it will be seen that there is room for cutting a hole that would pass a cube of even larger dimensions. Therefore, the one through which I cut a hole was not, as the reader may have hastily supposed, the larger one, but the smaller! Consequently, the larger cube would obviously remain the heavier. This could not happen if the smaller were passed through the larger."

About the structure of the methane molecule, Linus Pauling wrote in *The Architecture of Molecules* (W. H. Freeman and Company, 1964), "In 1858 the Scottish chemist A. S. Couper invented valence-bond formulas for chemical compounds and wrote the formula

for methane. Then in 1874 the 22-year-old Dutch chemist Jacobus Hendricus van't Hoff pointed out that some properties of substances could be simply explained by the assumption that the four bonds formed by a carbon atom are directed toward the corners of a tetrahedron, with the carbon atom at its center. . . . The structure theory of chemistry has been based on the tetrahedral carbon atom ever since. Spectroscopic studies of methane have shown that the . . . six H–C–H bond angles have the value 109.5° that is characteristic of the regular tetrahedron." Pauling also comments on the fact that a tetrahedron can be inscribed in a cube helps in the evaluation of some of its dimensional properties. (This is explored in the "Divided Cube" exercises of the Chapter 15 review.)

Before Rubik's cube, there was the Soma cube. Its inventor, Piet Hein, was a Danish poet and scientist with many interests. He thought of the idea of the Soma cube while attending a lecture by Werner Heisenberg on quantum mechanics! About the "impossible" wall of exercises 52 through 54, Martin Gardner wrote in *Knotted Doughnuts and Other Mathematical Entertainments* (W. H. Freeman and Company, 1986): "It *is* possible . . . to build a wall that from the front looks exactly like the one in the illustration. If the wall is viewed from behind, however, the hidden corner [at the bottom center] is missing, and an extra cube protrudes at some other spot." With a set of Soma cubes in the classroom, some students might enjoy trying to build the figures of exercises 49 through 51 as well as the "impossible" wall.

An Inside Job.

29. Alice lined the brick up with two edges of the top of the table and moved it over one length. She then measured the distance between the corner of the table top and the upper left corner of the brick.

•30. 6 cm. $(21 = \sqrt{18^2 + 9^2 + x^2}, 441 = 405 + x^2, x^2 = 36, x = 6.)$

31. Ollie's head!

Diagonals.

•32. 45°. (ΔBAC is an isosceles right triangle.)

•33. 60°. (If CE were drawn, ΔCAE would be an equilateral triangle.)

34. 90°. (CD is perpendicular to the plane of ΔABC.)

•35. About 35°. (Letting e be the edge of the cube, $\tan \angle CAD = \dfrac{CD}{AC} = \dfrac{e}{e\sqrt{2}} = \dfrac{1}{\sqrt{2}}$, $\angle CAD \approx 35°$.)

36. About 55°. (Because CD ∥ AF, $\angle DAF = \angle ADC$. In right ΔACD, $\angle CAD \approx 35°$; so $\angle ADC \approx 55°$.)

37. No. $\angle CAD = \angle DAE \approx 35°$; so $\angle CAD + \angle DAE \approx 70°$, but $\angle CAE = 60°$. (Ray AD is not between rays AC and AE, because the three rays are not coplanar.)

Hole Through a Cube.

38. 16 cm². $(4^2.)$

•39. $3\sqrt{2}$ cm. (AD is the hypotenuse of isosceles right ΔADE.)

40. 18 cm². $[(3\sqrt{2})^2.]$

41. 4.24 cm. $(3\sqrt{2}.)$

42. Yes. We have just proved this possible with the results of exercises 38 through 41.

Bond Angles.

43. AB = $e\sqrt{2}$. (AB is the diagonal of a square whose side is e.)

•44. AO = $\dfrac{e}{2}\sqrt{3}$. (AO is half the diagonal of the cube. The diagonal of the cube is $e\sqrt{3}$.)

45. AP = $\dfrac{e}{2}\sqrt{2}$. (AP = $\dfrac{1}{2}$AB.)

•46. 54.74°. $(\sin \angle AOP = \dfrac{AP}{AO} = \dfrac{\frac{e}{2}\sqrt{2}}{\frac{e}{2}\sqrt{3}} = \dfrac{\sqrt{2}}{\sqrt{3}}$, $\angle AOP \approx 54.74°.)$

47. 109.5°. $(\angle AOB = 2\angle AOP \approx 109.5°.)$

Soma Puzzles.

48. 27. (The first piece contains 3 cubes and the other six pieces each contain 4 cubes.)

•49. One. The structure appears to contain $27 + 1 = 28$ cubes.

50. Three. The structure appears to contain $3 \times 10 = 30$ cubes.

51. Three. The structure appears to contain $1 + 4 + 9 + 16 = 30$ cubes.

52. Ten.

53. Nine. (Five pieces can each provide one corner cube and the other two pieces can each provide two corner cubes.)

54. The pieces can provide only nine corner cubes, but the wall contains ten.

Flattened Cube.

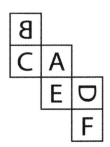

Chapter 15, Lesson 3

Set I (pages 635–637)

Chemists have synthesized several hydrocarbon compounds whose names are based on their polyhedral structures. In addition to *prismane* (a hexane in the shape of a triangular prism), there are *cubane*, *housane* (a pentagonal prism), and *dodecahedrane*. More spectacularly there are the *buckyballs* and their relatives—large cages formed by carbon atoms. These compounds have been studied intensively in recent years.

3-D Lettering.

- •1. L, I, and E.

  2. L is a hexagon, I is a quadrilateral (rectangle), and E is a dodecagon (12-gon).
- •3. Prisms.

Prismane.

  4. Because it is shaped like a prism.
- •5. Two.
- •6. Three.
  7. Six.
  8. Nine.
  9. 60° and 90°.

Polyhedral Nets.

- •10. The bases.
  11. The lateral faces.
- •12. *n*.
- •13. *n* + 2.
  14. Its lateral area.

  15. A cube (or, a rectangular solid).

Inside a Prism.

  16. The lateral faces.
  17. 837 ft² (27 × 31.)
- •18. 1,624 ft². [2(27 × 14) + 2(31 × 14).]

A Hexagonal Surprise.

- •19. A parallelogram.
- •20. BAB'C.
  21. CBC'D.
  22. DCD'E.
- •23. EDE'F.
  24. FAF'E.
  25. A parallelogram.
- •26. They appear to be parallel and equal.
  27. They appear to be congruent (and, if the green figure is viewed as three-dimensional, to lie in parallel planes).
  28. A triangular prism.

Set II (pages 637–638)

In *Crystals and Crystal Growing* (Anchor Books, 1960), Alan Holden and Phylis Singer report that "there are six styles of crystalline architecture, each employing a different sort of building block; and every crystal belongs to one of these six 'crystal systems.'" All six systems are prisms. Three are rectangular solids (*cubic, tetragonal, and orthorhombic*), named according to whether the edges have 1, 2, or 3 different lengths. The other three include two oblique prisms with parallelogram faces (*monoclinic* and *triclinic*) and a right hexagonal prism (*hexagonal*).

Concerning the Euler characteristic, Howard Eves remarks in *An Introduction to the History of Mathematics* (Saunders, 1990), "One of the most interesting theorems relating to any convex (or more generally any *simply connected*) polyhedron, is that $v - e + f = 2$. This may have been known to Archimedes (ca. 225 B.C.), and was very nearly stated by Descartes about 1635. Since Euler later independently announced it in 1752, the result is often referred to as the Euler-Descartes formula." We will return to the Euler characteristic in later lessons of this chapter.

Long House.

•29. Pentagons.

30. Rectangles.

31. Five.

32. 1,215 ft². (15 × 81.)

•33. 1,458 ft². [2(9 × 81).]

34. *Example figure:*

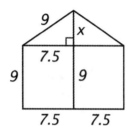

Approximately 14 ft. (Other roundings are acceptable.) ($x^2 + 7.5^2 = 9^2$, $x^2 = 24.75$, $x \approx 5$, $9 + x \approx 14$.)

Crystal Forms.

•35. They are rectangles.

36. The lateral edges are oblique to the planes of the bases.

37. They are parallelograms because their opposite sides are equal.

Feeding Trough.

•38. Yes. The bases of *every* prism are congruent.

39. Yes. AB = GH. AB = CD (given) and CD = GH (the opposite sides of a parallelogram are equal); so AB = GH (substitution).

40. Yes. AE ∥ CG and AE = CG. AE and CG are lateral edges of a prism; so they are parallel. AE = CG because they are both equal to BF (or DH); the opposite sides of a parallelogram are equal.

41. ABFE and DCGH are rectangles because they are lateral faces of a right prism. (They are also congruent because all of their corresponding parts are equal.)

Magic Box.

42. A rectangle.

43. SAS. (AE = DH, ∠AEX = ∠DHY, and EX = HY.)

•44. AX > AE.

45. The area of AXYD is greater than the area of AEHD. (AD = AD, but AX > AE; AD·AX > AD·AE.)

•46. A right triangular prism.

47. ΔAXE and ΔDYH.

48. A right trapezoidal prism.

49. ABCD and XFGY; ABFX and DCGY.

Euler's Discovery.

•50. $2n$.

51. $n + 2$.

52. $3n$.

53. 2. [$V + F - E = 2n + (n + 2) - 3n = 2$.]

Set III (page 639)

This exercise is based on the work of the eighteenth-century French crystallographer René Just Haüy described in Elizabeth A. Wood's *Crystals and Light* (Van Nostrand, 1964). Wood writes: "It is in the nature of naturalists to describe in detail the odd things that they find, and the early naturalists wrote full descriptions of the crystals they found, recording carefully the angles between their faces. One of them [Haüy] noticed the special angular relationships between the faces of a crystal. . . . It was this small-whole-number relationship between the tangents of the interfacial angles of crystals that convinced the early naturalists that crystals were made up of building blocks." In *Crystals and Crystal Growing* (Anchor Books, 1960), Alan Holden and Phylis Singer elaborate on this idea: "Crystals . . . can always be imagined to be made of little building blocks. . . . The building block is somewhat fictitious, of course. Nobody supposes that the ultimate fine structure of crystals looks like a collection of brick-shaped particles. The assembly of identical building blocks is merely a way of portraying the crystal's repetitive orderliness. Each building block represents a grouping of atoms or molecules which is regularly repeated throughout the crystal."

Topaz Angles.

1. $\tan \angle 1 = \dfrac{a}{b}$ and $\tan \angle 2 = \dfrac{2a}{b}$.

2. $\tan \angle 2 = 2 \tan \angle 1$.

3. 0.4557.
   ($\tan \angle 1 = \tan 24.5° \approx 0.4557$.)

4. 0.9114.
   [$\tan \angle 2 = 2 \tan \angle 1 \approx 2(0.4557) = 0.9114$.]

5. 42.3°.

6. Yes.

Chapter 15, Lesson 4

Set I (pages 642–644)

Tables of squares and cubes of numbers have been discovered on Babylonian tablets thought to date to 2100 B.C. The Pythagoreans may have been the first to associate these numbers with geometric figures.

According to Genesis 6:14–16, the ark had three stories and was divided into various rooms. Taking 18 inches as the length of a cubit (as in the *Scofield Reference Bible*), the average height of each story would have been 15 feet. The *Answers in Genesis* website suggests that the cubit may have been longer, about 20.4 inches.

Exercise 26 reveals that Cavalieri's Principle had been discovered long before the man for whom it was named was born. A good source of information on Cavalieri and his work is the article "Slicing It Thin" by Howard Eves included in *The Mathematical Gardner*, edited by David A. Klarner (Prindle, Weber & Schmidt, 1981).

The four candy bars weigh about the same: 3 Musketeers, 2.13 oz; Snickers, 2.07 oz; Butterfinger, 2.1 oz; Milky Way, 2.05 oz.

Squares and Cubes.

1. Its area.

•2. "*x* cubed."

3.

•4. Its volume.

•5. 27.

6. 1,728.

•7. 36.

•8. 1,296. (36^2.)

9. 46,656. (36^3.)

Largest Suitcase.

10. 2,772 in³. ($22 \times 14 \times 9$.)

11. 1,728. (12^3.)

•12. About 1.6 ft³. ($\dfrac{2,772}{1,728}$.)

Noah's Ark.

13. 450,000 cubic cubits. ($300 \times 50 \times 30$.)

•14. 1,518,750 ft³. ($450,000 \times 1.5^3$.)

15. Approximately 500 boxcars.
    ($\dfrac{1,518,750}{3,000} = 506.25$.)

Weathering Rock.

•16. 1 m³. (1^3.)

•17. 6 m². (6×1^2.)

18. 8.

19. 1 m³.

•20. 12 m². [$8(6 \times 0.5^2)$.]

21. 64.

22. 1 m³.

23. 24 m². [$64(6 \times 0.25^2)$.]

24. It stays the same.

25. It increases (doubling at each stage).

Keng-chih's Principle.

26. Cavalieri's Principle.

•27. A right hexagonal prism and an oblique hexagonal prism.

28. That the prisms have equal volumes.

Hole in the Ground.

29. *(Student answer.)* (Alice probably got 1,860,867 ft^3 because $123^3 = 1,860,867$.)

30. Because Ollie saw Alice using her calculator. If she had realized that a hole in the ground doesn't have any dirt in it, she wouldn't have been doing that.

Prism Volumes.

•31. 108 cubic units.
$[V = Bh = \frac{1}{2}(4 \times 6)9 = 12 \times 9 = 108.]$

32. 440 cubic units.
$[V = Bh = (8 \times 5)11 = 40 \times 11 = 440.]$

•33. 288 cubic units.
$\{V = Bh = [\frac{1}{2}(4)(5 + 11)]9 = 32 \times 9 = 288.\}$

Candy Bars.

•34. 79 cm^3.
$(V = lwh = 3.0 \times 2.1 \times 12.5 = 78.75 \approx 79.)$

35. 66 cm^3.
$(V = lwh = 3.3 \times 2.0 \times 10.0 = 66.)$

•36. 55 cm^3. $\{V = Bh = [\frac{1}{2}(1.3)(2.3 + 3.5)]14.5 = 54.665 \approx 55.\}$

37. The 3 Musketeers bar. No; weight is not the same thing as volume.

Set II (pages 644–646)

Exercises 41 through 46 develop the formula for the cube of a binomial. A good class exercise would be to confirm the result algebraically.

Exercises 57 through 60 on the Toyota are based on material in A. K. Dewdney's *200% of Nothing* (Wiley, 1993). Dr. Dewdney wrote the "Mathematical Recreations" column in *Scientific American* for 8 years. As the exercises reveal, the commercial seems not only to confuse length with volume but also to have gotten the figures wrong.

Ream of Paper.

•38. 0.004 in. $(\frac{2}{500}.)$

39. 0.374 in^3. $[V = Bh = (8.5 \times 11)(0.004) = 0.374.]$

•40. 187.156 in^2.
$[A = 2(8.5 \times 11) + 2(8.5 \times 0.004) + 2(11 \times 0.004) = 187.156.]$

Binomial Cube.

•41. $(a + b)^3$.

42. Eight.

•43. Two of them.

•44. a^2b.

45. ab^2.

46. $a^3 + 3a^2b + 3ab^2 + b^3$.

Chinese Problem.

47. A right triangular prism.

48. $\frac{1}{2}abc$. $[V = Bh = \frac{1}{2}(ab)c.]$

•49. $ab + ac + bc + c\sqrt{a^2 + b^2}$.
$[A = 2(\frac{1}{2}ab) + ac + bc + c\sqrt{a^2 + b^2}.]$

SAT Problem.

50.

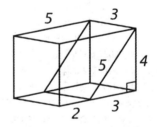

•51. 3 cm. $(5x = 15, x = 3.)$

52. 18 cm^3. $[V = Bh = \frac{1}{2}(3 \times 4)3 = 18.]$

53. 42 cm^3. $(5 \times 4 \times 3 = 60; 60 - 18 = 42.)$

Filling a Pool.

•54. 236.5 ft^2. $[A = \frac{1}{2}(3 + 5)16.5 + \frac{1}{2}(5 + 9)11.5 + \frac{1}{2}(9 + 7.5)9 + \frac{1}{2}(7.5 + 3)3 = 66 + 80.5 + 74.25 + 15.75 = 236.5.]$

•55. 4,730 ft³. ($V = Bh = 236.5 \times 20 = 4,730$.)

56. A little under 40 hours.
($\frac{4,730}{2} = 2,365$ minutes ≈ 39.4 hours.)

Advertising Claim.

•57. 9 ft³. ($0.75 \times 4 \times 3$.)

58. 13.5 ft³. ($6 \times 0.75 \times 3$.)

59. 18 ft³. ($6 \times 4 \times 0.75$.)

60. The extra room on the inside surely refers to volume, not length, so the wording should say "over two *cubic* feet more room." It is also strange that the commercial uses the number "two" when the volume should be increased by at least 9 cubic feet.

Measuring a Toad.

61. Its volume.

62. Push the toad under the water in a full measuring cup that can hold it comfortably. See how much water the toad displaces (the water will flow out).

Set III (page 646)

Benjamin Franklin's experiment in which he poured a spoonful of oil on the surface of a pond took place in 1765. He became curious about the effect of oil on water as a result of observing that the wakes of ships moving through greasy water were remarkably smooth. Although Franklin did not do the actual calculations, his experiment has been called a landmark in science in that it made possible the first rational measurement of the size of a molecule. A recreation of Franklin's experiment is pictured and described on pages 199–202 of Philip and Phylis Morrison's *The Ring of Truth* (Random House, 1987).

Molecule Experiment.

1. 0.0000002 cm. [$V = Bh$; so $h = \frac{V}{B}$.

$V = 4$ cm³ and $B = \frac{1}{2}(4,000 \text{ m}^2) = 2,000 \text{ m}^2 =$

$2,000 \times 100^2$ cm² $= 20,000,000$ cm².

$h = \frac{4}{20,000,000} = \frac{2}{10,000,000} =$

0.0000002 cm.]

2. 5×10^{20} (or 500 quintillion). [If the edge of the cube is 0.0000002 cm $= 2 \times 10^{-7}$ cm, its volume is (2×10^{-7} cm)³ $= 8 \times 10^{-21}$ cm³. If the cubes are packed together with no spaces between them, there would be as many as $\frac{4 \text{ cm}^3}{8 \times 10^{-21} \text{ cm}^3} = 0.5 \times 10^{21} =$ 5×10^{20} molecules.]

Chapter 15, Lesson 5

Set I (pages 649–650)

Ferrocene was first synthesized in 1951. At room temperature, it consists of yellow orange crystals that smell like camphor. It is used as a fuel additive, in rocket propellant, and in the production of antibiotics.

Leonardo da Vinci's sketch of a parachute appeared in a part of his *Codex Atlanticus* in which he explored mechanical flight. It was not until three centuries later that Louis Lenormand of France is credited for inventing the parachute. Its first trial took place in 1785, when a dog in a basket with a parachute was dropped from a balloon at a high altitude. Da Vinci's design was tested for the first time in 2000. According to BBC News, "a British man, Adrian Nicholas, dropped from a hot air balloon 3,000 meters (10,000 feet) above the ground, after ignoring expert advice that the canvas and wood contraption would not fly. . . . In the wider open spaces of Mpumalanga, South Africa, Mr. Nicholas safely floated down, saying the ride was smoother than with modern parachutes."

X-Ray Beam.

•1. A pyramid.

•2. Its apex.

3. Its base.

•4. A cross section.

•5. A lateral face.

Ferrocene.

6. Pentagonal (or regular pentagons).

7. At the apex of each pyramid.

8. The lateral edges.

9. Isosceles.

Pyramid Volumes.

•10. 128 cubic units.

$$[V = \frac{1}{3}Bh = \frac{1}{3}(8^2)6 = 128.]$$

11. 400 cubic units.

$$[V = \frac{1}{3}Bh = \frac{1}{3}(10 \times 15)8 = 400.]$$

•12. 70 cubic units.

$$\{V = \frac{1}{3}Bh = \frac{1}{3}[\frac{1}{2}(5 \times 12)]7 = 70.\}$$

A Regular Pyramid.

•13. They are equal. (They are radii of a regular polygon.)

14. They are right angles (and so are equal). (Because PO is the altitude, it is perpendicular to the plane of the pyramid's base; if a line is perpendicular to a plane, it is perpendicular to every line in the plane that passes through the point of intersection.)

15. They are right triangles. They are congruent (by SAS).

16. They are equal. (Corresponding parts of congruent triangles are equal.)

17. They are isosceles triangles. They are congruent (by SSS).

Leonardo's Claim.

18. A parachute (in the shape of a square pyramid).

•19. 576 yd³. $[V = \frac{1}{3}Bh = \frac{1}{3}(12^2)12 = 576.]$

20. No. The man looks too big. He looks as tall as the pyramid above him, which has an altitude of 12 yd = 36 ft.

Triangle and Pyramid Compared.

21. $A = \frac{1}{2}bh.$

•22. $V = \frac{1}{3}Bh.$

•23. 55.125 square units.

$$[A = \frac{1}{2}(10.5)(10.5) = 55.125.]$$

24. 55. $(1 + 2 + 3 + \cdots + 10 = 55.)$

•25. 385.875 cubic units.

$$[V = \frac{1}{3}(10.5)^2(10.5) = 385.875.]$$

26. 385. $(1^2 + 2^2 + 3^2 + \cdots + 10^2 = 385.)$

Set II (pages 651–653)

Surprising as it might seem, blindness is less a disability in mathematics than it is in many other areas of study. The blind mathematician Nicholas Saunderson was mentioned earlier in the text (Chapter 4, Lesson 1) with regard to his invention of the pin-board. In his *Penguin Book of Curious and Interesting Mathematics* (Penguin, 1997), David Wells quotes French author Denis Diderot concerning Saunderson: "Wondrous things are told about him, and his progress in literature and in the mathematical sciences lends credence to all of them. He was the author of a very perfect book of its kind, the *Elements of Algebra*, in which the only clue to his blindness is the occasional eccentricity of his demonstrations, which would perhaps not have been thought up by a sighted person. To him belongs the division of the cube into six equal pyramids having their vertices at the center of the cube and the six faces as their bases; this is used for an elegant proof that a pyramid is one-third of a prism having the same base and height."

David Burton, in his *History of Mathematics— An Introduction* (Allyn & Bacon, 1985), observes that, if the dimensions of the Great Pyramid were chosen so that the square of its altitude is equal to the area of one of its lateral faces, then, given that a is the altitude of a face triangle and $2b$ is the length of a side of the base, $\frac{a}{b}$ is the golden ratio.

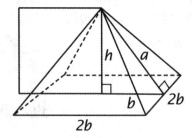

Expressing the claim algebraically, we have $h^2 = \frac{1}{2}(2b \cdot a) = ab$. Applying the Pythagorean Theorem to the figure above, we get $a^2 = h^2 + b^2$; so $h^2 = a^2 - b^2$. Substituting for h^2 gives

$a^2 - b^2 = ab$, from which we have $a^2 - ab - b^2 = 0$. Dividing by b^2 gives $(\frac{a}{b})^2 - (\frac{a}{b}) - 1 = 0$.

Letting $x = \frac{a}{b}$, we have $x^2 - x - 1 = 0$. Using the quadratic formula to solve for x gives

$$x = \frac{1 \pm \sqrt{(-1)^2 - 4(1)(-1)}}{2(1)} = \frac{1 \pm \sqrt{5}}{2}.$$ Because

$x = \frac{a}{b} > 0$, we have $\frac{a}{b} = \frac{1 + \sqrt{5}}{2} \approx 1.618 \ldots$, the golden ratio.

The way in which the African salt funnel is used is described by Paulus Gerdes in *Geometry from Africa* (Mathematical Association of America, 1999): "The funnel is hung . . . and salt containing earth is put in it. A bowl is placed beneath the funnel, hot water is poured on the earth in the funnel, and saltwater is caught in the bowl. After evaporation, salt remains in the bowl." The method for making a model of the funnel from a square piece of paper is taken from Gerdes' book, which also includes a description of how African artisans weave the funnel.

Thomas Banchoff in his book titled *Beyond the Third Dimension* (Scientific American Library, 1990) calls the Egyptian formula for finding the volume of the frustrum of a square pyramid "a high point in the geometry of the ancient world." Two strange aspects of this formula are that: (1) it is suggested by Problem 14 of the Moscow Papyrus (ca. 1890 B.C.), a manuscript that makes no mention of how to find the volume of a complete pyramid, and (2) the method for finding the volume of a complete pyramid does not appear until the work of Democritus in the fifth century B.C., more than a thousand years later! It is interesting to note that a frustrum of a pyramid is pictured on the back of the U. S. dollar bill.

The Divided Cube.

27. x^3.

•28. $\frac{1}{6}x^3$. (The cube is divided into six congruent pyramids.)

29. x^2.

•30. $\frac{1}{2}x$.

31. $\frac{1}{2}x^3$. $[(x^2)(\frac{1}{2}x).]$

32. $V = \frac{1}{3}Bh$ (because $\frac{1}{6}x^3 = \frac{1}{3}(\frac{1}{2}x^3)$).

Pyramid Numbers.

•33. 231,000 ft². ($481^2 = 231,361$.)

•34. 612 ft. $[PM^2 = PO^2 + OM^2 =$
$481^2 + (\frac{756}{2})^2 = 374,245;$
$PM = \sqrt{374,245} \approx 612.]$

•35. 231,000 ft².
$[A = \frac{1}{2}bh \approx \frac{1}{2}(756)(612) \approx 231,336.]$

36. Yes. (The answers to exercises 33 and 35 are about the same.)

•37. 92,000,000 ft³.
$[V = \frac{1}{3}Bh = \frac{1}{3}(756)^2(481) = 91,636,272.]$

More on the Euler Characteristic.

•38. $n + 1$.

39. $n + 1$.

40. $2n$.

41. The Euler characteristic for pyramids is 2 because $V + F - E = (n + 1) + (n + 1) - 2n = 2$.

Salt Funnel.

42. *Student model of funnel:*

•43. It is a pyramid.

44. An equilateral triangle.

•45. Isosceles right triangles.

•46. $\frac{3}{4}s^2$.

•47. $\frac{\sqrt{2}}{24}s^3$. $[V = \frac{1}{3}Bh = \frac{1}{3}(\frac{s^2}{4})(\frac{s\sqrt{2}}{2}) = \frac{\sqrt{2}}{24}s^3.]$

The Frustum of a Pyramid.

• 48. $\frac{1}{3}a^2(x + h)$.

49. $\frac{1}{3}b^2x$.

50. $\frac{1}{3}a^2(x + h) - \frac{1}{3}b^2x$, or $\frac{1}{3}(a^2x + a^2h - b^2x)$.

• 51. $OM = \frac{1}{2}a$ and $PN = \frac{1}{2}b$.

• 52. Corresponding sides of similar triangles are proportional.

• 53. $\dfrac{x+h}{x} = \dfrac{\frac{1}{2}a}{\frac{1}{2}b} = \dfrac{a}{b}$.

54. $x = \dfrac{bh}{a-b}$. [$xa = xb + bh$, $xa - xb = bh$,

$x(a - b) = bh$, $x = \dfrac{bh}{a-b}$.]

55. $\frac{1}{3}(a^2x + a^2h - b^2x) = \frac{1}{3}[(a^2 - b^2)x + a^2h] =$

$\frac{1}{3}[(a^2 - b^2)\dfrac{bh}{a-b} + a^2h] = \frac{1}{3}[(a + b)bh + a^2h] =$

$\frac{1}{3}(abh + b^2h + a^2h) = \frac{1}{3}(a^2 + ab + b^2)h$.

56. 504 cubic units. [$\frac{1}{3}(12^2 + 12 \times 3 + 3^2)8 = 504$.]

57. No. Trying the numbers for exercise 56 in the Babylonian expression gives

$\frac{1}{2}(12^2 + 3^2)8 = 612 \neq 504$.

Set III (page 653)

How Many Faces?

1. Answer C (seven). Pyramid ABCD has four faces and pyramid EFGHI has five faces. When they are put together, the two faces that coincide are lost so that one would expect the resulting solid to have $4 + 5 - 2 = 7$ exposed faces.

2.

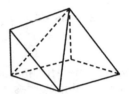

Answer A (five). When the pyramids are put together in the way described, two pairs of faces lie in common planes, so only two faces are formed rather than four. (This can be proved with elementary geometry by showing that the dihedral angles of the solids are supplementary.)

Chapter 15, Lesson 6

Set I (pages 656–658)

Finding volumes of solids of revolution is an important application of integration in the calculus. The formulas for the shell and disk methods are based on right circular cylinders. Some of your students might be interested in seeing the disk equation $V = \int_a^b \pi[f(x)]^2 dx$ and comparing it with the geometric formula for the volume of a cylinder.

The Channel tunnel consists of two large tunnels each containing a single track railway line and a smaller service tunnel midway between them. The tunnels are 50 kilometers long and were bored at an average depth of 45 meters below the ocean floor. More than 5 million cubic meters of rock were removed in building them, a volume twice that of the Great Pyramid.

Primitive Perspective.

1. A cone, a cylinder, and a pyramid.

2. *Example figures:*

Italian into English.

3. Volume, area, base.

4. Cylinder.

5. Cone.

•6. Product.

7. A third.

8. Altitude.

9. The second.

Solids of Revolution.

•10. Its axis.

•11. The altitude.

•12. The lateral surface.

•13. The bases.

14. It is perpendicular to them.

•15. 45π in³. $[V = Bh = (\pi3^2)5 = 45\pi.]$

16. No. Its volume is 75π in³.
   $[V = Bh = (\pi5^2)3 = 75\pi.]$

17. A right triangle.

18. No. If it were revolved about the hypotenuse, the resulting solid would consist of two cones with a common base.

Hockey Puck.

•19. 7 in². $[A = \pi r^2 = \pi(1.5)^2 = 2.25\pi \approx 7.]$

20. 7 in³. $[V = Bh = 2.25\pi(1) \approx 7.]$

Cylindrical Tunnel.

21. 64 m². $[A = \pi r^2 = \pi(4.5)^2 = 20.25\pi \approx 64.]$

•22. 4,800 m³. $[V = Bh = 20.25\pi(75) \approx 4,800.]$

Sliced Solids.

23. Cones.

24. *Example figures:*

•25. They have equal volumes.

•26. Cavalieri's Principle.

Sausages in a Can.

•27. 35π cm³. $[7Bh = 7(\pi1^2)5 = 35\pi.]$

•28. 110 cm³.

29. 54π cm³. $[Bh = (\pi3^2)6 = 54\pi.]$

30. 170 cm³.

•31. 65%. $(\dfrac{35\pi}{54\pi} \approx \dfrac{110}{170} \approx 0.65.)$

Triangles or Cones.

32. The triangles have equal areas. (They are both composed of two 3-4-5 right triangles, or $\dfrac{1}{2}bh = \dfrac{1}{2}8 \cdot 3 = \dfrac{1}{2}6 \cdot 4 = 12.$)

33. The cone with the base of radius 4 has a larger volume. Its volume is $\dfrac{1}{3}Bh = \dfrac{1}{3}(\pi4^2)3 = 16\pi$, whereas the volume of the cone with the base of radius 3 is $\dfrac{1}{3}Bh = \dfrac{1}{3}(\pi3^2)4 = 12\pi.$

Set II (pages 658–661)

The pipes for exercises 34 and 35 are drawn to scale, making it possible to guess without making any calculations that the two pipes in arrangement B can carry the most water.

Pipe Puzzle.

•34. That they have the same length.

35. Arrangement B. The areas of the cross sections of the pipes are: A, $\pi4^2 = 16\pi$; B, $2(\pi3^2) = 18\pi$; C, $3(\pi2^2) = 12\pi$.

Cake Pans.

36. The round pan has a volume of 95 in³ and the square pan has a volume of 96 in³. [$V_1 = Bh = (\pi 4.5^2)1.5 \approx 95$ and $V_2 = Bh = (8^2)1.5 = 96$.]

37. They would be *about* the same, although in theory the height of the round cake would be greater. The same amount of batter is used in each pan, but the area of the base of the round pan is slightly smaller.

Engine Displacement.

•38. The diameter of its base.

39. Its altitude.

40. $V = Bh = \pi r^2 h$. Letting the bore = $d = 2r$ and the stroke = h, we have $r = \dfrac{d}{2}$; so $V = \pi(\dfrac{d}{2})^2 h = \dfrac{\pi}{4}d^2 h$.

•41. 352 in³. [$8V = 8(\dfrac{\pi}{4})(4.0)^2(3.5) \approx 352$.]

Stage Lighting.

•42. πh^2. ($A = \pi r^2 = \pi h^2$.)

43. $\dfrac{1}{3}\pi h^3$. [$V = \dfrac{1}{3}Bh = \dfrac{1}{3}(\pi h^2)h = \dfrac{1}{3}\pi h^3$.]

•44. It is multiplied by four. [$A = \pi r^2 = \pi(2h)^2 = 4\pi h^2$.]

45. It is multiplied by eight. {$V = \dfrac{1}{3}Bh = \dfrac{1}{3}[\pi(2h)^2]2h = \dfrac{8}{3}\pi h^3$.}

Paper Towels.

46. 176. [80.6 ft² = (80.6)(12²) in² = 11,606.4 in²; $\dfrac{11,606.4}{(11)(6)} \approx 175.9$.]

47. 311 in³. [$V = Bh = (\pi 3^2)11 = 99\pi \approx 311$.]

•48. 285 in³. [$99\pi - \pi(\dfrac{1.75}{2})^2 11 \approx 285$.]

49. 1.6 in³. ($\dfrac{285}{176} \approx 1.6$.)

Cylinder and Cone Areas.

50. A rectangle.

51. $2\pi r$.

•52. $2\pi rh$. ($A = bh = 2\pi rh$.)

53. Its total area. [$2\pi r(r + h) = 2\pi r^2 + 2\pi rh$, or the areas of the two bases plus the lateral area.]

•54. A sector of a circle.

•55. $2\pi r$. (The length of arc CD is the circumference of the base of the cone.)

56. $\dfrac{r}{l} \cdot (\dfrac{\text{lateral area}}{\text{area of circle O}} = \dfrac{\text{length of arc CD}}{\text{circumference of circle O}} = \dfrac{2\pi r}{2\pi l} = \dfrac{r}{l}$.)

•57. πrl. ($\dfrac{\text{lateral area}}{\pi l^2} = \dfrac{r}{l}$; so lateral area $= \dfrac{r}{l}\pi l^2 = \pi rl$.)

58. $\pi r\sqrt{r^2 + h^2}$. ($l^2 = r^2 + h^2$; so $l = \sqrt{r^2 + h^2}$.)

59. Its total area. [$\pi r(r + l) = \pi r^2 + \pi rl$, or the area of its base plus its lateral area.]

Coin Wrappers.

60. Because the width of the flat wrapper is 34 mm, its circumference when it is formed into a cylinder is 2(34) = 68 mm. Because $c = \pi d$, $d = \dfrac{c}{\pi} = \dfrac{68}{\pi} \approx 21.6$. The diameter of the cylinder is about 21.6 mm, just large enough for nickels with a diameter of 21 mm.

61. About 29 mm. ($c = \pi d = 18\pi$, $w = \dfrac{c}{2} = 9\pi \approx 28.3$.)

62. Slightly more than $\dfrac{\pi x}{2}$ mm. ($w = \dfrac{c}{2} = \dfrac{\pi x}{2}$.)

Rolling Cones.

63. Its apex, V, stays in place. The points on the edge of the base of the cone sweep out a circle with center V and radius l.

64. $\angle PVO = 30°$. If the cone returns to the same spot after two rotations, $2c = 2(2\pi r) = 2\pi l$; so $l = 2r$, and so $\angle PVO = 30°$.

Set III (page 661)

The picture of the cone of iron ore was taken by photographer Alex MacLean, the founder of Landslides, a Boston-based firm that specializes in lower-altitude aerial photography. The cone was located near the loading docks of Ashtabula, Ohio, ready to be taken by train to steel mills in southern Ohio and West Virginia. MacLean's photographs have been published in two beautiful books: *Look at the Land—Aerial Reflections of America*, with text by Bill McKibben (Rizzoli, 1993) and *Taking Measures Across the American Landscape*, with text by James Corner (Yale University Press, 1996).

Iron Ore.

1. About 190 ft. [From the width of the train cars, we know that 4 mm corresponds to 10 feet; so 1 mm corresponds to 2.5 feet. The diameter of the pile is 75 mm, or 75(2.5) = 187.5 ≈ 190 ft.]

2. About 64 ft.

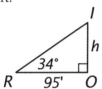

($\tan 34° = \dfrac{h}{95}$, $h = 95 \tan 34° ≈ 64$.)

3. About 28,000 ft². [$A = \pi r^2 ≈ \pi(95)^2 ≈ 28{,}000$.]

4. About 600,000 ft³.

[$V = \dfrac{1}{3}Bh ≈ \dfrac{1}{3}\pi(95)^2(64) ≈ 600{,}000$.]

5. About 137. ($\dfrac{600{,}000 \text{ ft}^3}{4{,}400 \text{ ft}^3} ≈ 137$.)

Chapter 15, Lesson 7

Set I (pages 665–667)

The earth obviously has a shadow *cone*, rather than the shadow cylinder depicted in Anno's whimsical painting. We have evidence of this shadow cone every time we observe an eclipse of the moon.

For our sun to become a black hole, its mass would have to be compressed into a sphere with a radius of about two-thirds of a mile (1 kilometer). If the earth were to become a black hole, it would have to be compressed to about the size of a marble.

Although Euclid considers spheres in the *Elements*, he does not show how to find their surface areas or volumes. He does establish in the last theorem (Proposition 18) of Book XII, however, that the ratio of the volumes of two spheres is equal to the cube of the ratio of their diameters. It was Archimedes who developed the formulas for the volume and area of a sphere in his work *On the Sphere and Cylinder*, which contains 53 theorems on these solids.

In *Beyond the Third Dimension* (Scientific American Library, 1990), Thomas Banchoff describes *Flatland* as "simultaneously a social satire and an introduction to the idea of higher dimensions. . . . [Its author, Edwin] Abbott used dimensional analogies to great effect in raising questions about the way we see the world, especially when we come into contact with the truly transcendental. For over a century, mathematicians and others have speculated about the nature of higher dimensions, and in our day the concept of dimension has begun to play a larger and larger role in our conception of a whole range of activities."

Your students may be surprised to learn that a baseball floats on water. In his book titled *Sport Science* (Simon & Schuster, 1984), Peter J. Brancazio observes that, in a list of balls commonly used in sports, including the football, baseball, golf ball, handball, soccer ball, basketball, and volleyball, only the golf ball and handball sink in water!

The fact established in exercises 24 and 25 that the two boxes contain the same amount of chocolate is contrary to many people's intuition, but, if we think of each sphere as inscribed in a cube, the result is obvious. The ratio of the volumes of the sphere and cube is a constant and does not depend on how large they might be.

Earth Shadow.

•1. They are equal.

•2. It is four times as large.

Black Hole.

•3. About 12.6 km².
[$A = 4\pi r^2 = 4\pi(1)^2 = 4\pi ≈ 12.6$.]

•4. About 4.2 km³.
[$V = \dfrac{4}{3}\pi r^3 = \dfrac{4}{3}\pi(1)^3 = \dfrac{4}{3}\pi ≈ 4.2$.]

5. 3 km. ($4\pi r^2 = \frac{4}{3}\pi r^3$, $12\pi r^2 = 4\pi r^3$, $3 = r$.)

6. 36π, or about 113 km^2, and 36π, or about 113 km^3. [$4\pi(3)^2 = 36\pi \approx 113$; $\frac{4}{3}\pi(3)^3 = 36\pi$.]

Archimedes' Discoveries.

•7. $\frac{1}{3}\pi x^3$. ($V = \frac{1}{3}\pi x^2 x = \frac{1}{3}\pi x^3$.)

8. $\frac{2}{3}\pi x^3$. ($V = \frac{1}{2} \cdot \frac{4}{3}\pi x^3 = \frac{2}{3}\pi x^3$.)

9. πx^3. ($V = \pi x^2 x = \pi x^3$.)

10. The volume of the hemisphere is twice that of the cone, and the volume of the cylinder is three times that of the cone. The sum of the volumes of the cone and hemisphere is equal to the volume of the cylinder.

Flatland.

•11. All radii of a sphere are equal.

•12. If a line is perpendicular to a plane, it is perpendicular to every line in the plane that passes through the point of intersection.

13. HL.

14. Corresponding sides of congruent triangles are equal.

15. It implies that the curve is a circle. The set of all points in a plane that are equidistant from a given point in the plane is a circle.

Sound Wave.

16. At its center.

17. Its area.

•18. It decreases.

Volume and Density.

•19. 212 cm^3. [$V = \frac{4}{3}\pi(3.7)^3 \approx 212$.]

20. 42 cm^3. [$V = \frac{4}{3}\pi(2.15)^3 \approx 42$.]

•21. 0.7 gram/cm^3.
($d = \frac{m}{v} = \frac{145 \text{ g}}{212 \text{ cm}^3} \approx 0.7 \text{ g/cm}^3$.)

22. A baseball will float on water.

23. It is more than 42 grams.
($d = \frac{m}{v} > 1$, $\frac{m}{42} > 1$, $m > 42$.)

Chocolate Packing.

24. (Student answer.)

25. The volume of chocolate in each box is $\frac{8}{3}\pi d^3$; so they contain the same amount of chocolate.

First box: $16[\frac{4}{3}\pi(\frac{d}{2})^3] = \frac{64}{3}\pi\frac{d^3}{8} = \frac{8}{3}\pi d^3$.

Second box: $2(\frac{4}{3}\pi d^3) = \frac{8}{3}\pi d^3$.

26. (Student answer.)

27. The first box contains more wrapping paper. The amounts are related to (but not the same as) the surface areas of the chocolates. If the wrappings are similar to one another, they will be proportional to the squares of the radii (or diameters) of the spherical chocolates that they enclose. Therefore, the box that contains more wrapping paper can be found by comparing the total areas of the chocolates in each box.

First box: $16(4\pi(\frac{d}{2})^2) = 64\pi\frac{d^2}{4} = 16\pi d^2$.

Second box: $2(4\pi d^2) = 8\pi d^2$.

Set II (pages 667–669)

Martin Gardner observed in one of his "Mathematical Games" columns for *Scientific American* (reprinted in *Fractal Music, Hypercards and More . . .* (W. H. Freeman and Company, 1992) that Max Bill's *Half Sphere Around Two Axes* "is based on an old folk method of quickly slicing an apple into congruent halves" and that "it is not as easy as it looks." Some of your students might enjoy the challenge of trying to cut an apple in this way.

In his "Games" column in the October 1983 issue of *Omni* magazine Scot Morris features the "Hollow Earth," telling the story of Cyrus Teed and his strange cult and commenting on the geometry explored in exercises 32 through 37. Morris wrote: "What's most infuriating is that a little mathematical fiddling turns this crazy theory into a proposition that is virtually impossible to refute. The trick is done by inversion, a purely geometric transformation that lets a mathematician turn shapes inside-out. When a sphere is inverted, every point outside is mapped to a corresponding point inside, and vice versa." Morris asked the great geometer H. S. M. Coxeter if there were "any way to prove that we *aren't* inside a hollow earth" and Coxeter couldn't think of any. Morris observes that "just as the geometry of space inverts, so do all the laws of physics. Toward the center of a hollow Earth, light slows down and everything shrinks—atoms, astronauts, spaceships, and measuring rods." More on this topic can be found at various sites on the internet.

Of the two methods for finding the volume of water in exercises 44 and 45, the first method is exact. The second method, treating the spherical shell as if it were a right prism wrapped around the sphere, provides a good approximation only if h, the thickness of the shell, is very small in comparison with r, the radius of the inner sphere.

Exercise 46 illustrates the fact that an expression that measures area must produce *square* units, whereas an expression that measures volume must produce *cubic* units.

This is even more evident in the "volume" formulas given for exercises 50 through 54 for spheres in different dimensions. Because our experience is limited to three dimensions, we have special names for only *linear*, *square*, and *cubic* units. The formulas in the table are derived from general formulas given by Clifford Pickover in *Surfing Through Hyperspace* (Oxford University Press, 1999). As the powers of π in them suggest, there are two different general formulas, depending on whether the number of dimensions is odd or even. The volume of a k-dimensional sphere where k is even is

$$V = \frac{\pi^{\frac{k}{2}} r^k}{(\frac{k}{2})!}$$

whereas, if k is odd, the volume is

$$V = \frac{\pi^{\frac{k-1}{2}} [\frac{k+1}{2}]! 2^{k+1} r^k}{(k+1)!}.$$

(I am tempted to put an exclamation point at the end of the second formula, but that would only cause confusion!)

Half Sphere.

•28. $\frac{2}{3}\pi r^3$. $[\frac{1}{2}(\frac{4}{3}\pi r^3) = \frac{2}{3}\pi r^3.]$

29. $2\pi r^2$. $[\frac{1}{2}(4\pi r^2) = 2\pi r^2.]$

•30. $\frac{3}{2}\pi r^2$. $[3(\frac{1}{2}\pi r^2) = \frac{3}{2}\pi r^2.]$

31. $3.5\pi r^2$. $(2\pi r^2 + \frac{3}{2}\pi r^2 = 3.5\pi r^2.)$

Hollow Earth.

32. $\triangle$OTP is a right triangle because, if a line is tangent to a circle, it is perpendicular to the radius drawn to the point of contact.

•33. Either leg of a right triangle is the geometric mean between the hypotenuse and its projection on the hypotenuse (or by similar triangles).

34. 65 mi. $[\frac{240,000}{3,960} = \frac{3,960}{OP'},$
$240,000\, OP' = (3,960)^2, OP' \approx 65.]$

•35. 3,895 mi.
$(AP' = AO - OP' = 3,960 - 65 = 3,895.)$

36. 900 ft. $[\frac{93,000,000}{3,960} = \frac{3,960}{OP'},$
$93,000,000\, OP' = (3,960)^2,$
$OP' \approx 0.17$ mi ≈ 900 ft.]

37. Near the center of the earth.

Melting Snowball.

•38. About 15 hours.
$[V = \frac{4}{3}\pi r^3 = \frac{4}{3}\pi(6\text{ cm})^3 \approx 905\text{ cm}^3;$
905 minutes $\approx$ 15 hours.]

•39. About 13 hours.
$[V = \frac{4}{3}\pi(3\text{ cm})^3 \approx 113\text{ cm}^3;$
$905\text{ cm}^3 - 113\text{ cm}^3 = 792\text{ cm}^3;$
792 minutes $\approx$ 13 hours.]

40. About 452 cm². $[A = 4\pi(6)^2 \approx 452.]$

41. About 113 cm². $[A = 4\pi(3)^2 \approx 113.]$

42. No. It takes about 13 hours for the diameter of the snowball to decrease 6 inches and only 2 hours for it to decrease the remaining 6 inches. (The diameter decreases at an increasing rate.)

43. No. The surface area of the snowball decreases about $452 - 113 = 339$ cm^2 in 13 hours (an average of 26 cm^2 per hour) and 113 cm^2 in 2 hours (about 56 cm^2 per hour). (The surface area also decreases at an increasing rate.)

Earth Under Water.

•44. 325,000,000 mi^3.
$(V = \frac{4}{3}\pi(3,960 + 1.65)^3 - \frac{4}{3}\pi(3,960)^3 =$
$\frac{4}{3}\pi(3,961.65^3 - 3,960^3) \approx 325,000,000.)$

45. 325,000,000 mi^3.
$(V \approx 4\pi r^2 h = 4\pi(3,960)^2(1.65) \approx 325,000,000.)$

Doughnuts.

46. The first expression must give its volume because (a units)(b units)$^2 = ab^2$ cubic units; the second expression must give its area because (a units)(b units) = ab square units.

•47. The second expression.

48. It would be doubled.
$[2\pi^2(2a)b^2 = 4\pi^2ab^2 = 2(2\pi^2ab^2).]$

49. It would be multiplied by four.
$[2\pi^2a(2b)^2 = 8\pi^2ab^2 = 4(2\pi^2ab^2).]$

Spheres in Different Dimensions.

50. A circle.

51. Its area.

52. Its diameter.

53.

Dimensions	Volume
1	2
2	3.14
3	4.19
4	4.93
5	5.26
6	5.17
7	4.72

54. As the number of dimensions increases, the volume of the sphere at first increases and then decreases.

Set III (page 669)

Frank Marshall, the producer of *Raiders of the Lost Ark,* said concerning the rolling-boulder sequence in the movie: "A huge boulder comes crashing down through a chute and chases after Indy. We had to make it look like it rolled free, but we also had to be able to control it. The boulder was on this contraption like an arm; it was free-spinning and the arm was hidden. The boulder tumbled down the chute, then it had to be taken back if we wanted to do another take. But we couldn't do it again very quickly because we had to put in the stalactites that got broken off as it rolled out."

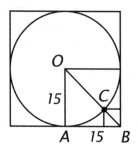

1. You could lie down along the edge of the tunnel. In the figure above, $\triangle OAB$ is an isosceles right triangle; so $OB = 15\sqrt{2}$ and $CB = OB - OC = 15\sqrt{2} - 15 = 15(\sqrt{2} - 1) \approx$ 6.2 feet. The sides of the small square are about $\frac{6.2}{\sqrt{2}} \approx 4.4$ feet; so there is plenty of room.

2. *(Student answer.)* (It might not work if the corners of the passage were rounded.)

Chapter 15, Lesson 8

Set I (pages 672–674)

About the Statue of Liberty, Robert Osserman in his book titled *Poetry of the Universe* (Anchor Books, 1995) explains: "When the Statue of Liberty was built, the first step was the construction of a small-scale model. Building the statue to scale from the model in one step would have been a very risky operation. The sculptor decided to construct a number of intermediate models, scaling the statue up by a modest amount at each step. One reason for caution is that each time the scale is doubled, the volume—and therefore the weight—goes up by a factor of eight. On the other hand, the strength of the supports, which depends on cross-sectional area, is multiplied only by a factor of four. That leads to the inevitable conclusion that whatever shape you design, if built on a large enough scale, will collapse under its own weight."

A similar lesson on this subject appears in J. E. Gordon's *The Science of Structures and Materials* (Scientific American Library, 1988) under the title "The Danger of Scaling Up": "Engineering design practice is based to a large extent upon experience and precedent. One successful project leads to another, and things generally get bigger and bigger. The tendency is to scale up a successful design, keeping the working stresses the same. Not very long ago the author had a head-on collision over the conference table with NASA about its wind-energy program. A number of safe and successful windmills are 175 feet in diameter. NASA proposed, in effect, to scale up this design to a diameter of 400 feet. I had to point out that, for reasons of modern fracture mechanics, extensive redesign would be needed if the bigger windmills were to operate safely."

The difficulty of comparing the volumes of solids from two-dimensional drawings of them is demonstrated by exercises 22 through 26 on olive sizes. The volume of a supercolossal olive is almost twice that of a jumbo olive, something that might be hard to guess from the side-view drawings of exercises 19 and 21.

The three pyramids at Giza were built in a period of about 150 years (ca. 2650 B.C.–2500 B.C.) and are named after the kings who built them.

Similar or Not?

1. Two cones are not always similar.
    Example figures:

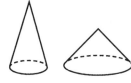

•2. Two cubes are always similar.

3. Two triangular prisms are not always similar.
    Example figures:

4. Two spheres are always similar.

5. Two square pyramids are not always similar.
    Example figures:

Scaling Up.

•6. It is multiplied by 9. (3^2.)

•7. It is multiplied by 27. (3^3.)

8. It is multiplied by 400. (20^2.)

9. It is multiplied by 8,000. (20^3.)

•10. It is multiplied by n^2.

•11. It is multiplied by n^3.

Giant Windmill.

12. 3.2. ($\frac{400}{125}$.)

•13. 10.24. (3.2^2.)

•14. 32.768. (3.2^3.)

15. Approximately 82,000 pounds (or 41 tons!) ($32.768 \times 2,500 = 81,920$.)

Gulliver in Lilliput.

16. By cubing 12. (They were basing the amount of food that Gulliver needed on his volume.)

•17. That Gulliver's body was similar to their bodies.

18. 144 times as much. (Basing the amount of cloth on surface areas gives $12^2 = 144$.)

Earth and Mars.

19. About 1.9 times. ($\frac{3,960}{2,100} = 1.8857\ldots$)

•20. About 3.6 times. [$(1.8857\ldots)^2 \approx 3.6$.]

21. About 6.7 times. [$(1.8857\ldots)^3 \approx 6.7$.]

Olive Sizes.

22. *Exact size:*

•23. About 22 mm. ($\frac{x}{18} = \frac{28}{23}$, $23x = 504$, $x \approx 22$.)

24. *Exact size:*

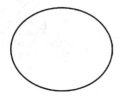

•25. About 1.8 times. [$(\frac{28}{23})^3 \approx 1.8$.]

26. About 9 grams. (1.8×5.)

Three Pyramids.

27. Cheops, 0.62; Chephren, 0.67; Mycerinus, 0.61.

28. Cheops and Mycerinus, because the ratio of their height to their base edge is closest to being the same.

Set II (pages 674–676)

There are two stories associated with the problem of the duplication of the cube, one of the three famous problems of antiquity. In *Science Awakening* (Oxford University Press, 1961), B. L. van der Waerden quotes from a supposed letter from Eratosthenes to king Ptolemy:

"It is said that one of the ancient tragic poets brought Minos on the scene, who had a tomb built for Glaucus. When he heard that the tomb was a hundred feet long in every direction, he said:

'You have made the royal residence too small, it should be twice as great. Quickly double each side of the tomb, without spoiling the beautiful shape.'

He seems to have made a mistake. For when the sides are doubled, the area is enlarged fourfold and the volume eightfold. The geometers then started to investigate how to double a given body, without changing its shape; and this problem was called the duplication of the cube, since they started with a cube and tried to double it. After they had looked for a solution in vain for a long time, Hippocrates of Chios observed that, if only one could find two geometric means between two line segments, of which the larger one is double the smaller, then the cube would be duplicated. This transformed the difficulty into another one, not less great."

The difficulty is in obtaining the length of the edge of the larger cube solely by geometric construction. Without this restriction, the problem can be easily solved. Exercises 37 through 43 present a simple solution, but, unfortunately, the figure used cannot be constructed by straightedge and compass.

Exercises 47 and 48 comparing the shape of liquid in a conical cup to its container has an interesting biological connection. In *Cat's Paws and Catapults* (Norton, 1998), Steven Vogel wrote: "The need for growth without loss of function can impose severe geometrical limitations. Consider the possible shapes of mollusks, a widespread and diverse group that includes scallops, slugs, snails, and squid. Mollusk shell doesn't grow, so a shell cannot enlarge except by adding incremental additions to edges and inner surfaces. For most shapes [such as cylinders and rectangular solids], such incremental additions would quickly lead to awkward changes in proportions. . . . Hollow cones are much better than cylinders or most other shapes if what matters is growth in size without simultaneous change in shape. And mollusk shells are basically cones. . . ."

The iron-doughnut problem is a classic puzzle in physics. Martin Gardner observes that opticians take advantage of the fact that a ring remains similar to itself as it expands when they heat the frame of a pair of glasses in order to remove a lens.

Doubling a Cube.

29. $6x^2$.

•30. $24x^2$. [$6(2x)^2 = 24x^2$, or $2^2 \cdot 6x^2 = 24x^2$.]

31. x^3.

•32. $8x^3$. [$(2x)^3 = 8x^3$, or $2^3 \cdot x^3 = 8x^3$.]

33. 4. ($\frac{24x^2}{6x^2} = 4$, or 2^2.)

34. 8. ($\frac{8x^3}{x^3} = 8$, or 2^3.)

•35. The square.

36. The cube.

Plato's Method.

37. The altitude to the hypotenuse of a right triangle is the geometric mean between the segments into which it divides the hypotenuse.

•38. $x = \sqrt[3]{2}$; $y = \sqrt[3]{4}$. [$\frac{1}{x} = \frac{x}{y}$ and $\frac{x}{y} = \frac{y}{2}$, $y = x^2$ and $y^2 = 2x$, $(x^2)^2 = 2x$, $x^3 = 2$, $x = \sqrt[3]{2}$; $y = (\sqrt[3]{2})^2 = \sqrt[3]{4}$.]

•39. PA, 1; PB, 2; PC, 4; PD, 8.

•40. It is twice as large.

41. $x = \sqrt[3]{3}$; $y = \sqrt[3]{9}$. [$\frac{1}{x} = \frac{x}{y}$ and $\frac{x}{y} = \frac{y}{3}$, $y = x^2$ and $y^2 = 3x$, $(x^2)^2 = 3x$, $x^3 = 3$, $x = \sqrt[3]{3}$; $y = (\sqrt[3]{3})^2 = \sqrt[3]{9}$.]

42. PA, 1; PB, 3; PC, 9; PD, 27.

43. It is three times as large.

Rubber Bands.

44. No. Although the corresponding lengths and widths of the rubber bands are proportional, we don't know whether their corresponding thicknesses are proportional.

•45. Eight. ($\frac{2\frac{1}{2}}{1\frac{1}{4}} = \frac{\frac{1}{8}}{\frac{1}{16}} = 2$; $2^3 = 8$. If the bands were made from the same material, their weights would have the same ratio as their volumes.)

46. The fact that the ratio $\frac{5,780}{1,420} \approx 4.1$ indicates that the weight, and hence volume, of a #31 band is about four times that of a #10 band. The rubber bands are evidently not similar in shape and apparently have approximately the same thickness.

Two Cups.

47. Yes. $\triangle ABE \sim \triangle CDE$ by AA because $\angle A = \angle DCE$ and $\angle AEB = \angle CED$.

48. Yes. Because $\triangle ABE \sim \triangle CDE$, $\frac{AB}{CD} = \frac{AE}{CE}$; so the corresponding dimensions of the cones are proportional.

49. No. The rectangles are *not* similar because $\frac{AB}{CD} = 1$, whereas $\frac{BF}{DF} > 1$. Alternatively, the rectangles are not similar, because they do not have the same shape.

50. No.

Height and Weight.

•51. About 47 pounds. [$(\frac{4}{6})^3 \times 160 \approx 47$.]

•52. About 93 pounds. [$(\frac{5}{6})^3 \times 160 \approx 93$.]

53. About 254 pounds. [$(\frac{7}{6})^3 \times 160 \approx 254$.]

54. About 379 pounds. [$(\frac{8}{6})^3 \times 160 \approx 379$.]

Heated Ring.

55. *(Student answer.)*

56. The diameter of its hole gets larger. Because it is given that the shape stays the same, *all* dimensions must increase in order to remain proportional.

Teddy Bear Trouble.

57. No. She would need $2^2 = 4$ times as much fur to cover the surface, $2^3 = 8$ times as much kapok to fill the volume, 2 times as much length of ribbon and the same number of eyes as for the smaller bears.

58. The kapok.

59. Five. (She has enough stuffing to make 40 small bears; $\frac{40}{8} = 5$.)

60. The ribbon.

According to the *Guinness Book of World Records*, a clam weighing 734 pounds was found off Ishigaki Island, Okinawa, Japan, in 1956.

Colossal Clam.

We will assume that the clams are similar and that the ratio of their weights is equal to the ratio of their volumes.

Let x = the weight of the larger clam in ounces:

$$\frac{x}{1} = (\frac{48}{2.4})^3 \; ;$$

so $x = 20^3 = 8,000$. The larger clam would weigh 8,000 ounces, or 500 pounds.

Chapter 15, Lesson 9

Set I (pages 678–680)

The prefixes used in naming the five regular polyhedra are of Greek origin: *tetra, hexa, okta, dodeka,* and *eikosa.* Surprisingly, however, of the terms referring to "four" most commonly used in mathematics, only "tetrahedron" comes from Greek. Terms such as "quadrilateral," "quadrant," "quadratic," "quadrillion," and "quadruple" come from Latin. It is also interesting to note that, although the names of the five regular polyhedra refer to *all* polyhedra having those numbers of faces, they are used almost exclusively to mean the *regular* polyhedra. Terms such as "pentahedron" or "decahedron," for which no regular polyhedrons exist, are almost never used.

Exercises 6 through 11 consider two- and three-dimensional versions of the same idea. It seems appropriate that the sides of the "midpoint triangle" in two dimensions are $\frac{1}{2}$ the sides of the original triangle, whereas the edges of the "midpoint tetrahedron" in three dimensions are $\frac{1}{3}$ the edges of the original tetrahedron. A rather surprising difference between the two situations, however, is that, although the large triangle can be "filled" with four of the smaller triangles, the large tetrahedron cannot be "filled" with *any* number of the smaller tetrahedrons. In *Martin Gardner's Sixth Book of Mathematical Games from Scientific American* (W.H. Freeman and Company, 1971), Gardner explains: "Because equilateral triangles also tile a plane, one might

suppose that regular and congruent tetrahedrons would also pack snugly to fill space. This seems so intuitively evident that even Aristotle, in his work *On the Heavens,* declared it to be the case. The fact is that among the Platonic Solids the cube alone has this property. If the tetrahedron also had it, it would long ago have rivaled the cube in popularity for packaging."

In *The Penguin Book of Curious and Interesting Mathematics* (Penguin, 1997), David Wells reports, in a questionnaire of mathematicians regarding the "most beautiful theorems" of mathematics, the fact that there are five regular polyhedra ranked near the top of the list. Benno Artmann (*Euclid—The Creation of Mathematics,* Springer, 1999) points out that "Euclid's treatment of the regular polyhedra is especially important for the history of mathematics because it contains the first example of a major classification theorem. Such theorems start with a definition (or axiomatic description) and end with a list of objects satisfying the description. There is no official definition of a regular solid in the list of definitions at the beginning of Book XI. However, we find one in the very last theorem of the *Elements,* which asserts the completeness of the preceding list of polyhedra." Exercises 30 through 36 suggest the form of Euclid's argument.

Numbers and Names.

- •1. Eight.

  2. 12.

- •3. Quadrilateral.

  4. Cube.

  5. An icosahedron is a regular polyhedron that has 20 faces. An icosagon is a polygon that has 20 sides.

Two Triangles.

- •6. Yes. (One explanation: The sides of the smaller triangle are midsegments of the larger triangle; so they are parallel to its sides. From the parallel lines, we know that AEDF, BDEF, and CDFE are parallelograms. The opposite angles of a parallelogram are equal, so the angles of the smaller triangle are equal to the angles of the larger triangle; the triangles are similar by AA.)

  7. They are twice as long. (This also follows from the Midsegment Theorem.)

8. The perimeter of the larger triangle is twice the perimeter of the smaller one. (The ratio of the perimeters of two similar polygons is equal to the ratio of a pair of corresponding sides.)

•9. The area of the larger triangle is four times the area of the smaller one. (The ratio of the areas of two similar polygons is equal to the square of the ratio of a pair of corresponding sides.)

Two Tetrahedrons.

•10. The surface area of the larger tetrahedron is $3^2 = 9$ times the surface area of the smaller one.

11. The volume of the larger tetrahedron is $3^3 = 27$ times the volume of the smaller one.

da Vinci Polyhedron.

12. A dodecahedron.

13. Regular pentagons.

•14. 12.

15. Five.

•16. Three.

17. Two.

•18. 20.

19. 30.

Escher Box.

•20. An icosahedron.

21. Equilateral triangles.

22. 20.

23. Three.

24. Five.

25. Two.

•26. 12. ($\frac{20 \times 3}{5}$.)

27. 30. ($\frac{20 \times 3}{2}$.)

Euclid's Claim.

28. Regular polygons.

•29. That they are congruent.

•30. 180°. ($3 \times 60°$.)

31. 240°. ($4 \times 60°$.)

32. 300°. ($5 \times 60°$.)

33. Because, if six equilateral triangles surround a point, they must lie in a plane (because the sum of the six angles is 360°).

•34. 270°. ($3 \times 90°$.)

35. 324°. ($3 \times 108°$.)

36. The figure is flat if the sum is 360°; it is solid if the sum is less than 360°.

Set II (pages 680–682)

The golden ratio appears as the ratio of the lengths of the edges of the dodecahedron and its inscribed cube in exercise 45. The exercise reveals that this ratio comes from the relation of the length of a diagonal of a regular pentagon to one of its sides.

Exercises 47 through 60 reveal something rather surprising. As the number of sides of a regular polygon inscribed in a circle increases, the perimeter and area of the polygon increase. As the number of faces of a regular polyhedron inscribed in a sphere increases, its surface area and volume do not necessarily increase! Howard Eves wrote in the first volume of *In Mathematical Circles* (Prindle, Weber & Schmidt, 1969): "The uninitiated will almost always intuitively believe that of the regular dodecahedron (a solid having 12 faces) and a regular icosahedron (a solid having 20 faces) inscribed in the same sphere, the icosahedron has the greater volume. . . . The reverse is actually the case." Not only is the inscribed dodecahedron closer than the icosahedron to the sphere in volume, it is also closer to it in surface area. If these solids are *circumscribed* about a sphere rather than *inscribed* in one, however, it is the icosahedron whose surface area and volume are closer to those of the sphere. Another remarkable fact about the solids inscribed in the sphere is that the ratio of their volumes is equal to the ratio of their surface areas.

Octahedron and Cube.

37.

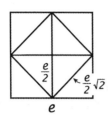

•38. $\frac{e}{2}\sqrt{2}$. (From the isosceles right triangle.)

39. $\frac{1}{2}e^2$. $[A = (\frac{e}{2}\sqrt{2})^2 = \frac{1}{2}e^2.]$

40. $\frac{e}{2}$.

•41. $\frac{1}{6}e^3$. $\{V = 2(\frac{1}{3}Ah) = 2[\frac{1}{3}(\frac{1}{2}e^2)(\frac{e}{2})] = \frac{1}{6}e^3.\}$

42. It is one-sixth of the volume of the cube.

Dodecahedron Construction.

43. By HL. (ΔABP and ΔCBP are right triangles, AB = BC because ABCDE is regular, and BP = BP.)

•44. ∠ABP = 54° and ∠BAP = 36°.
   (∠ABP = ∠CBP; so
   ∠ABP = $\frac{1}{2}$∠ABC = $\frac{1}{2}$108° = 54°;
   ∠BAP = 90° − 54° = 36°.)

•45. 1.618. (In right ΔABP,

   sin ∠ABP [or cos ∠BAP] = $\frac{AP}{AB}$; AC = 2AP;

   so $\frac{AC}{AB} = \frac{2AP}{AB}$ = 2 sin 54° = 1.618. This is
   the golden ratio.)

•46. About 1.8. [The volume of the cube is
   $1^3 = 1$ and the volume of the dodecahedron
   is 1 + 6(0.135) = 1.81.]

Round About.

47. That there exists a circle that contains all of its vertices.

48. That there exists a sphere that contains all of a regular polyhedron's vertices.

49. The polygon with 20 sides.

50. *(Student answer.)* (The polyhedron with 20 faces seems like a reasonable guess.)

51. 12.6 square units.
   $[A = 4\pi r^2 = 4\pi(1)^2 = 4\pi \approx 12.6.]$

•52. 0.714 unit.

•53. 10.5 square units.

54. 1.051 units.

55. 9.6 square units!

56. *(Student answer.)* (Again the icosahedron seems like a reasonable guess.)

57. 4.2 cubic units.
   $[V = \frac{4}{3}\pi r^3 = \frac{4}{3}\pi(1)^3 = \frac{4}{3}\pi \approx 4.2.]$

•58. 2.8 cubic units. $[\frac{1}{4}(0.714)^3(15 + 7\sqrt{5}) \approx 2.8.]$

59. 2.5 cubic units. $[\frac{5}{12}(1.051)^3(3 + \sqrt{5}) \approx 2.5.]$

60. The dodecahedron!

Set III (page 682)

About this "Pyramid Puzzle," Jerry Slocum and Jack Botermans wrote in *Puzzles Old and New* (University of Washington Press, 1986): "This put-together puzzle consists of two identical pieces and at first sight it seems too easy to include in a serious puzzle book. However, appearances can be deceptive and this apparently simple puzzle has frustrated many a self-confident individual." The puzzle has been marketed in various versions, including one in which the two pieces have each been cut in half so that there are four pieces and another one with four pieces, each piece consisting of four connected spheres.

Tetrahedron Puzzle.

The solution is shown at the right above.

Set I (pages 684–686)

In *Visual Intelligence* (Norton, 1998), Donald Hoffman considers the ambiguity of seeing depth: "The image at the eye has two dimensions; therefore it has countless interpretations in three dimensions. . . . *Anytime* you see depth you construct it, not just when you view strange drawings . . . but also in everyday life. . . . Quite simply, your visual system is biased. It constructs only those 3D worlds that conform to its rules. Most others it simply ignores. These rules are powerful. They prune the possible depths you can see from infinity down to one or two. This is useful, for if you saw all options then you would see depth flip or distort with your every glance." Vasarely's sculpture of the "flipping cubes" is a dramatic example of the problems that we can have in viewing a two-dimensional picture as three-dimensional.

The two figures by Gaetano Kanizsa are almost identical yet are interpreted very differently. The second, seen as three-dimensional, has the same depth flip: Is the top of the box the quadrilateral at the upper left or the one at the upper right?

A *cord* is an unusual unit of volume for several reasons: it is only used to measure *wood*, it is associated with a specific *shape*, and that shape is *not a cube*. *The American Heritage Dictionary of the English Language* defines it as: "A unit of quantity for cut fuel wood, equal to 128 cubic feet in a stack measuring 4 by 4 by 8 feet."

The building considered in exercises 13 through 16 is the Baptistery of the Cathedral of Santa Maria del Fiore in Florence, Italy. It dates from the eleventh century, and its octagonal shape was evidently inspired by a building that had *previously* existed on the site.

The "inverse square law" considered for light intensity in exercises 17 through 19 appears in Newton's law of universal gravitation and in Coulomb's law for electric charges. The authors of *Project Physics* (Holt Rinehart Winston, 1981) remark that "these two great laws arise from completely different sets of observations and apply to completely different kinds of phenomena. Why they should match so exactly is to this day a fascinating puzzle."

Exercises 20 through 24 are based on one of the methods used by astronomers to estimate the number of stars in our galaxy, the Milky Way. It is assumed in this method that the density of stars throughout the galaxy is the same as that in our "neighborhood," 10 stars in a sphere with a radius of 10 light-years. The Milky Way itself is not spherical, of course, but shaped somewhat like a pancake with a bulge in the middle. Current estimates suggest that it may contain as many as 400 billion stars.

A simple method for determining the volume of a regular tetrahedron in terms of the length of its edge is explored in exercises 25 through 31. A more challenging approach would be to treat it as a triangular pyramid so that the length of one of its altitudes has to be considered.

Flipping Cubes.

- •1. 4 and 8.

- 2. A line.

- •3. They appear to be perpendicular.

- 4. 1, 7, and 11.

- 5. It depends on how the figure (an optical illusion) is seen. (For example, if 3 is seen as the top of a cube and 9 as the bottom of a cube, the planes do not appear to be parallel. If 3 is seen as the top of a cube and 9 is seen as the floor of a cubical space, the planes do appear to be parallel.)

- 6. That they do not intersect.

Seeing Things.

- 7. (*Student answer.*) (The most likely interpretation is that this figure is two-dimensional: an irregular hexagon with three concurrent diagonals.)

- 8. (*Student answer.*) (The most likely interpretation is that this figure is three-dimensional: a polyhedron with six quadrilateral faces.)

Cord of Wood.

- 9. 128. ($V = lwh = 8 \cdot 4 \cdot 4 = 128$.)

- •10. 16 ft. ($128 = 4 \cdot 2 \cdot x$, $8x = 128$, $x = 16$.)

Box Pattern.

- 11. The hexagons.

- 12. A right hexagonal prism.

Octagonal Building.

- •13. A prism and a pyramid.

14. 1,080,000 ft³.
    ($V = Bh = 12,000 \times 90 = 1,080,000$.)

•15. 80,000 ft³. ($V = \frac{1}{3}Bh = \frac{1}{3}12,000 \times 20 = 80,000$.)

16. 1,160,000 ft³. (1,080,000 + 80,000.)

Screen Illumination.

17. They are twice as long.

•18. It is four times as large. ($2^2 = 4$.)

19. The same amount of light is spread over an area four times as large; so the light is one-fourth as bright.

Stars.

20. 500. ($\frac{5,000}{10}$.)

•21. 250,000. (500^2.)

22. 125,000,000. (500^3.)

•23. 1,250,000,000. ($10 \times 125,000,000$.)

24. 125,000,000,000. ($100 \times 1,250,000,000$.)

Divided Cube.

•25. $e\sqrt{2}$.

26. They are equilateral and congruent.

•27. A regular tetrahedron.

28. Five. (The regular tetrahedron and the four pyramids on its faces.)

•29. $\frac{1}{6}e^3$. [$V = \frac{1}{3}Bh = \frac{1}{3}(\frac{1}{2}e^2)e = \frac{1}{6}e^3$.]

•30. $\frac{1}{3}e^3$. [$e^3 - 4(\frac{1}{6}e^3) = \frac{1}{3}e^3$.]

31. One-third.

Set II (pages 686–688)

More from Archimedes' *On the Sphere and Cylinder* is included in exercises 32 through 42. About 150 years after Archimedes' death, the great Roman statesman Cicero found his tombstone. He wrote: "When I was quaestor [administrator] in Sicily I managed to track down Archimedes' grave. . . . I remembered having heard of some simple lines of verse which had been inscribed on his tomb, referring to a sphere and cylinder

modelled in stone on top of the grave. . . . Finally, I noticed a little column, just visible above the scrub: it was surmounted by a sphere and cylinder. . . . The verses were still visible, though approximately the second half of each line had been worn away." Although the tomb no longer exists, Archimedes' discoveries of the remarkable connections between the area and the volume of a sphere and its circumscribed cylinder are part of his lasting memorial.

The "twisted ring" of exercises 43 through 45 is equivalent to a Möbius strip. In one of his "Mathematical Games" columns for *Scientific American*, Martin Gardner wrote: "The cross section of a Möbius band actually is a rectangle, very much longer than it is wide. The band itself may be regarded as a four-sided prism, twisted so that one end has rotated 180 degrees before the two ends are joined. Viewed in this way it is a solid ring with two distinct 'faces.'" More concerning Möbius strips as twisted prisms can be found in Gardner's article titled "Twisted Prismatic Rings" in *Fractal Music, Hypercards and More* . . . (W. H. Freeman and Company, 1992).

About Cavalieri's method for calculating the volume of a wine barrel, John Pottage comments in *Geometrical Investigations* (Addison-Wesley, 1983): "Cavalieri's barrel rule may well have been the source of an important sequence of mathematical development. . . . It was probably this formula for the capacity of a cask of wine which suggested to James Gregory . . . the more general formula for approximate integration, often written in the form

$$\int_0^{2h} y\,dx = \frac{1}{3}h(y_0 + 4y_1 + y_2)$$

which, in its turn, led to the outstanding generalizations due to Newton and Cotes, and subsequently to the minor generalization known as 'Simpson's rule.'"

About the Schlegel diagrams of polyhedra, Thomas Banchoff wrote in *Beyond the Third Dimension* (Scientific American Library, 1990): "The central projections of regular polyhedra form beautiful patterns which reveal the structure and symmetries of these objects. If a regular polyhedron has all of its vertices on a sphere in three-space, then we may use central projection from the north pole to obtain an image on the horizontal plane at the south pole. . . . The images of the edges of a polyhedron are said to form a *Schlegel diagram* of the polyhedron, named for Viktor Schlegel, the German mathematician who invented this type of diagram in 1883." The diagrams are used in exercises 57 through 61 to verify that the Euler characteristic of the regular polyhedra is the same number that we proved for prisms and pyramids: 2. (See the commentary for exercises 50 through 53 of Chapter 15, Lesson 3.)

According to Archimedes.

32. A circle whose radius is equal to the radius of the sphere.

33. If the radius of a great circle and the sphere is r, the area of a great circle is πr^2 and the surface area of the sphere is $4\pi r^2$.

34. The volume of the cone is
$$\frac{1}{3}Bh = \frac{1}{3}\pi r^2 r = \frac{1}{3}\pi r^3 \text{ and the volume of the}$$
sphere is $\frac{4}{3}\pi r^3$; so the theorem is true.

35. $\frac{4}{3}\pi x^3$.

•36. $2\pi x^3$. $[V = Bh = \pi x^2(2x) = 2\pi x^3.]$

•37. $\frac{2}{3}$. $\left(\dfrac{\frac{4}{3}\pi x^3}{2\pi x^3} = \dfrac{2}{3}.\right)$

38. $4\pi x^2$.

39. $4\pi x^2$. $[A = bh = 2\pi x(2x) = 4\pi x^2.]$

•40. $6\pi x^2$. $(4\pi x^2 + 2\pi x^2 = 6\pi x^2.)$

41. The two areas are equal.

42. $\frac{2}{3}$. $\left(\dfrac{4\pi x^2}{6\pi x^2} = \dfrac{2}{3}.\right)$

Twisted Ring.

43. Two.

44. 10 cubic units. $(10 \times 1^3.)$

•45. 40 square units. (Each cube has four exposed faces; so each cube has an exposed surface area of $4 \times 1^2 = 4$ square units. $10 \times 4 = 40.$)

SAT Problem.

46.

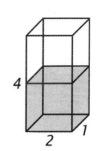

47.

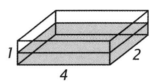

48. 2 feet and 0.5 foot, respectively. $(1 \times 2 \times d_1 = 4, d_1 = 2; \ 2 \times 4 \times d_2 = 4, d_2 = 0.5.)$

Wine Barrel.

49. The area of its base.

•50. The area of the horizontal cross section halfway up the barrel.

•51. $\pi r^2 h$. $[\frac{1}{3}(2\pi r^2 + \pi r^2)h = \frac{1}{3}(3\pi r^2)h = \pi r^2 h.]$

52. A cylinder.

53. $\frac{1}{3}\pi r^2 h$. $[\frac{1}{3}(2\pi 0^2 + \pi r^2)h = \frac{1}{3}\pi r^2 h.]$

54. A cone. [If $R = 0$, the wine-barrel figure becomes an "hour glass" of two cones with a common apex, bases of radius r, and altitudes of $\frac{1}{2}h$. $2(\frac{1}{3}\pi r^2)(\frac{1}{2}h) = \frac{1}{3}\pi r^2 h$; so the two cones are together equal in volume to a single cone with the same base and altitude h.]

55. $\frac{4}{3}\pi R^3$. $[\frac{1}{3}(2\pi R^2 + \pi 0^2)2R = \frac{4}{3}\pi R^3.]$

56. A sphere.

Euler Characteristic.

57. Yes. (The three small triangles and the triangle containing them.)

•58. Four.

•59. Six.

60. | Polyhedron | F | V | E |
| --- | --- | --- | --- |
| Tetrahedron | 4 | 4 | 6 |
| Cube | 6 | 8 | 12 |
| Octahedron | 8 | 6 | 12 |
| Dodecahedron | 12 | 20 | 30 |
| Icosahedron | 20 | 12 | 30 |

61. It is 2 for all of them.

Chapter 16, Lesson 1

Set I (pages 692–693)

The beach ball of exercises 16 through 24 is a "spherical dodecahedron," the result of projecting the edges of a regular dodecahedron on the surface of its circumscribing sphere. Because of the duality of the regular dodecahedron and regular icosahedron, the ball is also a "spherical icosahedron." The sphere is divided into 120 congruent right triangles, of which 6 are contained in each equilateral triangle of the icosahedron and 10 are contained in each regular pentagon of the dodecahedron. More on the mathematics of this topic can be found in chapter 1 of *Spherical Models,* by Magnus J. Wenninger (Cambridge University Press, 1979).

Garden Sphere.

- •1. At the center of the sphere.

- •2. A radius of the sphere.

- •3. Two hemispheres.

- 4. A diameter.

Basic Differences.

- 5. Both geometries.

- 6. Euclidean geometry.

- •7. Euclidean geometry.

- 8. Sphere geometry.

- 9. Euclidean geometry.

- •10. Both geometries.

- 11. Euclidean geometry.

Shortest Routes.

- 12. Arc 3.

- 13. It is the largest.

- 14. The length of the minor arc of the great circle that contains them.

- 15. Because the shortest route between two points is along a great circle.

Beach Ball.

- •16. $186°$. $(36° + 60° + 90°.)$

- 17. No. In these triangles, their sum is more than $90°$.

- •18. $\angle A = 2(36°) = 72°$, $\angle B = \angle C = 60°$; $\angle A + \angle B + \angle C = 192°$.

- 19. $\angle A = 36°$, $\angle E = 90°$, $\angle F = 2(36°) = 72°$; $\angle A + \angle E + \angle F = 198°$.

- 20. $\angle C = 60°$, $\angle D = 36°$, $\angle F = 3(36°) = 108°$; $\angle C + \angle D + \angle F = 204°$.

- 21. $\angle A = \angle D = \angle F = 2(36°) = 72°$; $\angle A + \angle D + \angle F = 216°$.

- •22. Yes.

- 23. The greater the area of the triangle, the greater the sum of its angles seems to be.

- 24. A very small triangle.

Set II (pages 693–695)

Exercises 25 through 28 describe some of the content of *Sphaerica,* the major work of Menelaus. The following summary of the first book of *Sphaerica* is from Morris Kline's *Mathematical Thought from Ancient to Modern Times* (Oxford University Press, 1972): "In the first book, on spherical geometry, we find the concept of a spherical triangle, that is, the figure formed by three arcs of great circles on a sphere, each arc being less than a semicircle. The object of the book is to prove theorems for spherical triangles analogous to what Euclid proved for plane triangles. Thus the sum of two sides of a spherical triangle is greater than the third side and the sum of the angles of a triangle is greater than two right angles. Equal sides subtend equal angles. Then Menelaus proves the theorem that has no analogue in plane triangles, namely, that if the angles of one spherical triangle equal respectively the angles of another, the triangles are congruent. He also has other congruence theorems and theorems about isosceles triangles."

The bowling-ball exercises were inspired by some remarks on geodesics by David Henderson in *Experiencing Geometry in Euclidean, Spherical and Hyperbolic Spaces* (Prentice Hall, 2001). In chapter 2, titled "Straightness on Spheres", Henderson describes some activities to help in visualizing great circles as "the only straight lines on the surface of a sphere." He suggests: (1) stretching a rubber band on a slippery sphere—it will stay in place only on a great circle, (2) rolling a ball on a straight chalk line—the chalk will form a great circle on the ball, (3) putting a stiff strip of paper "flat" on a sphere—it will lie properly only along a great circle, (4) using a small toy car whose wheels are fixed so that it rolls along a straight line on a plane—the car will roll around a great circle on a sphere but not around other curves. Henderson's book is a valuable resource for anyone who teaches geometry.

Spherical Triangles.

•25. Arcs of great circles.

26. The sum of any two sides of a triangle is greater than the third side.

27. No. In Euclidean geometry, the triangles would be similar but not necessarily congruent.

28. The greater the area of the triangle, the greater the sum of its angles.

Seemingly Parallel.

29. They do not intersect.

•30. Both curves cannot correspond to lines. (Lines are great circles and great circles divide the sphere into two equal hemispheres. If one of these curves divides the sphere into two equal hemispheres, the other one clearly does not.)

Euclidean and Sphere Geometries.

•31. One.

•32. No.

33. That they are parallel.

34. No. The corresponding angles formed by lines AB and AC with line BC are equal because they are right angles, but the lines intersect in point A.

35. That they are parallel.

36. No.

37. It is greater than either remote interior angle (and equal to their sum).

38. No. In △ABC, for instance, the exterior angle at B is equal to the remote interior angle at C.

39. It is 180°.

•40. No. The sum of the angles of △ABC is greater than 180°.

Wet Paint.

41. It is finite in length, has a definite width and thickness, and so on.

42. It makes sense if the "line" on the ball is a great circle.

43. The radius (or diameter) of the sphere.

Equilateral and Right.

44. It has 90° angles rather than 60° angles. Also, the sum of its angles is 270°, not 180°.

•45. It has three right angles rather than one.

•46. It is four times as long. (The two perpendicular lines through point A divide the line through points B and C into four equal parts.)

47. 4π units. [$c = 2\pi(2) = 4\pi$.]

48. 3π units.

49. 16π square units. [$A = 4\pi r^2 = 4\pi(2)^2 = 16\pi$.]

50. 2π square units. (△ABC is one of eight congruent triangles into which the lines containing its sides divide the sphere.)

Set III (page 695)

The following material is from the "Games" column of *Omni* magazine, reprinted in *The Next Book of Omni Games,* by Scot Morris (New American Library, 1988):

"Ross Eckler, editor of *Word Ways,* sent us an interesting question based on his geographical research. As you move from place to place during your life, including visits to foreign countries, visualize the corresponding paths of your antipodal points—the point traced exactly through the center of the earth to the opposite side of the globe.

What fraction of the time, would you say, is your antipodal trace on water, and what fraction of the time is it on land?

Most people confronted with this problem allow for the fact that about two thirds of the globe is covered with water. So they answer that their antipodal point would be on water anywhere from 65 to 90 percent of the time. The truth of matter is that for most Americans the answer is likely to be closer to 100 percent.

The reason for this is the extremely nonrandom distribution of land and water. For example, there are only a few square miles of the continental United States that map onto land: two specks in Southeast Colorado that correspond to St. Paul and Amsterdam Islands, in the Indian Ocean, and a larger speck on the Montana–Canada border that corresponds to Kerguelen Island, also in the Indian Ocean. None of these points are near major highways, however, so most people have never been there. Western Europe and Israel—common tourist destinations—also map into water, with the exception of parts of Spain.

Interestingly, the Hawaiian islands are totally land-antipodal, mapping into Botswana; most of Alaska, however, maps into water, except for a region near Point Barrow that is opposite of Antarctica.

Have you taken a trip to South America? If so, your antipodal point might have dried off. Bogota, Quito, Lima, and much of Chile and Argentina are opposite China, Sumatra, and neighboring lands.

A visit to Australia does no good whatever. The entire continent neatly fits into the Atlantic Ocean, with Perth missing Bermuda by only a few miles.

The chances are very high, therefore, that at no time during your life could you have dug a hole through the center of the earth and struck land on the other side."

Antipodal Points.

1. The sphere of the earth has been mapped twice so that each pair of antipodal points coincide.

2. For most points of the earth that are on land, the corresponding antipodal points are in an ocean.

Chapter 16, Lesson 2

Set I (pages 698–699)

The western and eastern borders of Wyoming are each about 280 miles in length. The southern and northern borders are about 370 miles and 350 miles, respectively. A nice class exercise might be to use the latitudes of the southern and northern borders (41°N and 45°N) to calculate the ratio of the circumferences of the circles of latitude and show that it is in reasonable agreement with the ratio of these lengths. Students who know their geography may enjoy naming the states that border Wyoming. (Montana on the north, Utah and Colorado on the south, Idaho and Utah on

the west, and South Dakota and Nebraska on the east.) Another question of interest would be to name other states whose borders are Saccheri quadrilaterals. (There is only one: Colorado.)

Theorem 85, that the summit angles of a Saccheri quadrilateral are equal, is the first theorem proved by Saccheri in his *Euclid Freed of Every Flaw*. The first English edition of Saccheri's book, titled *Girolamo Saccheri's Euclides Vindicatus* and edited and translated by George Bruce Halsted with the original Latin on the facing pages, was published by Open Court in 1920. It was republished with added notes in 1986 by the Chelsea Publishing Company.

Wyoming.

- •1. Its legs.

- •2. The base angles.

- •3. The summit.

- 4. The summit angles.

- •5. WY and MO.

- •6. ∠Y and ∠O.

Theorem 87.

- •7. Two points determine a line.

- 8. The legs of a Saccheri quadrilateral are equal.

- •9. All right angles are equal.

- 10. SAS.

- 11. Corresponding parts of congruent triangles are equal.

- 12. SSS.

- 13. Corresponding parts of congruent triangles are equal.

Theorem 89.

- •14. The Ruler Postulate.

- 15. Two points determine a line.

- •16. It is a birectangular quadrilateral whose legs are equal.

- •17. The summit angles of a Saccheri quadrilateral are equal.

18. The "whole greater than part" theorem.

19. Substitution.

•20. An exterior angle of a triangle is greater than either remote interior angle.

21. The transitive property.

Theorem 90.

22. The "three possibilities" property.

•23. A Saccheri quadrilateral.

24. The summit angles of a Saccheri quadrilateral are equal.

•25. The fact that ∠D > ∠C.

•26. If the legs of a birectangular quadrilateral are unequal, the summit angles opposite them are unequal in the same order.

27. The fact that ∠D > ∠C.

28. CB > DA.

Set II (pages 699–700)

Of Nasir Eddin (1201–1274), Howard Eves wrote in *An Introduction to the History of Mathematics* (Saunders, 1990): "He wrote the first work on plane and spherical trigonometry considered independently of astronomy. Saccheri (1667–1733) started his work on non-Euclidean geometry through a knowledge of Nasir Eddin's writings on Euclid's parallel postulate. His was the only attempt to prove this postulate in the period from the ancient Greeks up to the Renaissance." An outline of Nasir Eddin's approach to the postulate is included on pages 208–210 of the first volume of *The Thirteen Books of Euclid's Elements*, with introduction and commentary by Sir Thomas L. Heath (Dover, 1956).

Theorem 86 is the second theorem proved by Saccheri in his *Euclid Freed of Every Flaw*.

A curious fact about the flag of Poland: if it is flown upside down, it becomes the flag of both Monaco and Indonesia!

Nasir Eddin.

•29. CD.

30. ABDC is a birectangular quadrilateral in which ∠2 > ∠3 (∠2 is obtuse and ∠3 is acute); so CD > AB. If the summit angles of a birectangular quadrilateral are unequal, the legs opposite them are unequal in the same order.

Theorem 88.

•31. DM = CM because ΔADM ≅ ΔBCM (SAS).

32. The midpoint of a line segment divides it into two equal segments.

•33. In a plane, two points each equidistant from the endpoints of a line segment determine the perpendicular bisector of the line segment.

34. ∠D = ∠C because the summit angles of a Saccheri quadrilateral are equal; so ΔADN ≅ ΔBCN (SAS); AN = BN because corresponding parts of congruent triangles are equal.

35. In a plane, two points each equidistant from the endpoints of a line segment determine the perpendicular bisector of the line segment.

Polish Flag.

36.
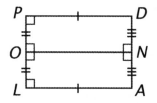

37. A Saccheri quadrilateral.

•38. DA.

39. The summit angles of a Saccheri quadrilateral are equal.

•40. The line segment connecting the midpoints of the base and summit of a Saccheri quadrilateral is perpendicular to both of them.

41. Birectangular.

Alter Ego.

•42. In a plane, two lines perpendicular to a third line are parallel.

43. A quadrilateral is a parallelogram if two opposite sides are both parallel and equal.

•44. The opposite angles of a parallelogram are equal.

45. Because a quadrilateral each of whose angles is a right angle is a rectangle.

46. It is equal to and parallel to the base.

Irregular Lots.

47. AF > CD. ABEF is a birectangular quadrilateral in which ∠1 > ∠F, and so AF > BE; BCDE is a birectangular quadrilateral in which ∠D > ∠2, and so BE > CD. The conclusion that AF > CD follows from the transitive property.

48. ∠D > ∠F. ACDF is a birectangular quadrilateral in which AF > CD. If the legs of a birectangular quadrilateral are unequal, the summit angles opposite them are unequal in the same order.

Set III (page 701)

Accompanying the photograph by Alex S. MacLean of the jog in the road in Castleton, North Dakota, from *Taking Measures Across the American Landscape* (Yale University Press, 1996) is this caption by Hildegard Binder Johnson: "It did not take long for legislators to understand that a township could not be exactly six miles on each side if the north-south lines were to follow the lines of longitude, which converged, or narrowed, to the north. The grid was, therefore, corrected every four townships to maintain equal allocations of land. These slippages, measured every twenty-four miles (or every thirty minutes if one was driving), poignantly register the spherical and magnetic condition of our planet, doglegging to counter the diminishing distances as one moves north."

Offsets.

1. No. The sides of the townships lie along map directions; so the lines containing their legs all intersect at the North (and South) Pole.

2. They are shorter than the bases. It is given that AB = DF = 6 miles; DE < DF, and so DE < AB.

3. They become shorter and shorter. (If a township had its summit very close to the North Pole, it would look very much like an isosceles triangle.)

Set I (page 704)

Lobachevskian Theorem 1, that the summit of a Saccheri quadrilateral is longer than its base, is part of the third theorem proved by Saccheri in his *Euclid Freed of Every Flaw*. (Saccheri covers all three cases, stating that the summit is equal to, less than, or greater than the base, depending on whether the summit angles are right, obtuse, or acute.)

Lobachevskian Theorem 1.

•1. The summit angles of a Saccheri quadrilateral are acute in Lobachevskian geometry.

2. An acute angle is less than 90°.

•3. Substitution.

•4. A line segment has exactly one midpoint.

5. Two points determine a line.

•6. The line segment connecting the midpoints of the base and summit of a Saccheri quadrilateral is perpendicular to both of them.

•7. A quadrilateral that has two sides perpendicular to a third side is birectangular.

8. If the summit angles of a birectangular quadrilateral are unequal, the legs opposite them are unequal in the same order.

•9. The Addition Theorem of Inequality.

10. Substitution.

Lobachevskian Theorem 2.

11. △ADM ≅ △CPM by SAS (AM = MC, DM = PM, and ∠AMD = ∠CMP). △BEN ≅ △CPN for the same reason.

12. Corresponding parts of congruent triangles are equal.

13. Substitution.

•14. A birectangular quadrilateral whose legs are equal is a Saccheri quadrilateral.

•15. The summit of a Saccheri quadrilateral is longer than its base in Lobachevskian geometry.

16. Substitution.

17. Substitution.

18. Multiplication (or division).

Set II (pages 705–706)

Although our study of non-Euclidean geometry is limited to two dimensions, exercises 37 through 46 illustrate some of its strange consequences in three dimensions. The simple pattern of equilateral triangles that can be folded to form a regular tetrahedron in Euclidean geometry results in an irregular tetrahedron in Lobachevskian geometry. The pattern of squares that can be folded to form an open box in the shape of a cube in Euclidean geometry does not exist in Lobachevskian geometry. The Set III exercise of Lesson 2 illustrated the difficulty of trying to subdivide the surface of a sphere (the Riemannian plane) into squares and the same is true of the Lobachevskian plane. The existence of squares and cubes depends on the Euclidean Parallel Postulate; so areas and volumes in non-Euclidean geometry are not measured in square or cubic units. The existence of a square in a geometry forces that geometry to be Euclidean because the square can tile the plane, and subdividing the resulting squares produces a Cartesian grid that forces a Pythagorean pattern on the entire plane.

Lobachevsky and His Geometry.

19. *(Student answer.)* (Below the portrait of Lobachevsky is his name. The last word is "geometry.")

•20. Through a point not on a line, there is more than one line parallel to the line.

21. The summit angles of a Saccheri quadrilateral are acute.

22. The summit of a Saccheri quadrilateral is longer than its base.

23. A midsegment of a triangle is less than half as long as the third side.

Lobachevskian Quadrilateral.

24.

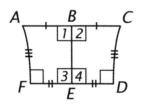

25. A Saccheri quadrilateral.

26. Acute. The summit angles of a Saccheri quadrilateral are acute in Lobachevskian geometry.

27. Right. The line segment connecting the midpoints of the base and summit of a Saccheri quadrilateral is perpendicular to both of them.

•28. Birectangular.

29. If the summit angles of a birectangular quadrilateral are unequal, the legs opposite them are unequal in the same order.

30. In Lobachevskian geometry, the line segment that connects the midpoints of the summit and base of a Saccheri quadrilateral is shorter than either leg.

Riemannian Quadrilateral.

31. Obtuse.

32. *(Student answer.)* (Yes.)

33. DC < AB.

•34. In Riemannian geometry, the summit of a Saccheri quadrilateral is shorter than its base.

Riemannian Triangle.

35. In Riemannian geometry, a midsegment of a triangle is more than half as long as the third side.

36. No. There are no parallel lines in Riemannian geometry.

Triangular Pyramid.

•37. They are all half as long.

38. They are equilateral and congruent.

39. The edges of the base are less than half as long as the sides of △ACE.

40. Its lateral faces are congruent isosceles triangles in which the base is shorter than the legs. Its base is an equilateral triangle.

Open Box.

41. It would have the shape of a cube.

•42. They are congruent squares.

43. Saccheri quadrilaterals.

44. They would be longer than the sides of CFIL. (In Lobachevskian geometry, the summit of a Saccheri quadrilateral is longer than its base.)

45. A square.

46. A square is also a Saccheri quadrilateral but, in Lobachevskian geometry, the summit of a Saccheri quadrilateral is longer than its base. (Also, in Lobachevskian geometry, the summit angles of a Saccheri quadrilateral are acute, not right.)

Set III (page 706)

In the section on non-Euclidean geometries in *Mathematics—The Science of Patterns* (Scientific American Library, 1994), Keith Devlin observes that, with adjustments such as those considered in Lesson 1, the surface of a sphere is a good model for representing Riemannian geometry. He then asks, what surface would be a good model for representing Lobachevskian geometry? Devlin writes: "The answer turns out to involve a pattern that is familiar to every parent. Watch a child walking along, pulling a toy attached to a string. If the child makes an abrupt left turn, the toy will trail behind, not making a sharp corner, but curving around until it is almost behind the child once again. This curve is called a tractrix." A pseudosphere is a surface produced by rotating a tractrix about its asymptote; in contrast with a sphere, which has a constant positive curvature, a pseudosphere has a constant negative curvature. Remarkably, a pseudosphere matched with the appropriate sphere has the same volume and surface area. For more on this subject, see *A Book of Curves*, by E. H. Lockwood (Cambridge University Press, 1963).

Directions for crocheting a pseudosphere are given by David Henderson in *Experiencing Geometry in Euclidean, Spherical, and Hyperbolic Spaces* (Prentice Hall, 2001).

As with geometry on the surface of a sphere, lines on a pseudosphere are represented by geodesics, the shortest paths on the surface between pairs of points. These paths are easily pictured on a sphere because they are along arcs of great circles. Unfortunately, for the pseudosphere, the only simple geodesics are the meridians, illustrated by lines AB and DC in the figure illustrating the Saccheri quadrilateral, and

the circles orthogonal to them, of which AD and BC in the figure are arcs. H. S. M. Coxeter in *Introduction to Geometry* (Wiley, 1969) points out that, because of this shortcoming, the pseudosphere is "utterly useless as a means for drawing significant hyperbolic figures."

Pseudosphere.

From the figure, it appears that AD < BC; so BC is the summit of the Saccheri quadrilateral. It follows that AB = DC because they are the legs. Also, ∠BAD and ∠ADC are right angles and ∠ABC and ∠DCB are acute angles because they are, respectively, the base angles and the summit angles of the Saccheri quadrilateral. Lines AB and DC are parallel because, in a plane, two lines perpendicular to a third line are parallel (this theorem comes before the Parallel Postulate in Euclidean geometry).

Chapter 16, Lesson 4

Set I (pages 709–710)

Lobachevskian Theorem 3, that the sum of the angles of a triangle is less than 180°, is part of Proposition 15 proved by Saccheri in his *Euclid Freed of Every Flaw*. Saccheri states it thus: "By any triangle ABC, of which the three angles are equal to, or greater, or less than two right angles, is established respectively the hypothesis of right angle, or obtuse angle, or acute angle."

Saccheri's Proposition 16 includes the corollary about the sum of the angles of a quadrilateral being less than 360°.

English mathematician John Wallis (1616–1703) tried to prove Euclid's Fifth Postulate but assumed in doing so that, for every triangle, there exists a similar triangle of any arbitrary size. Saccheri was aware of Wallis's attempt and discusses it in his book. The proof of Lobachevskian Theorem 4 outlined in exercises 14 through 20 is, in fact, the one used by Saccheri in establishing that Wallis's assumption and Euclid's Fifth Postulate are logically equivalent.

Lobachevskian Theorem 3.

1. ΔADM ≅ ΔCPM by SAS (AM = MC, DM = PM, and ∠AMD = ∠CMP). ΔBEN ≅ ΔCPN for the same reason.

•2. ∠D and ∠E.

3. AD and BE.

- 4. A Saccheri quadrilateral.

- 5. The summit angles of a Saccheri quadrilateral are acute in Lobachevskian geometry.

  6. ∠DAB + ∠EBA < 180°.

  7. ∠1 = ∠5 and ∠4 = ∠6.

- 8. Substitution.

  9. Substitution.

Corollary to Lobachevskian Theorem 3.

- 10. The sum of the angles of a triangle is less than 180° in Lobachevskian geometry.

  11. Addition.

  12. Betweenness of Rays Theorem.

  13. Substitution.

Lobachevskian Theorem 4.

- 14. Corresponding angles of similar triangles are equal.

  15. SAS.

  16. Substitution. (∠D = ∠A and ∠1 = ∠A; ∠F = ∠C and ∠2 = ∠C.)

- 17. The angles in a linear pair are supplementary.

  18. Addition.

  19. Substitution.

  20. The sum of the angles of a quadrilateral is less than 360° in Lobachevskian geometry.

Set II (pages 710–712)

Exercises 31 and 32 lead to the theorem that, in Lobachevskian geometry, an exterior angle of a triangle is greater than the sum of the remote interior angles. Exercises 33 and 34 reveal that, in Lobachevskian geometry, an angle inscribed in a semicircle is acute. Both theorems are considered by Saccheri, the first as a corollary to his Proposition 15 and the second as his Proposition 18.

The following excerpt is from the preface to the second English edition of *Euclid Freed from Every Flaw* (Chelsea, 1986): "In this book Girolamo Saccheri set forth in 1733, for the first time ever, what amounts to the axiom systems of non-

Euclidean geometry. It is in this book that, for the first time in history, theorem after theorem of [Lobachevskian] non-Euclidean geometry is stated and proved. . . . Why—one cannot help asking—did Saccheri not evaluate correctly what he had achieved, why did he not claim credit for the discovery of non-Euclidean geometry?. . . In his era, it would seem, the existence of a valid geometry alternative to Euclid's was, quite literally, unthinkable. Not impossible, not wrong, but *unthinkable*."

Exercises 35 through 50 show an example of an attempt to "double a triangle" in Lobachevskian geometry. For the right triangle considered, when the hypotenuse is doubled, one leg of the resulting triangle is less than doubled, whereas the other leg is more than doubled!

Lobachevskian Triangles.

- 21. No. The sums of the angles of the two triangles may not be equal.

  22. Their sum is less than 90° (or, they are not complementary).

  23. Each angle is less than 60°.

  24. The longer the sides, the smaller the angles.

Triangle on a Sphere.

  25. ∠B = ∠C.

- 26. ∠A + ∠B + ∠C > 180°.

- 27. Riemannian geometry.

Triangle on a Pseudosphere.

  28. ∠B = ∠C.

  29. ∠A + ∠B + ∠C < 180°.

  30. Lobachevskian geometry.

Exterior Angles.

- 31. ∠1 = ∠A + ∠B.

  32. The conclusion about ∠1 is not true in Lobachevskian geometry. ∠1 + ∠2 = 180° (the angles in a linear pair are supplementary) and ∠A + ∠B + ∠2 < 180° (the sum of the angles of a triangle is less than 180° in Lobachevskian geometry); so ∠A + ∠B + ∠2 < ∠1 + ∠2 (substitution). It follows that ∠A + ∠B < ∠1 (subtraction); so, in Lobachevskian geometry, ∠1 > ∠A + ∠B.

Angle Inscribed in a Semicircle.

33. It is a right angle.

34. The conclusion about ∠ABC is not true in Lobachevskian geometry.
∠1 + ∠ABC + ∠4 < 180° (the sum of the angles of a triangle is less than 180° in Lobachevskian geometry). OA = OB = OC (all radii of a circle are equal); so ∠1 = ∠2 and ∠3 = ∠4 (if two sides of a triangle are equal, the angles opposite them are equal). Also, ∠ABC = ∠2 + ∠3, and so
∠2 + (∠2 + ∠3) + ∠3 < 180° (substitution); so 2(∠2 + ∠3) < 180° and ∠2 + ∠3 < 90° (division). So ∠ABC < 90°. An angle inscribed in a semicircle in Lobachevskian geometry is acute.

Magnification and Distortion.

•35. ΔADE ~ ΔABC (AA).

36. ∠D = ∠ABC.

37. Each side of ΔADE is twice as long as the corresponding side of ΔABC.

38.

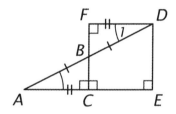

39. ∠DBF ≅ ΔABC. ∠1 = ∠A, BD = AB, and ∠FBD = ∠CBA; so they are congruent by ASA.

40. BF = BC.

•41. ∠F is a right angle. (∠F = ∠ACB.)

•42. CFDE is a birectangular quadrilateral (in two different ways).

•43. ∠FDE is acute. (The sum of the angles of a quadrilateral is less than 360° in Lobachevskian geometry. The other three angles of CFDE are right angles.)

44. DE > FC. If the summit angles of a birectangular quadrilateral are unequal, the legs opposite them are unequal in the same order.

45. FC = 2BC. (FB = BC because ΔDBF ≅ ΔABC.)

•46. DE > 2BC. (Substitution.)

47. CE < FD. (If the summit angles of a birectangular quadrilateral are unequal, the legs opposite them are unequal in the same order.)

48. CE < AC. (FD = AC; so CE < AC by substitution.)

49. AE < 2AC. [CE < AC, and so AC + CE < AC + AC (addition); so AE < 2AC.]

50. AD = 2AB, DE > 2BC (exercise 46), and AE < 2AC (exercise 49).

Set III (page 712)

Marvin Jay Greenberg, in a section of *Euclidean and Non-Euclidean Geometries—Development and History* (W. H. Freeman and Company, 1980) titled "What Is the Geometry of Physical Space?" wrote: "Certainly, engineering and architecture are evidence that Euclidean geometry is extremely useful for ordinary measurement of distances that are not too large. However, the representational accuracy of Euclidean geometry is less certain when we deal with larger distances. For example, let us interpret a 'line' physically as the path traveled by a light ray. We could then consider three widely separated light sources forming a physical triangle. We would want to measure the angles of this physical triangle in order to verify whether the sum is 180° or not (such an experiment would presumably settle the question of whether space is Euclidean or [non-Euclidean].

Gauss allegedly performed this experiment, using three mountain tops as the vertices of his triangle. The results were inconclusive. Why? Because any physical experiment involves experimental error. Our instruments are never completely accurate. Suppose the sum did turn out to be 180°. If the error in our measurement were at most 1/100 of a degree, we could conclude only that the sum was between 179.99° and 180.01°. We could never be sure that it actually was 180°.

Suppose, on the other hand, that measurement gave us a sum of 179°. Although we would conclude only that the sum was between 178.99° and 179.01°, we would be certain that the sum was less than 180°. In other words, the only conclusive result of such an experiment would be that space is [Lobachevskian]! . . .

To repeat the point: because of experimental error, a physical experiment can never prove conclusively that space is Euclidean—it can prove only that space is non-Euclidean."

Furthermore, the fact that the same quasar can be observed along *two different directions* indicates that light paths violate our assumptions about straight lines in Euclidean space.

Astronomical Triangle.

1. No. The difference could be due to errors in measurement.

2. It would be easier to prove that physical space is non-Euclidean. It is impossible to measure anything exactly; so it is impossible to know that the sum of the three angle measures is *exactly* 180°.

Chapter 16, Review

Set I (pages 714–715)

The beach ball of exercises 12 through 24 is another "spherical polyhedron," the result of projecting the edges of a regular polyhedron on the surface of its circumscribing sphere. Because of the duality of the cube and the regular octahedron, it is both a "spherical cube" and a "spherical octahedron." The sphere is divided into 48 congruent right triangles, of which 8 are contained in each square of the cube and 6 are contained in each equilateral triangle of the octahedron.

Sphere Geometry.

•1. A great circle.

•2. False.

3. False.

•4. True (because antipodal points are considered to be identical).

5. False.

6. False.

Saccheri Quadrilateral.

•7. AD.

8. AD > BC (or, AD is the longest side of ABCD).

9. ABCD cannot be a rhombus, because it is not equilateral.

10. Acute.

11. ABCD cannot be a rectangle, because ∠A and ∠D are not right angles. They are acute.

Beach Ball.

•12. 360°.

13. 360°.

14. 360°.

15.

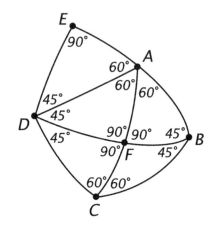

•16. ∠ABC = 90°; ∠BCD = 120°; ∠CDA = 90°; ∠DAB = 120°.

•17. The quadrilateral is a parallelogram.

18. No. There are no parallelograms in sphere geometry, because there are no parallel lines.

•19. It is a rhombus.

20. They are perpendicular to each other and they bisect each other. (It is given that all of the triangles are congruent.)

21. Yes.

22. ∠E = 90°; ∠EAF = 120°; ∠AFD = 90°; ∠FDE = 90°.

•23. No.

24. Yes.

Set II (pages 715–717)

The ladder-rung exercises demonstrate that, in Lobachevskian geometry, two parallel lines are not everywhere equidistant. The earliest known attempt to prove Euclid's Fifth Postulate was made by Posidonius in the first century B.C. Posidonius did it by changing Euclid's definition of parallel lines to say the two lines are parallel if the distance between them is always the same. Concerning this idea, Saccheri wrote in *Euclid Freed of Every Flaw*: "Thence now I may proceed to explain, why in the Preface to the Reader I have said: *not without a great sin against rigid logic two equidistant straight lines have been assumed by some as given*. Where I should point out that none of those is carped at, whom I have mentioned even indirectly in this book of mine, because they are truly great geometers, and verily free from this sin. But I say: *great sin against rigid logic*: for what else is it to assume as given *two equidistant straight lines*: unless either to assume; that every line equidistant in the same plane from a certain supposed straight line is itself also a straight line; or at least to suppose, that some one thus equidistant may be a straight line, as if therefore it were allowable to make assumption, whether by hypothesis, or by postulate, of any such distance of one from another? But it is certain neither of these can be made traffic of as if *per se* known."

It is probably no surprise that the Pythagorean Theorem does not hold in the non-Euclidean geometries, because geometric squares do not even exist in them. In terms of the lengths of the legs *a* and *b* and hypotenuse *c* of a right triangle, in Lobachevskian geometry, $c^2 > a^2 + b^2$, and in Riemannian geometry, $c^2 < a^2 + b^2$.

Maurits Escher created four Circle Limit prints based on Poincaré's model of Lobachevskian geometry. The figure of black and white regions devised by Poincaré appears in Chapter 12, Lesson 5. The geometer H. S. M. Coxeter had sent Escher a drawing of the Poincaré model in 1958 and Escher replied: "Since a long time I am interested in patterns with 'motives' getting smaller and smaller till they reach the limit of infinite smallness." Coxeter tells of this in *The Mathematical Gardner*, edited by David A. Klarner (Prindle, Weber & Schmidt, 1981), and describes the model: "the 'straight lines' of the [Lobachevskian] plane appear as arcs of circles orthogonal to a boundary circle drawn in the Euclidean plane. Angles are represented faithfully but distances are distorted, with the points of [the boundary circle] being infinitely far away. Although the [fish] get 'smaller gradually from the center towards the outside circle-limit', we enter the spirit of [Lobachevskian] geometry when we stretch our imagination in order to pretend that these [fish] are all . . . congruent."

Lobachevsky's Ladder.

25.

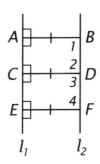

• 26. Saccheri quadrilaterals.

• 27. Acute.

28. $\angle 2 + \angle 3 < 180°$.

29. The angles in a linear pair are supplementary.

30. It proves that they cannot all be of equal lengths.

Pythagoras Meets Lobachevsky.

31. The Pythagorean Theorem.

• 32. Substitution.

33. $(\frac{1}{2}CA)^2 + (\frac{1}{2}CB)^2 = MN^2$, and so

$\frac{1}{4}CA^2 + \frac{1}{4}CB^2 = MN^2$; so

$\frac{1}{4}(CA^2 + CB^2) = MN^2$. If $CA^2 + CB^2 = AB^2$,

$\frac{1}{4}AB^2 = MN^2$ by substitution; so

$\frac{1}{2}AB = MN$.

• 34. A midsegment of a triangle is less than half as long as the third side.

35. The Pythagorean Theorem is not true in Lobachevskian geometry.

Lobachevsky Meets Escher.

36. Two circles are orthogonal iff they intersect in two points and their tangents at the points of intersection are perpendicular to each other.

• 37. Two points determine a line.

38. Through a point not on a line, there is more than one parallel to the line.

39. The summit angles of a Saccheri quadrilateral are acute (or, the summit of a Saccheri quadrilateral is longer than its base).

•40. The sum of the angles of a triangle is less than 180°.

41. A midsegment of a triangle is less than half as long as the third side.

The Geometry of the Universe.

42. It is finite. The volume of the space of the universe would be comparable to the area of the surface of a sphere, which is finite.

43. The astronaut would eventually return to the point from which he or she started. A straight line in space would be comparable to a line on the surface of a sphere, which is a great circle.

Final Review

Set I (pages 718–722)

German Terms.

1. Radius.

2. A diameter is a chord of a circle that contains its center.

3. A secant is a line that intersects a circle in two points; a tangent is a line that intersects a circle in exactly one point.

Regular Polygons.

4. A regular hexagon and a regular dodecagon.

5. They are convex, equilateral, equiangular, and cyclic.

6. 60° and 120°.

7. (Isosceles) trapezoids.

8. 720°. [Each angle has a measure of $2(60°) = 120°$; $6(120°) = 720°$.]

9. 1,800°. [Each angle has a measure of $60° + 90° = 150°$; $12(150°) = 1,800°$.]

What Follows?

10. Its base and altitude.

11. Is perpendicular to the chord.

12. Lies in the plane.

13. The triangle is a right triangle.

14. Have equal areas.

15. Are equal.

16. Tangent to the circle.

17. A line.

18. $e\sqrt{3}$.

19. In the same ratio.

20. The hypotenuse and its projection on the hypotenuse.

21. The area of its base and its altitude.

22. The ratio of any pair of corresponding sides.

23. Is –1.

24. Half the positive difference of its intercepted arcs.

Pythagorean Squares.

25. $w + x + y$.

26. The two areas are equal. Because the squares are from the sides of a right triangle, $(v + w) + (y + z) = (w + x + y)$; so $v + z = x$ by subtraction.

The Value of Pi.

27. $\triangle ABC$ is a right triangle. $\angle ACB$ is a right angle because it is inscribed in semicircle.

28. $\dfrac{AD}{CD} = \dfrac{CD}{DB}$. (The altitude to the hypotenuse of a right triangle is the geometric mean between the segments into which it divides the hypotenuse.)

29. $CD = \sqrt{10} \approx 3.16$. ($\dfrac{2}{CD} = \dfrac{CD}{5}$, $CD^2 = 10$, $CD = \sqrt{10}$.)

30. No. $\sqrt{10} > \pi$ because $\pi \approx 3.14$.

Euclid in Color.

31. The opposite angles of a cyclic quadrilateral are supplementary.

32. Inscribed angles that intercept the same arc are equal.

33. The sum of the angles of a triangle is 180°.

34. Two right angles, or 180°.

Largest Package.

35. 15,000 in³. ($10 \times 20 \times 75$.)

36. No. The length of its longest side is 75 in and the distance around its thickest part is $2(10 + 20) = 60$ in; $75 + 60 = 135 > 130$ in.

37. 26 in. ($e + 4e = 130$, $5e = 130$, $e = 26$.)

38. 17,576 in³. (26^3.)

Four Centers.

39. The orthocenter.

40. The incenter.

41. The centroid.

42. The circumcenter.

43. B (the incenter).

44. D (the circumcenter).

45. C (the centroid).

46. B (the incenter).

Formulas.

47. To find the circumference of a circle (using its diameter).

48. To find the length of a diagonal of a rectangular solid (using its length, width, and height).

49. To find the area of a trapezoid (using its altitude and two bases).

50. To find the altitude of an equilateral triangle (using its side).

Kindergarten Toy.

51. A, a cube; B, a cylinder; and C, a sphere.

52. $V = e^3$, $A = 6e^2$.

53. $V = \pi r^2 h$, $A = 2\pi rh + 2\pi r^2$ [or $A = 2\pi r(h + r)$].

54. $V = \dfrac{4}{3}\pi r^3$, $A = 4\pi r^2$.

Steep Roof.

55. $m\text{OA} = \dfrac{21-0}{12-0} = \dfrac{21}{12} = \dfrac{7}{4} = 1.75$;

$m\text{AB} = \dfrac{21-0}{12-24} = \dfrac{21}{-12} = -\dfrac{7}{4} = -1.75$.

56. 60°. ($\tan \angle AOB = \dfrac{21}{12} = 1.75$; $\angle AOB \approx 60.3°$.)

57. 24 ft. ($OA = AB = \sqrt{12^2 + 21^2} = \sqrt{585} \approx 24.2$.)

58. No. (It is not equilateral, because OB is slightly longer than the other two sides or, equivalently, because $\angle AOB$ is not 60°.)

Greek Crosses.

59. $12x$.

60. $5x^2$.

61. $10x^2$.

62. $x\sqrt{10}$.

63. $4x\sqrt{10}$.

Baseball in Orbit.

64. About 138 mi. [$c = 2\pi r = 2\pi(22) = 44\pi \approx 138$.]

65. About 83 minutes. ($\frac{138 \text{ mi}}{100 \text{ mi}/\text{hour}} =$

1.38 hours $\approx$ 82.8 minutes.)

Two Quadrilaterals.

66. $2w + 2x + 2y + 2z = 360°$ (the sum of the angles of a quadrilateral is 360°); so $2(w + x + y + z) = 360°$ and $w + x + y + z = 180°$ (division).

67. In $\triangle$AFD, $\angle F + w + z = 180°$ and, in $\triangle$BHC, $\angle H + x + y = 180°$ (the sum of the angles of a triangle is 180°); so $\angle F = 180° - w - z$ and $\angle H = 180° - x - y$ (subtraction).

68. $\angle F + \angle H = (180° - w - z) + (180° - x - y) = 360° - w - x - y - z = 360° - (w + x + y + z)$ (addition); so $\angle F + \angle H = 360° - 180° = 180°$ (substitution).

69. $\angle F$ and $\angle H$ are supplementary (if the sum of two angles is 180°, they are supplementary); so quadrilateral EFGH is cyclic (a quadrilateral is cyclic if a pair of its opposite angles are supplementary).

Long Shadows.

70. Approximately 5,500 ft.

($\tan 12.8° = \frac{1,250}{x}$, $x = \frac{1,250}{\tan 12.8°} \approx 5,500$.)

71. Approximately 26 ft. ($\tan 12.8° = \frac{6}{y}$,

$y = \frac{6}{\tan 12.8°} \approx 26.4$; or $\frac{1,250}{5,500} = \frac{6}{y}$, etc.)

Tree Geometry.

72. k.

73. k^2.

74. k^3.

75. k^3.

76. Those of the larger tree would be $10^2 = 100$ times the area of those of the smaller tree.

77. Those of the larger tree would be $10^3 = 1,000$ times the weight of those of the smaller tree.

78. The larger tree.

SAT Problem.

79. $3\pi s$ units. [$2(\frac{3}{4}2\pi s$.]

80. $\frac{3}{2}\pi s^2 + s^2$ [or $(\frac{3}{2}\pi + 1)s^2$] square units.

[$2(\frac{3}{4}\pi s^2) + s^2 = \frac{3}{2}\pi s^2 + s^2$.]

Sight Line.

81. About 23°. ($\tan \angle 1 = \frac{90}{210}$, $\angle 1 \approx 23°$.)

82. About 18°. ($\tan \angle 2 = \frac{70}{210}$, $\angle 2 \approx 18°$.)

Set II (pages 723–727)

Dog Crates.

83. 19 in. ($V = lwh$, $w = \frac{V}{lh} = \frac{12,540}{30 \times 22} = 19$.)

84. No. They are not similar, because their corresponding dimensions are not proportional: $\frac{30}{36} \neq \frac{22}{26}$.

($\frac{30}{36} \approx 0.83$ and $\frac{22}{26} \approx 0.85$.)

85. 51,364 in³.

[$V = (\frac{48}{30})^3(12,540) = 51,363.84$.]

Tangent Circles.

86. $\frac{AE}{EC} \cdot \frac{CD}{DB} \cdot \frac{BF}{FA} = 1$ (or any equivalent equation) by Ceva's Theorem.

87. Yes. If the radius of the three circles is r, $\frac{r}{r} \cdot \frac{r}{r} \cdot \frac{r}{r} = 1$.

88. Yes. Letting the radii of the three circles be r_1, r_2, and r_3, we have $\frac{r_1}{r_2} \cdot \frac{r_2}{r_3} \cdot \frac{r_3}{r_1} = 1$.

Squares on the Legs.

89. The altitude to the hypotenuse of a right triangle forms two triangles similar to it and to each other (or AA).

90. The ratio of the areas of two similar polygons is equal to the square of the ratio of the corresponding sides.

91. They are equal because $\alpha ABEF = AB^2$ and $\alpha BCHG = BC^2$.

92. Yes. DAFEB ~ DBGHC. (Their corresponding angles are equal and their corresponding sides are proportional.)

Formula Confusion.

93. (Student answer.) [Because $c = 2\pi r$, the formula $A = \frac{1}{2}rc$ is equivalent to $A = r(2\pi r) = \pi r^2$, which gives the area of the circle correctly.]

94. No. Because $A = 4\pi r^2$, the formula $V = \frac{1}{2}rA$ is equivalent to $V = \frac{1}{2}r(4\pi r^2) = 2\pi r^3$, but the volume of a sphere is $\frac{4}{3}\pi r^3$.

95. Yes. Because $A = 2\pi rh$, the formula $V = \frac{1}{2}rA$ is equivalent to $V = \frac{1}{2}r(2\pi rh) = \pi r^2 h$, which gives the volume of the cylinder correctly.

Regular 17-gon.

96. 3.12. ($17 \sin\frac{180}{17}$.)

97. 3.07. ($17 \sin\frac{180}{17} \cos\frac{180}{17}$.)

98. 62 cm. [$p = 2Nr \approx 2(3.12)(10) \approx 62$.]

99. 307 cm². [$A = Mr^2 \approx 3.07(100) = 307$.]

100. The corresponding measurements of the circle would be larger.

101. 63 cm and 314 cm².
    [$c = 2\pi r = 2\pi(10) = 20\pi \approx 63$.
    $A = \pi r^2 = \pi(10)^2 = 100\pi \approx 314$.]

Packing Circles.

102. $\sqrt{3}\,r^2$. [$\frac{\sqrt{3}}{4}(2r)^2 = \sqrt{3}\,r^2$.]

103. $\frac{1}{2}\pi r^2$. [$3(\frac{1}{6}\pi r^2) = \frac{1}{2}\pi r^2$.]

104. About 91%. ($\frac{\frac{1}{2}\pi r^2}{\sqrt{3}r^2} = \frac{\pi}{2\sqrt{3}} \approx 0.91$.)

Rain Gutter.

105.
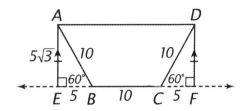

106. $\triangle AEB$ and $\triangle DFC$ are 30°-60° right triangles; $\triangle AEB \cong \triangle DFC$ (AAS).

107. In a plane, two lines perpendicular to a third line are parallel.

108. Corresponding parts of congruent triangles are equal.

109. ADFE is a parallelogram because AE ∥ DF and AE = DF (a quadrilateral is a parallelogram if two opposite sides are both parallel and equal); so AD ∥ EF (the opposite sides of a parallelogram are parallel).

110. $5\sqrt{3} \approx 8.66$ cm. ($\triangle ABE$ is a 30°-60° right triangle with hypotenuse AB = 10; so EB = 5 and AE = $5\sqrt{3}$.)

111. 130 cm². (ABCD is a trapezoid with BC = 10, AD = 20, and AE = $5\sqrt{3}$; so
    $\alpha ABCD = \frac{1}{2}5\sqrt{3}(10 + 20) = 75\sqrt{3} \approx 130$.)

Incircle.

112. The angle bisectors of a triangle are concurrent.

113. The incenter.

114. The tangent segments to a circle from an external point are equal.

115. If a line is tangent to a circle, it is perpendicular to the radius drawn to the point of contact.

116. In a plane, two lines perpendicular to a third line are parallel to each other.

117. ODCE is a parallelogram because both pairs of its opposite sides are parallel. OD = EC and OE = DC because the opposite sides of a parallelogram are equal.

118. AB = 29. ($\triangle ABC$ is a right triangle, and so $AB^2 = CA^2 + CB^2 = 20^2 + 21^2 = 400 + 441 = 841$; so AB = $\sqrt{841}$ = 29.)

119. $20 - r$. [ODCE is also a square; so CD = r. AD = AF (the tangent segments to a circle from an external point are equal) and AD = AC − CD = $20 - r$.]

120. $21 - r$.

121. AB = AF + FB; so $29 = (20 - r) + (21 - r)$, $29 = 41 - 2r$, $2r = 12$, $r = 6$.

Angles and Sides.

122. By the Law of Cosines.

123. $\cos C = \frac{1}{2}$. ($c^2 = a^2 + b^2 - 2ab \cos C$, $c^2 = c^2 + c^2 - 2c^2 \cos C$, $2c^2 \cos C = c^2$, $2 \cos C = 1$, $\cos C = \frac{1}{2}$.)

124. $\angle C = 60°$.

125. Yes. If $a = b = c$, ΔABC is equilateral; so each of its angles has a measure of 60°.

126. $\cos C = 0$. ($c^2 = a^2 + b^2 - 2ab \cos C$, $c^2 = c^2 - 2ab \cos C$, $2ab \cos C = 0$, $\cos C = 0$.)

127. $\angle C = 90°$.

128. Yes. If $c^2 = a^2 + b^2$, ΔABC is a right triangle with hypotenuse c, and so $\angle C = 90°$.

Circular Track.

129.

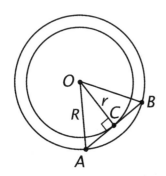

130. If a line is tangent to a circle, it is perpendicular to the radius drawn to the point of contact.

131. If a line through the center of a circle is perpendicular to a chord, it also bisects the chord.

132. ΔACO is a right triangle; so $AO^2 = AC^2 + CO^2$, $R^2 = 100^2 + r^2$, and so $R^2 - r^2 = 10,000$.

133. 31,416 m². [$A_{track} = \pi R^2 - \pi r^2 = \pi(R^2 - r^2) = 10,000\pi \approx 31,416$.]

da Vinci Problem.

134. $\frac{1}{4}\pi$. ($\frac{1}{4}\pi 1^2$.)

135. $\frac{1}{2}$. ($\frac{1}{2}1^2$.)

136. $\frac{1}{4}\pi - \frac{1}{2}$.

137. $2\pi - 4$. [$8(\frac{1}{4}\pi - \frac{1}{2})$.]

138. 2π. [In isosceles right ΔOPF, PF = 1; so OF = $\sqrt{2}$; $\pi(\sqrt{2})^2 = 2\pi$.]

139. 4. [$2\pi - (2\pi - 4)$.]

140. 1. ($\frac{1}{4}4$.)

A Problem from Ancient India.

141. 10 units. (FB = HA + AB. FB = $\sqrt{6^2 + (12 - x)^2}$, HA = 6, and AB = x. $\sqrt{36 + (144 - 24x + x^2)} = 6 + x$, $180 - 24x + x^2 = 36 + 12x + x^2$, $144 = 36x$, $x = 4$. FB = 6 + 4 = 10.)

Circumradius.

142. Every triangle is cyclic.

143. If a line through the center of a circle is perpendicular to a chord, it also bisects the chord.

144. ΔBOH ≅ ΔCOH (HL); so $\angle$BOH = $\angle$COH; so ray OH bisects $\angle$BOC; so $\angle$BOH = $\frac{1}{2}\angle$BOC.

145. A central angle is equal in measure to its intercepted arc.

146. An inscribed angle is equal in measure to half its intercepted arc.

147. $\angle$BOH = $\frac{1}{2}\angle$BOC = $\frac{1}{2}m\widehat{BC}$ = $\angle$A.

148. The sine of an acute angle in a right triangle is equal to the ratio of the opposite leg to the hypotenuse.

149. $\sin \angle BOH = \dfrac{\frac{a}{2}}{r}$ and $\angle BOH = \angle A$; so

$\sin A = \dfrac{\frac{a}{2}}{r}$, $r \sin A = \dfrac{a}{2}$, $a = 2r \sin A$,

$\dfrac{a}{\sin A} = 2r.$

150. By the Law of Sines.

151. 3 cm. ($\dfrac{a}{\sin A} = 2r$, $\dfrac{3}{\sin 30°} = 2r$, $6 = 2r$, $r = 3$.)

152. *Example figure:*

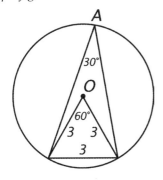

Conical Mountain.

153. 70 m. ($\triangle AFC \sim \triangle AGE$ by AA; so $\dfrac{AF}{AG} = \dfrac{FC}{GE}$,

$\dfrac{x}{x+100} = \dfrac{35}{85}$, $85x = 35x + 3{,}500$, and so

$50x = 3{,}500$, $x = 70$.)

154. 1,286,220 m³.
$[V = \dfrac{1}{3}Bh = \dfrac{1}{3}\pi(85)^2(170) \approx 1{,}286{,}220.]$

155. 89,797 m³. $[V = \dfrac{1}{3}Bh = \dfrac{1}{3}\pi(35)^2(70) \approx 89{,}797.]$

156. 89,797 m³. $[V = \dfrac{1}{2} \cdot \dfrac{4}{3}\pi(35)^3 \approx 89{,}797.]$

157. 1,100,000 m³.
(1,286,220 − 89,797 − 89,797 = 1,106,626.)

PRINCIPLES OF ALGEBRA 2

Learn algebra from a biblical perspective and use it to complete real-life tasks. *Algebra 2* uniquely provides different tools to help your student understand why algebra matters and how to use it to explore the wonder of God's creation.

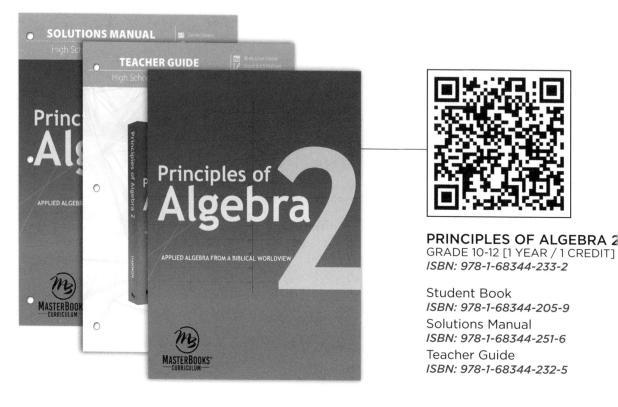

PRINCIPLES OF ALGEBRA 2
GRADE 10-12 [1 YEAR / 1 CREDIT]
ISBN: 978-1-68344-233-2

Student Book
ISBN: 978-1-68344-205-9
Solutions Manual
ISBN: 978-1-68344-251-6
Teacher Guide
ISBN: 978-1-68344-232-5

ENRICHMENT VIDEO COURSE ALSO AVAILABLE!

Visit MasterBooksAcademy.com for online video instruction that will help your student feel more confident and have a more independent learning experience.

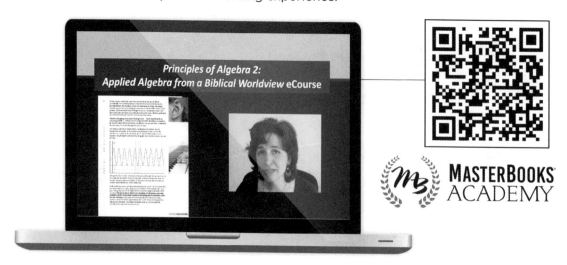

MASTERBOOKS ACADEMY

VISIT **MASTERBOOKS.COM** TO SEE OUR FULL LINE OF CURRICULUM